Engineering Physics

AF651592

Engineering Physics

(2nd Editon)

Dr. Prabir K Basu

Ph. D (NPL-New Delhi/BITS, Pilani)

Prof. Echelon Institute of Technology,
Faridabad

☆

Hrishikesh Dhasmana

M. Tech., IIT Mumbai

Asst. Prof. Lingaya's University,
Faridabad

Ane Books Pvt. Ltd.

New Delhi ♦ Chennai

Engineering Physics

Prabir K Basu and Hrishikesh Dhasmana

© Author

First Edition : 2009

Second Edition : 2010

Reprint : 2018 ,2019,2020

Published by

Ane Books Pvt. Ltd.
4821, Parwana Bhawan, 1st Floor, 24 Ansari Road,
Darya Ganj, **New Delhi** - 110 002, India
Tel.: +91(011) 23276843-44, Fax: +91(011) 23276863
e-mail: kapoor@anebooks.com,

Branch Office:

- Avantika Niwas, 1st Floor, 19 Doraiswamy Road, T. Nagar,
  Chennai - 600 017, Tel.: +91(044) 28141554, 28141209
  e-mail: anebookschennai@gmail.com, rathinam@anebooks.com

Disclaimer

Please be informed that the author and the publisher have put in their best efforts in producing this book. Every care has been taken to ensure the accuracy of the contents. However, we make no warranties for the same and therefore shall not be responsible or liable for any loss or any commercial damages accruing thereof. Neither the publisher nor the author is engaged in providing services of any professional nature and shall therefore not be responsible for any incidental, consequential, special or any other damages. Please do consult a professional where appropriate.

ISBN : 978-81-9083-224-3

All rights reserved. No part of this book may be reproduced in any form including photocopying, microfilms, photoprints, storage in any retrieval system, transmission in any permanent or temporary form, without the prior written consent of the publisher.

Printed at : Repro Knowledgecast Limited, Thane

Dedicated to

Our Parents

Roni, Piu, Saksham and all Our Well-wishers

Preface to the Second Edition

We feel great pleasure in bringing out the second edition of our book on Engineering Physics for engineering students. The authors feel the need of writing a book according to the use of engineering students in a simplest possible way. As we all know that physics is not only a subject, but it is a realization of phenomena happening around us. So this book is practically based on lecture notes which were applied to teach the students in such a way that it gives them an ample opportunity to understand the intricacies of different engineering aspects. Every effort has been made to make all the topics presentable in a lucid manner. The authors have consulted various books, journals and visited several internet sites to prepare the book. The whole syllabus dealt with several branches of physics applicable to engineering students. The philosophy of presentation and material content in the book are based on concept based approach towards the subject.

In this edition as per demand of our readers we have introduced solutions to every problem. Also some aspects of Ultrasonics have been introduced. We have tried to put hard efforts in placing contents in this book such that students feel comfortable with the syllabus in an interesting way.

We sincerely express our deep sense of gratitude to the chairman P. Bansal, director Dr. V.K. Agarwal, of Echelon Institute of Technology, Faridabad for their inspirational influence during teaching at EIT. We are grateful to our colleagues, friends, Mr. N. Udayakumar, Chairman, Udhaya Energy Photovoltaics Pvt Ltd, Coimbatore and Prof.(Dr.) Viresh Dutta, Center for Energy Studies, IIT-Delhi, Mr. Roop Singh and Prof.(Dr.) D.K.Thakur, Mr. R.M. Pujari, A.K. Shukla, P. Unijal, Dr. B.K. Sharma, Sunil Bist for their encouragement during preparation of the manuscript. We both are also very much thankful to all our family members for their heartiest cooperation during the preparation of the book. We are also grateful to the publishers, Ane Books, Pvt. Ltd., New Delhi, for all cooperation and help related to the timely publication of the book.

It will be kind enough for readers for any suggestions which can easily be accepted for further improvement in next edition.

Dr. P.K. Basu
Hrishikesh Dhasmana

Preface to the First Edition

We feel great pleasure in bringing out a book on Physics for engineering students. The authors feel the need of writing a book according to the use of engineering students in a simple possible way. As we all know that physics is not only a subject, but it is a realization of phenomena happening around us. So this book is practically based on lecture notes which were applied to teach the students in such a way that it gives them an ample opportunity to understand the nutty gritty of different engineering aspects. Every effort has been made to make all the topics presentable in a lucid manner. The authors have consulted various books, journals, libraries and visited several internet sites to prepare the book. The whole syllabus dealt with several branches of Physics applicable engineering students. The philosophy of presentation and material content in the book are based on concept based approach towards the subject. We have tried to put hard efforts in placing contents in this book such that students feel comfortable with the syllabus in an interesting way.

We sincerely express our deep sense of gratitude to Prof. G.V.K. Sinha, Chairman, Dr. P. Gadde and Prof. (Dr.) H.B. Singh, HOD, Applied Science and Humanities, Prof. (Dr.) S.V.A.V. Prasad, HOD, Department of Electronics, LIMAT, Faridabad, Mr. N. Udaya Kumar, Chairman, Udhaya Energy Photovoltaics Pvt. Ltd., Coimbatore, for their inspirational influence during our last years of teaching. This only pushed us to write the book. We are grateful to Prof. (Dr.) K.L. Moza, Venkatesh K, Dr. Ratna Sen and our other colleagues and friends for their encouragement during preparation of the manuscript. We both are also very much thankful to all our family members for their heartiest cooperation during the preparation of the book. We are also grateful to the publishers for timely publication of the book. It will be kind enough for readers for any suggestions which can easily be accepted for further improvement in next edition.

Dr. P.K. Basu

Contents

	Preface	*vii*
CHAPTER 1.	**Interference**	**1**
	1. Light Fundamentals	2
	2. Analytical Treatment for Interference	5
	3. Conditions for Interference of Light	7
	4. Types of Interference	7
	5. Theory of Interference Fringes by Division of Wavefront	7
	6. Interference by Division of Amplitude	14
	7. Newton's Rings	20
	8. Michelson's Interferometer	27
CHAPTER 2.	**Diffraction**	**33**
	1. (*a*) Fresnel's Diffraction	33
	1. (*b*) Fraunhofer's Diffraction	34
	2. Composition of Simple Harmonic Motions of Equal Amplitude and Period and Phases Increasing in Arithmetic Progression	35
	3. Fraunhofer's Diffraction Through a Single Slit	36
	4. Plane Diffraction Grating	41
	5. Applications	47
	6. Absent Spectra with a Grating	50
	7. Maximum Number of Orders Available with a Grating	50
	8. Dispersive Power of Grating	51
	9. Resolving Power	51
CHAPTER 3.	**Polarization**	**55**
	1. Polarization Fundamentals	56
	2. Doubly Refractive Crystal	60
	3. Difference between Positive and Negative Crystals	63
	4. Nicol's Prism	63
	5. Theory of Production of Plane, Circularly and Elliptically Polarized Lights	66
	6. Quarter Wave Plate	70
	7. Half Wave Plate	71
	8. Optical Activity	72
	9. Specific Rotation	73
	10. Lorentz Half Shade Polarimeter	74
	11. Bi-quartz Polarimeter	77
	12. Photoelasticity	78
CHAPTER 4.	**Laser**	**81**
	1. Coherence Properties of Laser	81
	2. Spontaneous and Stimulated Radiation	82

3. Principle of Laser Action 84
4. Ruby Laser 91
5. Helium Neon Laser (HE-NE) (Gas Laser) 92
6. Semiconductor Laser 95
7. Laser Beam Characteristics 97
8. Uses of Laser 98

CHAPTER 5. Fiber Optics 99

1. Light Ray Propagation Theory in Optical Fiber 101
2. Types of Optical Fiber 104
3. Applications 107
4. Advantages 109
5. Disadvantages 109

CHAPTER 6. Simple Harmonic Motion 111

1. Some Pavmeters of SHM 112
2. Forced Vibration 116
3. Resonance 118
4. Sharpness of Resonance 120
5. Examples 120
6. Energy of the Forced Oscillator 121
7. Damped Vibration 123
8. Energy of Damped Harmonic Oscillator 126
9. Logarithmic Decrement (λ) 127
10. Relaxation Time (τ) 128
11. Quality Factor (Q) 128

CHAPTER 7. Dielectrics 129

1. (a) Polar Dielectric 129
1. (b) Non-Polar Dielectric 130
2. Polarization 131
3. Different Relations between Electrical Parameters 132
4. Electric Susceptibility (X) 134
5. Gauss's Law in Presence of a Dielectric 136
6. Energy Stored in Dielectric 138
7. Types of Polarization 139
8. Frequency Dependence of Dielectric 141
9. Dielectric Losses 142

CHAPTER 8. Electromagnetic Theory 145

1. Basic Concepts 147
2. Maxwell's Equations 153
3. Derivation 154
4. Physical Significance of Maxwell's Equations 158
5. Plane Wave Equation in a Homogeneous Medium 160
6. Plane Wave Equation in Free Space 161
7. The Wave Equations for a Conducting Medium 162
8. Poynting's Theorem 163
9. Representation of Electromagnetic Wave 166

10. Waveguide 167
11. Co-axial Cable 168

Chapter 9. Special Theory of Relativity 169

1. Basic Concepts 169
2. Galilean Transformation (Old System) 170
3. Michelson – Morley Experiment 172
4. Failure of Galilean Transformation 176
5. Postulates of Special Theory of relativity 176
6. Lorentz Transformation (New System) 177
7. Variation of mass with velocity 185
8. Equivalence of Mass and Energy 187

Chapter 10. Nuclear Physics 189

1. Neutron Cross Section 189
2. Nuclear Fission 190
3. Chain Reaction and Reactor Criticality 192
4. Nuclear Fission Reactor 193
5. Nuclear Fusion 195
6. Interaction of radiation with matter 197
7. Absorption of gamma (γ) rays through matter 198
8. Nuclear Radiation Detectors 200

Chapter 11. Superconductivity 217

1. Introduction 217
2. Effect of Magnetic Field 219
3. Persistent current 219
4. Heat Capacity 220
5. Isotope Effect 220
3. Flux Exclusion : The Meissner Effect : 220
4. London Equations 221
5. Explanation of Meissner Effect and Flux Penetration from London Equations 223
6. Josephson Effect 225
7. Classifications of Superconductors (Based on their magnetic Field behavior) 226
8. Elements of BCS theory 227
9. Applications 228

Appendix 229

Problems with Solution 241

Questions for Assignment 289

Index 291

1

Interference

Light is a form of energy which provides our eyes the sensation of vision. The first thing any living thing in the world, whether be it plants or animals, requires is light. This form of energy is the basic building block of our cosmos. The transportation of energy can be possible in two ways, by physical transportation by some carriers or by a wave disturbance. The scientific study of the behaviour of light is called optics and covers reflection of light by a mirror or other object, refraction by a lens or prism, diffraction of light as it passes by the edge of an opaque object, and interference patterns resulting from diffraction and study of the polarization of light. Any successful theory of the nature of light must be able to explain these and other optical phenomena. The earliest scientific theories of the nature of light were proposed around the end of the 17th century in 1690, Christian Huygens proposed a theory that explained light as a wave phenomenon. However, a rival theory was offered by Sir Isaac Newton in 1704. Newton, who had discovered the visible spectrum in 1666, held that light is composed of tiny particles, or corpuscles, emitted by luminous bodies. By combining this corpuscular theory with his laws of mechanics, he was able to explain many optical phenomena. For more than 100 years, Newton's corpuscular theory of light was favored over the wave theory of light, partly because of Newton's great prestige and partly because not enough experimental evidence existed to provide an adequate basis of comparison between the two theories. Huygens (1629-1695) proposed the wave theory of light and explained the above phenomena. He considered the light waves as longitudinal waves like elastic waves. This created problem to explain the phenomenon of double refraction in polarization of light as we can not expect unsymmetrical behaviour of light about the axis of propagation. This concept of polarization was finally explained by Fresnel (1788-1829) by considering it as transverse waves. Young (1773-1829) also provided experimental evidence for wave theory of light. Finally, important experiments were done on the diffraction and interference of light by Thomas Young (1801) and A. J. Fresnel (1814–15) that could only be interpreted in terms of the wave theory. The polarization of light was still another phenomenon that could only be explained by the wave theory. Thus, in the 19th century the wave theory became the dominant theory of the nature of light.

1. LIGHT FUNDAMENTALS

Light is an electromagnetic wave. Its intensity is nothing but the amount of energy crossing across unit area perpendicular to the direction of propagation of light. When two waves having same frequency but with a constant initial phase difference travel in the same direction and in the same region, they influence each other. The final intensity in the region of superposition is different than the incident beam.

I. Wave

A wave is any disturbance flowing through space with time from the equilibrium position of generation. The transverse waves flow with crests and troughs traveling away from the center of disturbance. The longitudinal waves travel through the medium by compression and rarefaction of the medium. The transverse waves start from the center of disturbance and spread out in other regions. All particles located at crest, due to their position at that particular instant of time, will be at the same state of vibration, or in same phase. So for a wave surface phase of the wave is constant. This is true for trough positions also. The separation of two consecutive wave surfaces is of one wavelength.

II. Wavefront

According to wave theory, a point light source in an isotropic (same properties in all dirdirections) medium emits light in all directions. These waves spread out uniformly in form of ever increasing concentric spheres. The disturbance will reach simultaneously to all particles lying on the surface of a sphere with a point source at the centre.

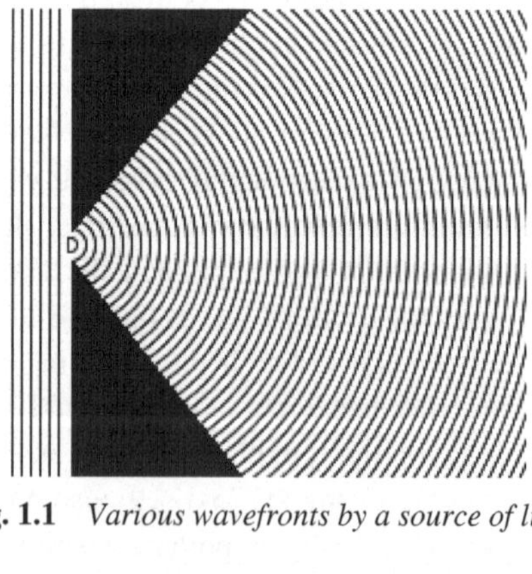

Fig. 1.1 *Various wavefronts by a source of light*

Thus, a wavefront at any time is defined as a continuous locus of all the neighbouring particles vibrating in the same phase. Fig.1.1 represents various wavefronts generated from a point light source propagating in a medium. The most distant wave surface from the source (loosely the front of the waves) is known as wavefront. A point source radiates waves uniformly in all directions and thus the wavefronts will be a family of concentric spheres. However, a linear source produces a family of coaxial cylindrical wavefronts. At very large distances from the source these waveforms can be assumed as plane wavefronts. The same logic applies to the wavefronts coming out from a narrow slit or large slits.

III. Rays

It is convenient for us to depict the waves in term of rays in place of wavefronts. Rays are basically imaginary lines drawn normal to the wavefronts. They are parallel lines for plane waves, and, radial lines for spherical waves. They represent the direction of the flow of light energy from one point to another. The intensity of the wave is the energy transferred (average) by the wave per unit time through a unit area of the surface perpendicular to the direction of propagation of the wave. Fig. 1.2 (*a*) shows spherical wavefronts coming out from a point source and Fig.1.2 (*b*) shows the presentation of rays from wavefronts.

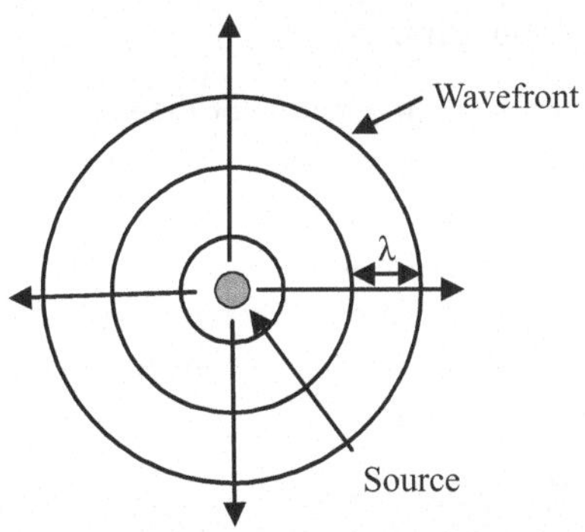

Fig. 1.2 (*a*) *Wavefronts from a point light source*

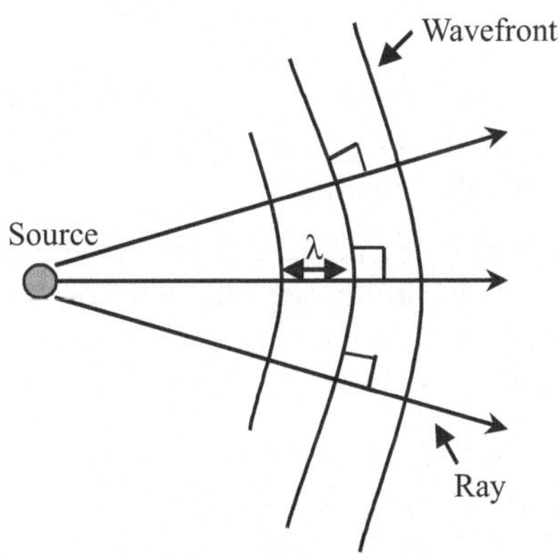

Fig. 1.2 (*b*) *Ray representation from wavefronts*

IV. Optical Medium

Optical medium is the material though which light propagates. Some materials like glass, air, water easily transmits light with very little scattering of light with their molecules or atoms, are known as 'transparent' medium. Some other materials transmit a part of light by making objects seeing through them less visible, are known as 'translucent' medium. 'Opaque' materials do not allow any light to pass through them.

While passing through any material medium the velocity of light reduces depending on the amount of scattering in the medium. This amount of scattering may be compared as the optical density of the medium. This is expressed as the 'refractive index' of the medium and can be defined as the ratio of the velocity of light in a vacuum to the velocity of light in the medium. This is a characteristic optical parameter for any material medium. The more the value of refractive index of the medium as compared to 1, the more optically dense the medium is.

V. Principle of Superposition

If two or more waves interact with each other the resultant displacement of the wave at any point at any time is determined by the principle of superposition. The principle of superposition states that the net displacement of a given point in space at any time due to two or more waves is the algebraic sum of the displacement produced at that point by all waves. Thus, if $y_1(x, t)$ and $y_2(x, t)$ are the wave functions characterizing two waves traveling in space along X-direction only then their resultant is given by

$$y(x, t) = y_1(x, t) + y_2(x, t)$$

This principle of superposition is applied to only small amplitude waveforms.

VI. Interference

The superposition of coherent waves traveling in the same direction in space results in interference. The modification in the distribution of intensity in the region of superposition is called Interference.

The resultant displacement of a particle of the medium acted upon by two or more waves simultaneously is the algebric sum of the displacements of the same particle due to individual waves in absence of the other. Mathematically, the resultant, y, of two waves, y_1 and y_2, is

$$y = y_1 \oplus y_2,$$

$\oplus$ sign means the algebric sum, *i.e.*, when y_1 and y_2 are in same direction, sign is positive, and when y_1 and y_2 are in opposite direction, sign is negative.

VII. Coherent Sources

For any interference to occur the phase difference between the interfering waves should be maintained constant. The sources which satisfy this condition of constant phase difference are known as 'Coherent Sources'. These are shown in Fig. 1.3.

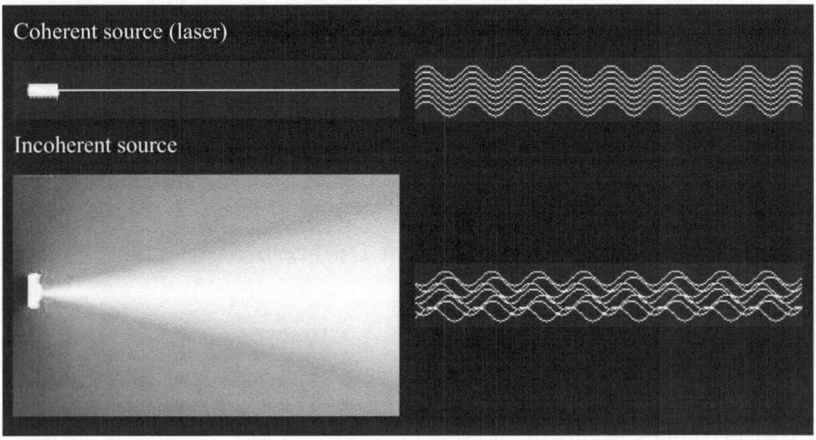

Fig. 1.3 *Examples of coherent and incoherent sources*

2. ANALYTICAL TREATMENT FOR INTERFERENCE

We consider the resultant, y, of two waves, y_1 and y_2, generated due to their interference. We also consider y_1 and y_2 are differing in phase by δ, have circular frequency ω with amplitudes a_1 and a_2 respectively, so that

$$y_1 = a_1 \sin \omega t, \text{ and } y_2 = a_2 \sin(\omega t + \delta).$$

Hence, $y = y_1 + y_2$

$\Rightarrow$

$$y = a_1 \sin \omega t + a_2 \sin(\omega t + \delta)$$
$$= [a_1 + a_2 \cos \delta] \sin \omega t + [a_2 \sin \delta] \cos \omega t$$
$$= (R \cos \theta) \sin \omega t + (R \sin \theta) \cos \omega t \qquad \text{...(1.1)}$$

where, $R \cos \theta = a_1 + a_2 \cos \delta$ and $R \sin \theta = a_2 \sin \delta$ ...(1.2)

Using eqn.(1.2) into (1.1), we have $y = R \sin(\omega t + \theta)$ ...(1.3)

Now, using eqn.(1.2), we get

$$(R \cos \theta)^2 + (R \sin \theta)^2 = [a_1 + a_2 \cos \delta]^2 + [a_2 \sin \delta]^2$$

$\Rightarrow$

$$R^2 = a_1^2 + a_2^2 + 2a_1 a_2 \cos \delta$$

The resultant intensity, I, is the square of resultant amplitude and is thus given by

$$I = R^2 = a_1^2 + a_2^2 + 2a_1 a_2 \cos \delta \qquad \text{...(1.4)}$$

Hence, the phase difference, δ, is given by

$$\delta = \left(\frac{2\pi}{\lambda}\right) \times \text{Path difference } (\Delta x) \qquad \text{...(1.5)}$$

When $\cos\delta = 1$, *i.e.*, $\delta = 2n\pi$, here $n = 0, 1, 2, \ldots$

path difference, $$\Delta x = \delta \times \frac{\lambda}{2\pi}$$

$$= 2n\pi \times \frac{\lambda}{2\pi}$$

$$= n\lambda$$

Then the maximum intensity is $I_{max} = (a_1 + a_2)^2$

When $\cos\delta = -1$, *i.e.*, $\delta = (2n+1)\pi$, here $n = 0, 1, 2, \ldots$

path difference, $$\Delta x = (2n+1)\,\pi \times \frac{\lambda}{2\pi}$$

$$= (2n+1)\lambda/2$$

Then the minimum intensity is $I_{min} = (a_1 - a_2)^2$

When the phase difference between the waves is zero, the waves are said to be in phase. If two waves of approximately equal amplitude are superimposed, when the crests coincide (that is, when they are in phase), the resulting amplitude will be equal to twice the individual amplitudes. This is called constructive interference. This is shown in Fig. 1.4.

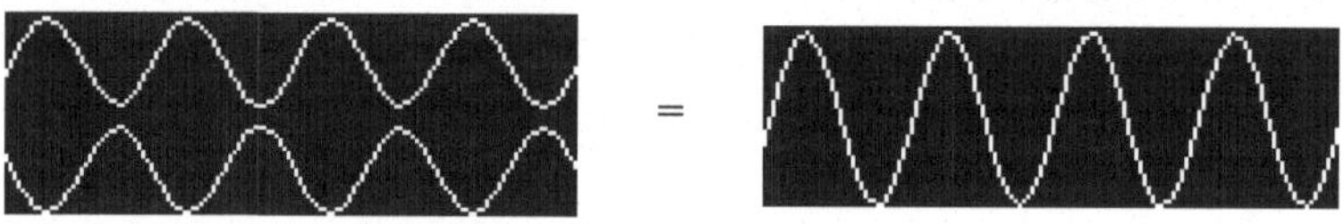

Fig. 1.4 *Constructive interference*

When the phase difference between the two waves is π, the waves are out of phase of 180° or in opposite phase. They are inverted with respect to each other everywhere. If the two waves are in antiphase (in other words, a crest and a trough coincide), the resultant amplitude will be zero. This is called destructive interference. This is shown in Fig. 1.5.

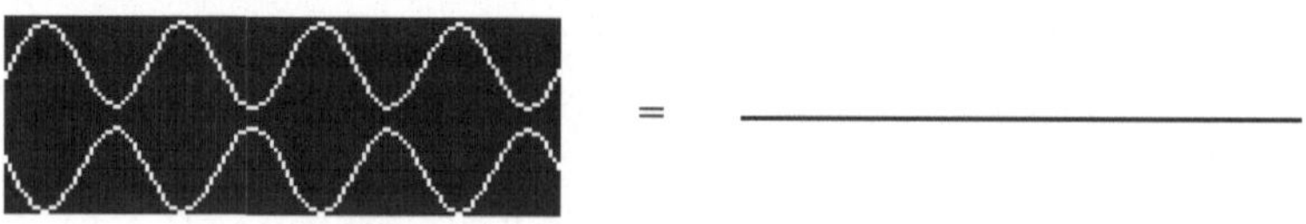

Fig. 1.5 *Destructive interference*

Interference occurs in longitudinal waves (*e.g.*, sound waves) also. The three wave parameters, amplitude, frequency and phase should fulfill their conditions to cause interference. The coherence condition should be maintained. If the waves differ in frequency, their phase difference fluctuates with time. This superposition of waves differing in frequency produces 'beats'.

3. CONDITIONS FOR INTERFERENCE OF LIGHT

(*i*) **Conditions for sustained interference**

(*a*) The sources should be coherent.

(*b*) The sources should emit continuous waves of the same wavelength and time period.

(*ii*) **Conditions for observations**

(*a*) The separation between the two sources ($2d$) should be small.

(*b*) The distance D between the two sources and the screen should be large.

(*c*) The background should be dark.

(*iii*) **Conditions for good contrast**

(*a*) The amplitudes of the interfering waves should be equal or nearly equal.

(*b*) The sources must be narrow, *i.e.,* they must be extremely small.

(*c*) The sources should be monochromatic.

4. TYPES OF INTERFERENCE

There are normally two methods for obtaining interference pattern.

I. Interference by Division of Wavefront

Here, the original wave is divided into two parts by using slits, prism or mirror, and the two wavefronts thus produced are brought together to produce the interference pattern. It requires a narrow source of light.

Examples : Fresnel Biprison, Young's Double slit experiment.

II. Interference by Division of Amplitude

In this case the original amplitude of the wave is divided into two parts by reflection or refraction and the divided parts are finally brought close together to produce the interference pattern. It requires a wide source of light.

Examples : Newton's Rings, Experiment Michelson Interferometer.

5. THEORY OF INTERFERENCE FRINGES BY DIVISION OF WAVEFRONT

I. Double Slit Interference

Thomas Young in 1801 first demonstrated experimentally the interference of waves.

Let us consider, a monochromatic source S as shown in Fig. 1.6 S_1 and S_2 are two pinholes equidistant from S. S_1 and S_2 are separated from each other by a distance of $2d$. XY is the screen placed at a distance D, parallel to S_1S_2. The point O on the screen is equidistant from S_1 and S_2. P is a point on XY at a distance of x from O. We have to find the status of fringe, either dark or bright, at P due to two monochromatic and coherent sources S_1 and S_2.

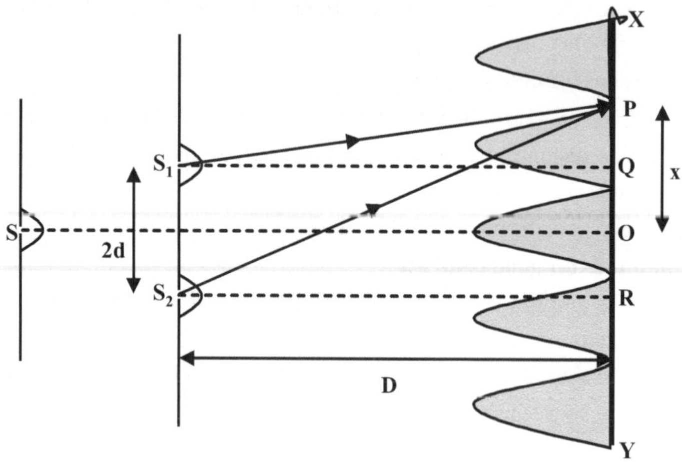

Fig. 1.6 *Schematic representation of double slit interference*

From ΔS_1QP,

$$S_1P^2 = S_1Q^2 + QP^2 = D^2 + (x-d)^2 \qquad \ldots(1.6)$$

From ΔS_2RP,

$$S_2P^2 = S_2R^2 + PR^2 = D^2 + (x+d)^2 \qquad \ldots(1.7)$$

Using eqns. (6) and (7),

$$S_2P^2 - S_1P^2 = \left[(x+d)^2 - (x-d)^2\right] = 4xd$$

$$\Rightarrow \qquad (S_2P - S_1P)(S_2P + S_1P) = 4xd \qquad \ldots(1.8)$$

However, $D >> 2d$ or $D > x$, hence, $(S_2P + S_1P) \approx 2D$ ...(1.9)

Using eqns. (1.8) and (1.9),

$$(S_2P - S_1P) = \frac{2xd}{D} \qquad \ldots(1.10)$$

(*a*) **For Bright Fringes :** The point P will be bright only when the path difference between the rays coming from S_1 and S_2 is a whole multiple of wavelength of monochromatic source, λ, *i.e.,*

$$(S_2P - S_1P) = n \times \lambda, \text{ where } n = 0, 1, 2, \ldots$$

So, using eqn.(10),

$$(S_2P - S_1P) = \frac{2xd}{D} = n\lambda \Rightarrow x = n \times \frac{D}{2d} \times \lambda$$

For n^{th} order for our convenience we can use x_n in place of x,

so, $$x_n = n \times \frac{D}{2d} \times \lambda$$

Using eqns. (1.8) and (1.9),

Here x_n is the position of nth maxima from center O on XY screen.

Hence, bright fringes will be formed at different values of x with $n = 0, 1, 2, \ldots$, and the distance between the consecutive fringes or the Fringe Width, β, is given by

$$x_2 - x_1 = \frac{2D\lambda}{2d} - \frac{D\lambda}{2d} = \frac{D\lambda}{2d}$$

$$\Rightarrow \qquad \beta = \frac{D\lambda}{2d}$$

(*b*) **For Dark Fringes :** The point P will be dark only when the path difference between the rays coming from S_1 and S_2 is an odd multiple of half wavelength of monochromatic source, λ, *i.e.*,

$$(S_2P - S_1P) = \left(\frac{2n+1}{2}\right) \times \lambda, \text{ where } n = 0, 1, 2, \ldots$$

So, using eqn.(10), we have

$$(S_2P - S_1P) = \frac{2x_n d}{D} = \left(\frac{2n+1}{2}\right)\lambda$$

$$\Rightarrow \qquad x_n = \left(\frac{2n+1}{2}\right) \times \frac{D}{2d} \times \lambda$$

Here x_n is the position of nth minima from center o on XY screen. Hence, dark fringes will be formed at different values of x_n with, and the distance between the consecutive fringes or the Fringe Width, β, is given by

$$x_2 - x_1 = \frac{5D\lambda}{4d} - \frac{3D\lambda}{4d} = \frac{D\lambda}{2d} \Rightarrow \beta = \frac{D\lambda}{2d}$$

Hence, fringe width β for both bright and dark fringes are equal and constant in an interference pattern as,

$$\beta = \frac{D\lambda}{2d} \qquad \ldots(1.11)$$

So, we found β is directly proportional to D, directly proportional to λ, and inversely proportional to d.

Application : Determination of the thickness of a thin sheet of transparent material (Glass, Mica, etc).

A transparent material of thickness t and refractive index μ is placed in the path of one of the interfering beam as shown in Fig.1.7. Here the fringe width β remains unchanged. P is a point on screen XY at a distance of x from O.

Now, light travels a distance $(S_1P - t)$ with a velocity c and a distance t with velocity c_g.

Here, $$\mu = \frac{c}{c_g} \Rightarrow c_g = \frac{c}{\mu}$$

So, time taken by light to go from S_1 to P is

$$= \frac{S_1P - t}{c} + \frac{t}{c/\mu} = \frac{S_1P + (\mu - 1)t}{c}.$$

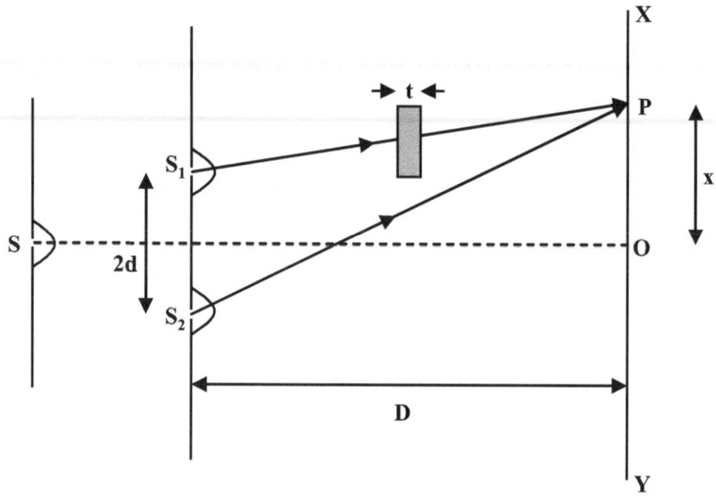

Fig. 1.7 *Double slit experiment with thin sheet of transparent material*

The path S_1 to P is equivalent to an air path of $(S_1P + (\mu - 1)t)$.

Therefore, the path difference of the two waves coming from coherent sources S_1 and S_2 at P is

$$(S_2P) - (S_1P + (\mu - 1)t) = [S_2P - S_1P] - (\mu - 1)t$$

And, $$[S_2P - S_1P] - (\mu - 1)t = \frac{2xd}{D} - (\mu - 1)t \quad \text{(Using eqn.(1.10))}$$

For n^{th} maxima, path difference $= n\lambda$

Hence, for n^{th} maxima,

$$\frac{2x_n d}{D} - (\mu - 1)t = n\lambda$$

$$\Rightarrow \quad x_n = \frac{D}{2d}\left[n\lambda + (\mu - 1)t\right]$$

This is the expression for x_n, the nth maxima position in the presence of the thin sheet. For no sheet plate, $t = 0$ and $x_n = \frac{Dn\lambda}{2d}$.

If S be the displacement of nth maxima by introducing the sheet, then

$$S = x_n\big|_{t=t} - x_n\big|_{t=0} = \frac{D}{2d}\left[n\lambda + (\mu - 1)t\right] - \frac{Dn\lambda}{2d}$$

$$\Rightarrow \qquad S = \frac{D}{2d}(\mu - 1)t$$

$$\Rightarrow \qquad t = \frac{S \times 2d}{D(\mu - 1)} \qquad \text{(here, } \mu \neq 1\text{)} \qquad \ldots(1.12)$$

So, thickness t can be calculated with

(*i*) the measurement of $2d$ and D by double slit experiment,

(*ii*) the measurement of S measured by replacing the monochromatic source by white light and cross wire is set on the white fringe. Reading is noted. Now monochromatic source and sheet is introduced, so white fringe is shifted. New position is noted. The difference between two readings is S.

By using $2d$, D and S, t can be calculated using eqn. (1.12).

II. Fresnel's Biprism (An Example of Division of Wavefront)

Fresnel used a biprism to obtain two coherent virtual sources to produce the interference pattern. In this Fresnel Biprism arrangement (as shown in Fig.1.8 below), the slit S is illuminated by a monochromatic light source. The light is passing through the biprism $ABCD$ having an angle $\angle ADC \approx 178° - 179°$. Edge D of the biprism $ABCD$ divides the incident wavefront into two parts.

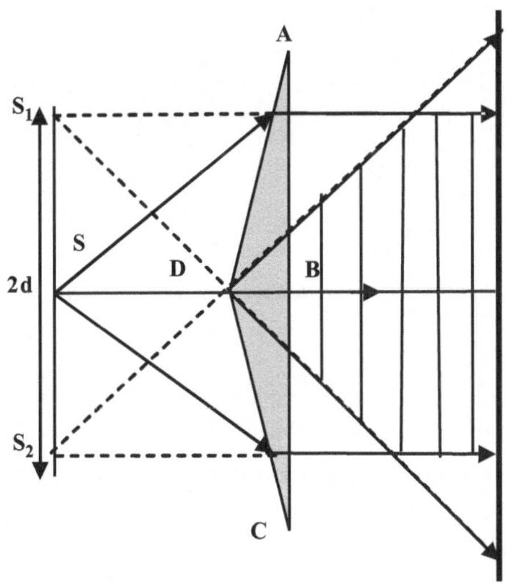

Fig. 1.8 *Wavefront division using Fresnel's Biprism*

1st part : This is passing through the upper half ABD and is deviated through small angle towards the lower half of the diagram. Thus it appears to diverge from the virtual image S_1.

2nd part : This is passing through the lower half *CBD* and is deviated through small angle towards the upper half of the diagram. It thus appears to diverge from the virtual image S_2.

As these two wavefronts arises from the same wavefront, S_1 and S_2 serve as coherent sources and the interference fringes are obtained in the overlapping region XY.

Applications: Determination of the wavelength of monochromatic light by if Fresnel's Biprism

Experimental adjustments steps

(*i*) Leveling of optical bench by spirit level is must.

(*ii*) Slit, biprism, eye piece all should be at same height.

(*iii*) Eye piece is focused on its cross wires.

(*iv*) Slit is made narrow, vertical and is illuminated by monochromatic light.

(*v*) The bi-prism is moved at right angles to the length of optical bench till overlapping region is brought in field of view.

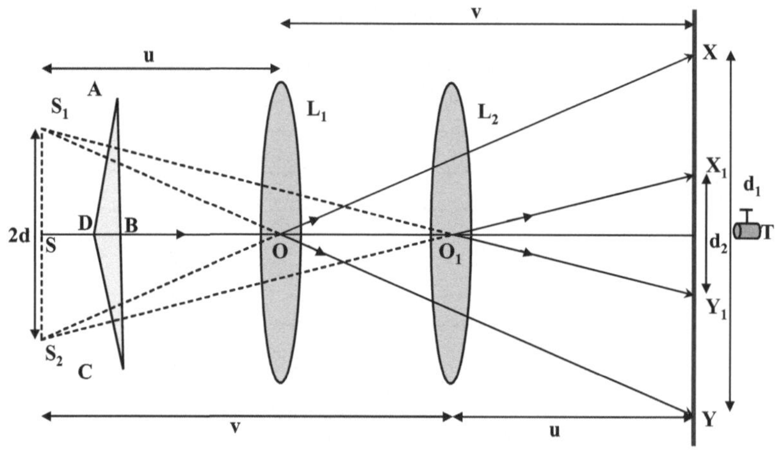

Fig. 1.9 *Schematic diagram for measurement of wavelength using Fresnel's biprism*

(*vi*) The tangent screw of slit is rotated so that the fringes appear.

(*vii*) The fringes are made clear by rotating the biprism in its own plane.

(*viii*) The eye piece is moved away from the biprism, the fringes often shift sideways causing Lateral shift. Remove lateral shift by the movement of eyepiece.

Theory

We already know from eqn.(11), $\lambda = \beta\left(\frac{2d}{D}\right)$

Procedure

(*i*) Measurement of β :

By adjusting cross wire on two consecutive bright images and measuring the distances between them we can get β.

(*ii*) Measurement of D :

By measuring directly the distance between slit and eye piece we can get D.

(*iii*) Measurement of d :

Now a lens with focal length less than one-fourth of the distance between bi-prism and eye-piece is mounted between bi-prism and eye-piece as shown in Fig. 1.9. The lens is adjusted in two positions L_1 and L_2 till sharp images of S_1 and S_2 are obtained in the field of view of eye piece. The distances d_1 and d_2 between the real images in two cases are measured.

Now, $\frac{d_1}{2d} = \frac{v}{u}$ (from similar triangles XOY and S_1OS_2)

Also, $\frac{d_2}{2d} = \frac{u}{v}$ (from similar triangles $X_1O_1Y_1$ and $S_1O_1S_2$).

So, $\frac{d_1}{2d} = \frac{1}{u/v} = 1\Big/\left(\frac{d_2}{2d}\right) \Rightarrow 2d = \sqrt{d_1 d_2}$

By using the values of $2d$, D and β in eqn.(6) we can easily get the value of λ.

White light fringes :

White light is a mixture of an infinite number of monochromatic sources of different λ. As per eqn.(1.11), each pair of monochromatic source of white light produces its own system of fringes with different β. The resultant interference

- shows zero order band and this is white,
- shows that on both sides of zero order band, few coloured fringes will be observed. Their inner and outer ends will be reddish and violet respectively,
- shows that after coloured fringes uniform illumination will be on both sides.

How to locate them

First, white light fringes are obtained and the cross wire is fixed on the zero order fringe (white one). Now, white light source is replaced by the monochromatic source. The fringe which is now coinciding with cross wire will be zero order fringe (central fringe) in monochromatic light system.

Relationship between d, x, α and μ for Fresnel's Biprism

As shown in Fig.1.10 the biprism is represented by *ABD*. For prism *ABD*, δ is the angle of deviation of beam *SA* after its incidence at the prism.

So, $$\mu = \frac{\text{Sin}\,\frac{\alpha+\delta}{2}}{\text{Sin}\left(\alpha/2\right)} \approx \frac{(\alpha+\delta)/2}{\left[\alpha/2\right]} \quad \text{[as } \alpha \text{ and } \delta \text{ are very small]}$$

$$\Rightarrow \quad \mu\alpha = \alpha + \delta$$

$$\Rightarrow \quad \delta = (\mu - 1)\alpha \qquad \ldots(1.13)$$

Now, from ΔASS_1, and using eqn (1.13), $[d = x\delta = x(\mu - 1)\alpha]$

$$\tan\delta = \frac{d}{x} \Rightarrow \delta \approx d/x \Rightarrow d = x\,\delta = x(\mu - 1)\alpha$$

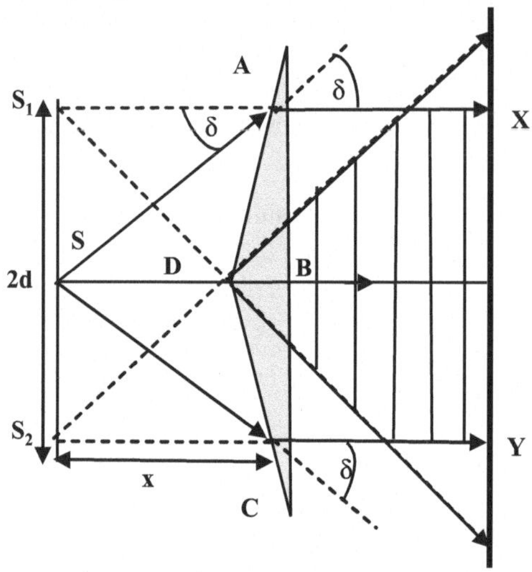

Fig. 1.10 *Schematic diagram of Feresnel's biprism to measure the virtual double slit separation*

6. INTERFERENCE BY DIVISION OF AMPLITUDE

I. Interference in Thin Films

The brilliant colours observed on soap bubble, a thin film of oil spread on the surface of water, etc., when they are exposed to an extended source of white light, are explained by Young on the basis of interference of light. However, this type of interference is completely different from interference discussed earlier. Young's double slit, Lioyd's mirror, Fresnel's Biprism are the examples where a pair of slits, a mirror, biprism divides the wave trains and thus known as division of wavefront. In case of soap bubbles etc., this type of interference is based on the interference of light of different intensities obtained generally by dividing light from same source by reflection. Hence, as intensity is divided and the divided reflected (transmitted) rays travel different path lengths and reunite (superpose) to form interference pattern. This type of interference is known as interference by the division of amplitude. A thin film should have the thickness of the order of wavelength of visible light. Otherwise we can consider the film as thick film.

II. Reflection Phase Change

Reflected light will experience a 180 degree phase change when it reflects from a medium of higher index of refraction and no phase change when it reflects from a medium of smaller refractive index. This is shown in Fig. 1.11 for an oil film floating on water. Fig. 1.12 shows reflection from a glass surface with its wave got inverted on reflection. This phase change is important in the interference which occurs in thin films, the design of anti-reflecting coatings, interference filters and thin film mirrors.

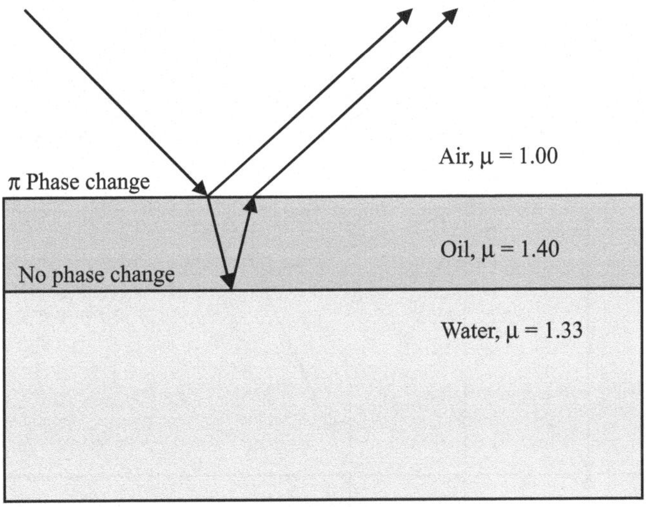

Fig. 1.11 *Example of a thin oil film*

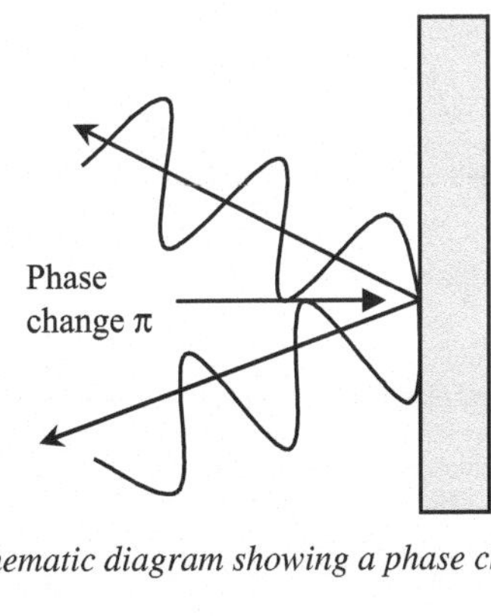

Fig. 1.12 *Schematic diagram showing a phase change of π on reflection*

III. Thin Film Interference for the Reflected Light

We here consider a uniform thickness transparent film of thickness t bounded by parallel surfaces both on top and bottom. An extended source is illuminating the film at an angle of i made along the normal to the surface.

Let GH and G_1H_1 be the two surfaces of a transparent film of thickness t and refractive index $\hat{\imath}$ as shown in Fig. 1.13 A ray of monochromatic light AB is incident on its upper surface GH. It is partly reflected along BR and refracted along BC. After one internal reflection at C, CD finally emerges out along DR_1 in air. The rays BE and DR_1 are coherent and can produce interference if they superpose on each other. Now $DR_1 \| BR$. We have to find out the effective path difference between rays BR and DR_1. We draw perpendiculars DE and BF on BR and CD respectively. We produce DC in backward direction which meets BQ produced at P. In Fig. 1.13 we have, the angle of incidence, i, is $\angle ABN$ and, the angle of refraction, r, is $\angle QBC$.

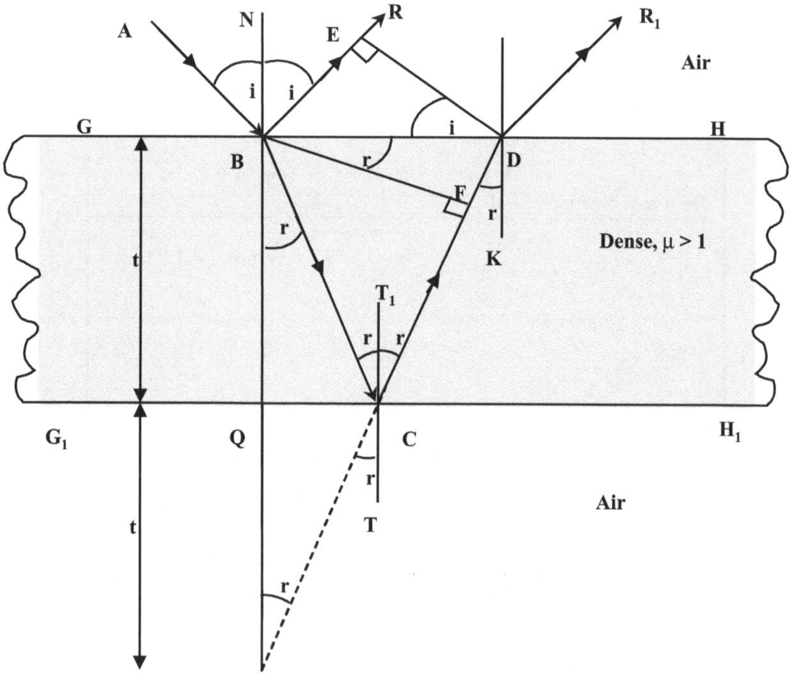

Fig. 1.13 *Diagram of thin film interference due to reflected light*

Now, $\angle BDE = 90° - \angle EBD = \angle NBE = i$,

and, $\angle QPC = 90° - \angle QCP = \angle TCP = \angle T_1CF = \angle T_1CB = \angle CBQ = r.$

So ΔPCB is isosceles.

The optical path difference Δ between two reflected rays BR and DR_1 is then

$$\Delta = \text{Path}(BC + CD)_{\text{Film}} - \text{Path}(BE)_{\text{Air}}$$

$$= \mu(BC + CD) - BE \quad \text{...(1.14)}$$

We know, $\mu = \dfrac{\sin i}{\sin r}$

$$= \frac{BE/BD}{FD/BD}\left[\frac{(from, \Delta BED)}{(from, \Delta BFD; \angle FBD = 90° - \angle FDB = \angle FDK = \angle FCT_1 = r)}\right]$$

$$\Rightarrow \quad \mu = \frac{BE}{FD} \Rightarrow BE = \mu FD \quad \text{...(1.15)}$$

Using eqns.(1.14) and (1.15), we have

$$D = \mu(BC + CD) - \mu FD$$

$$= \mu(BC + CF + FD) - \mu FD$$

$$= \mu(BC + CF) = \mu(PC + CF)$$

$= \mu PF$ [as ΔBCP is isosceles and $BC = PC$]

$= \mu\,(BP \cos r)$

[from right angled ΔBFP, cor $r = PF/BP$ and $PB = 2t$]

$$\Rightarrow \quad \Delta = 2\mu t \cos r \quad \text{...(1.16)}$$

A ray reflected at a surface backed by a denser medium suffers an abrupt change of phase by δ which is equivalent to a path difference of $\lambda/2$ [as phase difference $= \dfrac{2\pi}{\lambda} \times$ path difference].

Thus, the effective path difference Δ_{eff} becomes $2\mu t \cos r \pm \lambda/2$, *i.e.,*

$$\Delta_{\text{eff}} = 2\mu t \cos r \pm \lambda/2 \quad \text{...(1.17)}$$

Maxima : Maxima occurs when effective path difference, $\Delta_{\text{eff}} = n\lambda$. Here n is the order of interference.

So, using eqn.(1.17), $2\mu t \cos r \pm \lambda/2 = n\lambda$

$$\Rightarrow \quad 2\mu t \cos r = \left(n + \frac{1}{2}\right)\lambda \quad \text{...(1.18)}$$

Here the film will appear bright.

Minima : Minima occurs when effective path difference, $\Delta_{\text{eff}} = (2n \pm 1)\dfrac{\lambda}{2}$.

So, using eqn.(1.15), $2\mu t \cos r \pm \lambda/2 = (2n \pm 1)\dfrac{\lambda}{2}$

$$\Rightarrow \qquad 2\mu t \cos r = n\lambda \qquad \text{...(1.19)}$$

Here the film will appear dark.

Key Points

(*i*) If t is very small, *i.e.*, $t \ll \lambda$, the film will appear dark in reflected light.

(*ii*) If the incident monochromatic light is parallel, the whole film will appear uniformly dark or bright as t and r are constant..

IV. Due to Transmitted Light

We obtain two transmitted rays CT and ET_1 after simultaneous reflection and refraction as shown in Fig. 1.14. These rays originated from the same point source A. So they are in phase and produce sustain interference when combined. We draw perpendiculars CQ and EP on DE and CT respectively and produce ED to meet the produced CF at I. The effective path difference thus is given by,

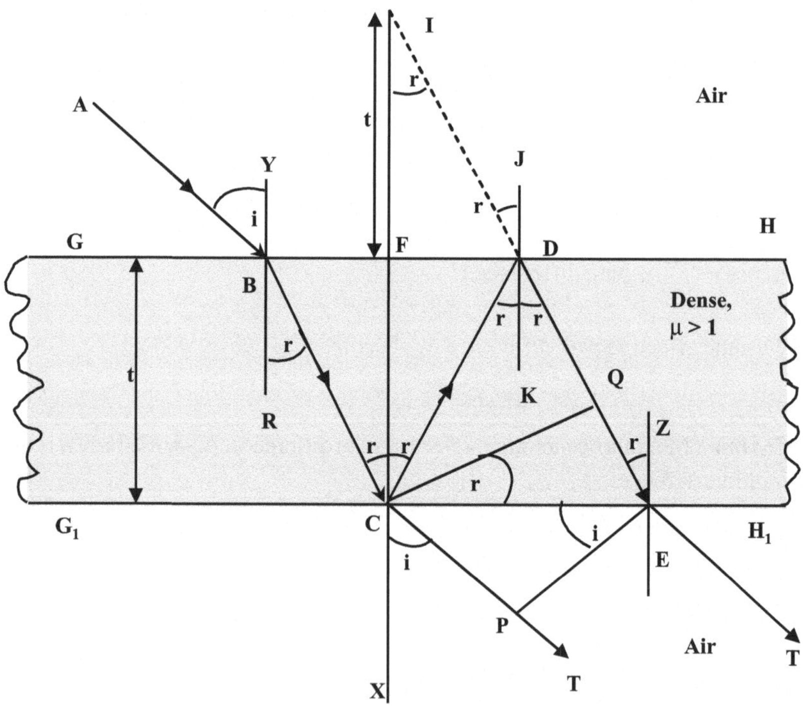

Fig. 1.14 *Diagram of thin film interference due to transmitted light*

$$\Delta = \mu\,(CD + DE) - PC \qquad \text{...(1.20)}$$

Extending ED & CF in backward direction, they meet at a point I.

Now, $$\mu = \frac{\sin i}{\sin r}$$

$$= \frac{PC/EC}{QC/EC}\left[\frac{(from, \Delta PCE; \angle PEC = 90° - \angle ECP = \angle PCX = \angle ABY = i)}{(from, \Delta QCE; \angle QCE = 90° - \angle QEC = \angle DEZ = \angle FCB = \angle RBC = r)}\right]$$

$$\Rightarrow \quad \mu = \frac{PC}{QE}$$

$$\Rightarrow \quad PC = \mu QE \qquad \text{...(1.21)}$$

Also, $\angle DIF = 90° - \angle FDI = \angle JDI = \angle KDQ = \angle KDC = \angle DCF = r$.

So, ΔDIC is isosceles.

Using eqns.(1.20) and (1.21), we have

$$D = \mu(CD + DQ + QE) - \mu QE = \mu(DI + DQ) = \mu QI$$

As $QI = CI.\cos r = 2t.\cos r$

$$\Rightarrow \quad \Delta = \mu.2t.\cos r \qquad \text{...(1.22)}$$

Here inside the film reflection at different points takes place at the surface backed by rarer medium, thus no abrupt change of μ takes place.

Maxima : Maxima occurs when effective path difference, $\Delta = n\lambda$.

So, using eqn.(1.22), $2\mu t \cos r = n\lambda$ ...(1.23)

Here the film will appear bright.

Minima : Minima occurs when effective path difference, $\Delta = (2n \pm 1)\frac{\lambda}{2}$.

So, using eqn.(1.22), $2\mu t \cos r = (2n \pm 1)\frac{\lambda}{2}$ ...(1.24)

Here the film will appear dark.

V. Colours of Thin Films

When a thin film is exposed to a white light beautiful colours are observed. The incident white light splits up by reflection at the top and bottom of the film. The splitted rays interfere and the interference of these rays is responsible for colours. The appearance of a reflected light depends on μ, t and r as the path difference is depending on them. For white light and a single film, t and r are constant, and μ varies with λ. At a particular position of the eye the interfering rays of only a certain λ will have a path difference satisfying the condition of bright fringe. Hence, only such λ will be present there. Other neighbouring λ will be present with diminished intensity. The colours for which the condition of minima is satisfied are absent. Similarly if the same point of the film is observed with different positions or different points of the film are observed with eye at the same position, a different set of colours is observed each time.

VI. Necessity of an Extended Source

For division by wavefront, the source used is narrow. These fringes can be obtained on the screen, or can be viewed through a eyepiece. However, in case of thin films, the narrow source put constraints on the visibility of the film. This is shown in Fig. 1.15. When the film is exposed by an extended source the portion A is seen by the reflected light originally coming from some point S_1 and B is seen coming from S_2. So each point of the film is seen by the reflected light originally coming from some point of the source. So the rays from different parts of the source are reflected from different points of the film so as to enter the eye placed in a suitable position. Therefore, we can see the entire film simultaneously.

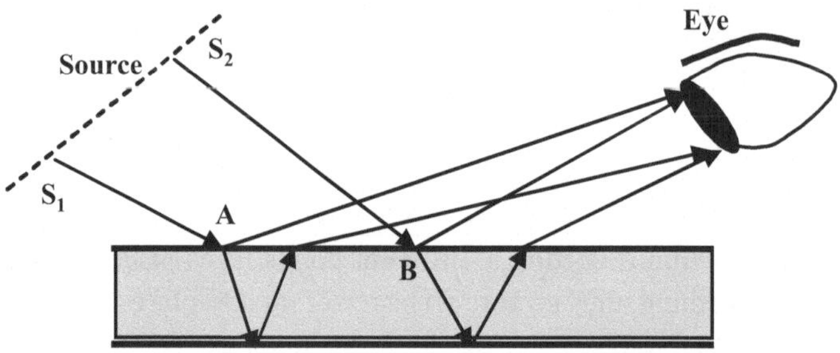

Fig. 1.15 *Diagram of light from an extended source reflected from a thin film*

7. NEWTON'S RINGS

The Newton's Ring method is an ideal example of interference by division of amplitude. Although Newton first observed these rings, it was again Young who explained the formation of fringes later on.

When a plano-convex lens having long focal length with its convex surface is placed on a plane glass plate, an air film of gradually increasing thickness is formed between the two. The thickness of the film at the point of contact is zero. If a monochromatic light is allowed to fall normally on this arrangement and the film is viewed in reflected light, alternate concentric dark and bright fringes around the point of contact are seen. The experimental arrangement is shown in Fig.1.16 (*a*).

Here, L is the plano-convex lens, P is the plane glass plate, C is the point of contact of L and P, G is the plane glass plate making an angle of 45° with vertical and M is the microscope. G reflects normally a part of incident light from sodium lamp towards the air film enclosed between L and P. A part of incident light is reflected by the curved surface of L and a part is transmitted which is reflected back from the plane surface of the plate. These two reflected rays interfere and give rise to an interference pattern in the form of concentric circular rings as in Fig. 1.16 (*b*).

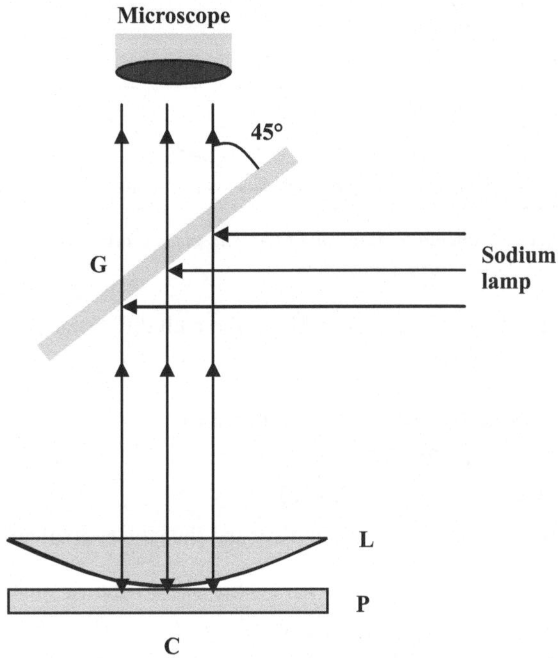

Fig. 1.16 (*a*) *Experimental arrangement of Newton's ring set up*

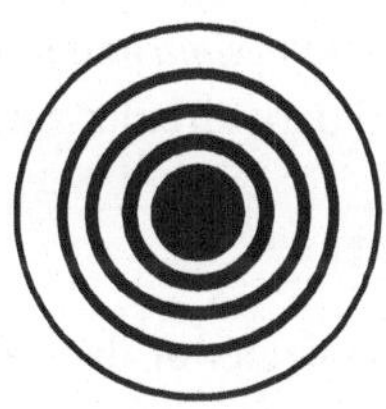

Fig. 1.16 (*b*) *Observed Newton's rings*

Explanation

As shown in Fig. 1.17 light ray *AB* falls on *L*. A part of it is reflected at *C* which goes out as ray 1 without any phase reversal.

Other part is refracted along *CD*. At *D* it is again reflected and goes out as ray 2 with a phase reversal of π. The reflected rays 1 and 2 produce interference fringes and hence fulfill the condition of interference.

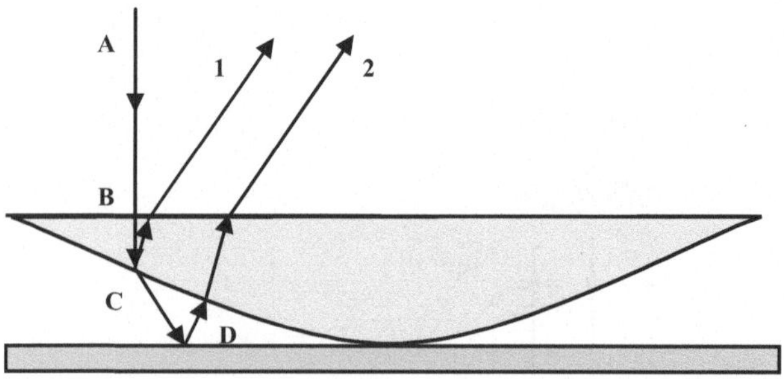

Fig. 1.17 *Schematic diagram of the lens-glass plate assembly for normal incidence*

As the rings are observed in reflected light the path difference is

$$\Delta = 2\mu t \cos r + \lambda/2$$

For air film, $\mu = 1$ and for normal incidence, $r = 0$.

$\Rightarrow$ $$\Delta = 2t + \lambda/2$$

At point of contact, $t = 0 \Rightarrow \Delta = \lambda/2$.

This is the condition of minimum intensity and thus the **central spot is dark.**

For n^{th} maximum, $\Rightarrow$ $2t + \lambda/2 = n\lambda$. This shows that a maximum of a particular order n will occur for a constant value of t. Since locus of all those points which are satisfying this condition produce circle. That's why the interference pattern is observed in the form of concentric circular rings.

For different values of t different *maxima* and *minima* will occur, which results into various concentric bright and dark circular rings

Theory (For Reflected Light)

Here we have to calculate the diameters of dark and bright rings. As shown in Fig.1.18 the lens *LOL* is placed on a glass plate *AB*. *LOL* is a part of spherical surface centered at *C* with radius of curvature *R*. *r* be the radius of Newton's ring corresponding to the air film thickness *t*. So, for the nth order bright ring, we have

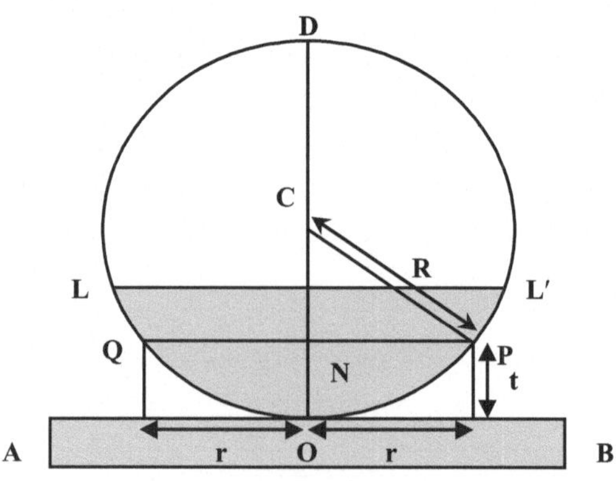

Fig. 1.18 *Schematic structure of lens-glass plate assembly to determine t*

$$2t + \lambda/2 = n\lambda$$

$\Rightarrow$ $$2t = (2n-1)\lambda/2 \qquad \text{...(1.25)} \quad [\text{Here}, n = 1, 2, 3, \ldots]$$

Also for the nth order dark ring

$\Rightarrow$ $$2t = n\lambda \qquad \text{...(1.26)} \quad [\text{Here}, n = 0, 1, 2, 3, \ldots]$$

From Fig.1.18 using the property of circle, $NP \times NQ = NO \times ND$ (as two chords OD and PQ intersect at N, OD is diameter)

$\Rightarrow$ $$r \times r = t \times (2R - t) = 2Rt - t^2 \approx 2Rt \qquad \text{(as } t \text{ is very small)}$$

$\Rightarrow$ $$t = \frac{r^2}{2R} \qquad \text{...(1.27)}$$

For Bright Ring

Using eqns.(1.25) and (1.27), $$2 \times \frac{r^2}{2R} = (2n-1)\lambda/2$$

$\Rightarrow$ $$r^2 = (2n-1) \times \frac{\lambda R}{2}$$

$\Rightarrow$ $$(d)^2 = 2(2n-1)\lambda R \qquad (\text{Diameter}, d = 2r)$$

$\Rightarrow$ $$d = \sqrt{2\lambda R\,(2n-1)}$$

So, $$d \propto \sqrt{2n-1}.$$

Hence, diameters of bright rings are proportional to the square root of odd natural numbers.

For Dark Ring

Using eqns.(1.24) and (1.27), $$2 \times \frac{r^2}{2R} = n\lambda \Rightarrow d = 2\sqrt{n\lambda R}.$$

So, $$d \propto \sqrt{n}$$

Hence, diameters of dark rings are proportional to the square root of natural numbers.

Key Points

(*i*) **Dark central spot :** At O, $t = 0$, and this is satisfied only by eqn.(1.26) for dark fringe with zeroth order fringe. This indicates that for reflected light the central zeroth order fringe is dark.

(*ii*) **Fringes of equal thickness :** Eqns.(1.23) and (1.24) also indicates that occurrence of alternate dark and bright fringes is due to the variation of t. Each minimum or maximum is a locus of constant film thickness. So these fringes are the fringes of equal thickness.

(*iii*) **Circular fringes :** The circular ring wedge of air film here bulging from the center of contact to the outer surface can be considered like an air-wedge rotated around an axis passing through the point of contact. The locus of these points for same thickness form a circle centered around O' and as the t is uniform at all points of any circle, the fringes are therefore circular.

(*iv*) As the order of pattern increases the pattern seems to be shrinking, *i.e.,* the consecutive difference between maxima or minima is decreasing as order n increases.

I. Applications

(*a*) **Determination of unknown wavelength (λ)**

We have for diameter of nth dark ring, D_n, as

$$D_n^2 = 4n\lambda R \qquad \ldots(1.28)$$

Also for diameter of $(n + p)$th dark ring, D_{n+p}, as

$$D_{n+p}^2 = 4(n + p)\lambda R \qquad \ldots(1.29)$$

From eqns.(1.28) and (1.29), $D_{n+p}^2 - D_n^2 = 4(n + p)\lambda R - 4n\lambda R = 4p\lambda R$

$$\Rightarrow \qquad \lambda = \frac{D_{n+p}^2 - D_n^2}{4pR}$$

This gives the value of unknown λ using Newton's Ring Experiment.

The diameters of successive dark rings are measured using a low power traveling microscope and a graph between D_n^2 with number of ring, n is drawn as shown in Fig. 1.19. The plot is a straight line with slope $4R\lambda$.

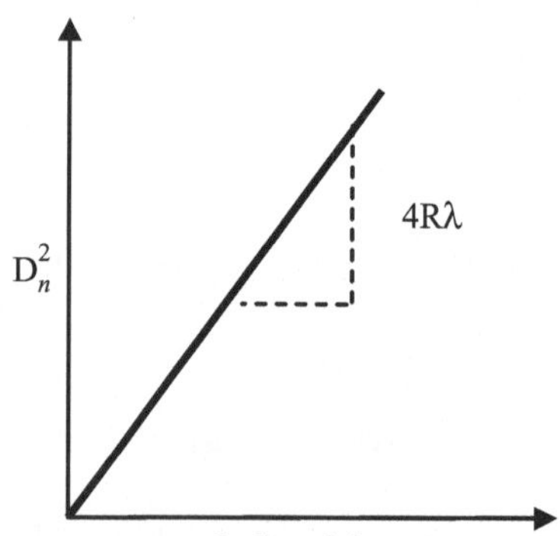

Fig. 1.19 *Plot of ring diameter versus order of rings*

Thus we have, $\lambda = \dfrac{slope}{4R}$

Using a spherometer, R is obtained and thus the value of λ is computed.

(*b*) **Determination of refractive index (μ) of a liquid**

When the air gap between lens and glass plate is replaced by a liquid of refractive index μ, the condition of interference for dark fringes becomes

$$2\mu t \cos r = m\lambda, m = 0, 1, 2, \ldots$$

For normal incidence ($r = 0°$) and so

$$\Rightarrow \qquad 2\mu t = m\lambda \quad \Rightarrow \quad t = \frac{m\lambda}{2\mu}$$

If the radius of mth dark fringe is r_m and its diameter, D_m, then,

$$r_m^2 = 2Rt = 2R \times \frac{m\lambda}{2\mu} = \frac{mR\lambda}{\mu}$$

$$\Rightarrow \qquad (D_m)_{liq}^2 = \frac{4mR\lambda}{\mu}$$

Thus relatively the interference pattern will shrink due to the presence of the liquid.

For $(m+p)^{\text{th}}$ dark fringe with diameter, D_{m+p}, then,

$$\left(D_{m+p}\right)_{liq}^2 = \frac{4(m+p)R\lambda}{\mu}$$

So,

$$\left(D_{m+p}\right)_{liq}^2 - (D_m)_{liq}^2 = \frac{4(m+p)R\lambda}{\mu} - \frac{4mR\lambda}{\mu} = \frac{4pR\lambda}{\mu}$$

$$\Rightarrow \qquad \lambda = \mu \times \frac{\left(D_{m+p}\right)_{liq}^2 - (D_m)_{liq}^2}{4pR} \qquad \ldots(1.30)$$

For air interface, $\mu = 1$, so eqn.(1.30) gives

$$\left(D_{m+p}\right)_{air}^2 - (D_m)_{air}^2 = 4p\lambda R \qquad \ldots(1.31)$$

So, by dividing eqn.(1.30) by eqn.(1.31), we have

$$\frac{\lambda}{4p\lambda R} = \frac{\mu \times \left[\dfrac{\left(D_{m+p}\right)_{liq}^2 - (D_m)_{liq}^2}{4pR}\right]}{\left(D_{m+p}\right)_{air}^2 - (D_m)_{air}^2}$$

$$\Rightarrow \qquad \mu = \frac{\left[\left(D_{m+p}\right)_{air}^2 - (D_m)_{air}^2\right]}{\left[\left(D_{m+p}\right)_{liq}^2 - (D_m)_{liq}^2\right]}$$

Thus by measuring diameter of interference rings of different order in presence of liquid and in absence of liquid, we can determine the refractive index of the liquid by using the above expression.

(*c*) **Newton's Rings formed by two curved surfaces**

Case-I: The lenses are plano-convex and plano-concave

Here a plano-convex lens of radius of curvature R_1 is placed on the plano-concave lens of radius of curvature R_2 at the point of contact O as shown in Fig. 1.20. A thin air film is formed between them and under exposure of light it will produce concentric circular rings. This can be viewed and measured in a microscope. Let nth dark ring passes through P and its radius is r_n. So at P, the film thickness is t.

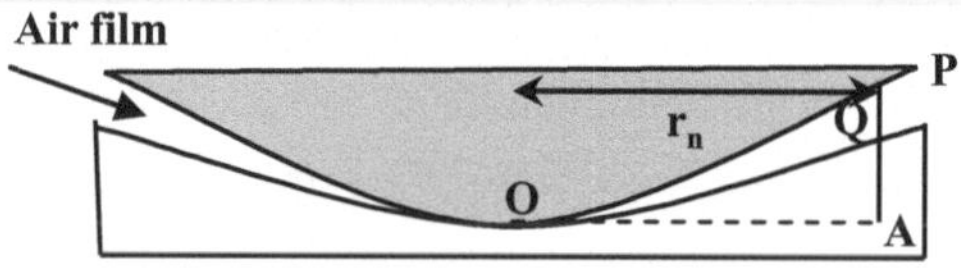

Fig. 1.20 *Schematic diagram of a plano-convex lens inside a plano-concave lens*

Hence, $$PQ = t = PA - QA \qquad ...(1.32)$$

However, from the properties of circles as in eqn. (1.27),

$$PA = \frac{r_n^2}{2R_1} \;\&\; QA = \frac{r_n^2}{2R_2} \qquad ...(1.33)$$

Hence, from eqns.(1.32) and (1.33), $t = \dfrac{r_n^2}{2}\left[\dfrac{1}{R_1} - \dfrac{1}{R_2}\right]$.

For air film with normal incidence and reflected light, path difference is $2t - \lambda/2$.

So, for dark rings, $2t = n\lambda$, where, $n = 1, 2, \ldots$

$$\Rightarrow \quad r_n^2\left[\frac{1}{R_1} - \frac{1}{R_2}\right] = n\lambda$$

For bright rings, $2t = \left(\dfrac{2n-1}{2}\right)\lambda$, where $n = 1, 2, \ldots$

$$\Rightarrow \quad r_n^2\left[\frac{1}{R_1} - \frac{1}{R_2}\right] = \left(\frac{2n-1}{2}\right)\lambda$$

Case-II: Both the lenses are plano-convex

Here a plano-convex lens of radius of curvature R_1 is placed on the plano-convex lens of radius of curvature R_2 at the point of contact O as shown in Fig. 1.21. A thin air film is formed between them and under exposure of light it will produce concentric circular rings. This can be viewed and measured in a microscope.

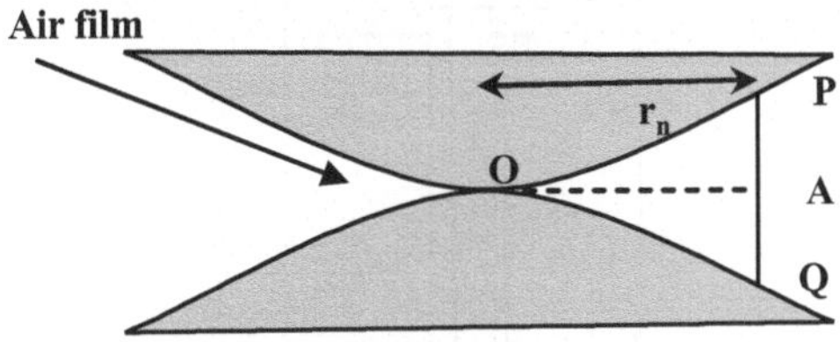

Fig. 1.21 *Schematic diagram of two a planoconvex lenses in contact*

Let n^{th} dark ring passes through P and its radius is r_n. So at P, the film thickness is t.

Hence, $$PQ = t = PA + QA \qquad \ldots(1.34)$$

However, from the properties of circles as in eqn.(1.27),

$$\text{PA} = \frac{r_n^2}{2R_1} \;\&\; QA = \frac{r_n^2}{2R_2} \qquad \ldots(1.35)$$

Hence, from eqns.(1.34) and (1.35), $t = \frac{r_n^2}{2}\left[\frac{1}{R_1} + \frac{1}{R_2}\right]$.

For air film with normal incidence and reflected light, path difference is $2t - \lambda/2$.

So, for dark rings, $2t = n\lambda$, where $n = 1, 2, \ldots$

$$\Rightarrow \quad r_n^2\left[\frac{1}{R_1} + \frac{1}{R_2}\right] = n\lambda$$

For bright rings, $2t = \left(\frac{2n-1}{2}\right)\lambda$, where $n = 1, 2, \ldots$

$$\Rightarrow \quad r_n^2\left[\frac{1}{R_1} + \frac{1}{R_2}\right] = \left(\frac{2n-1}{2}\right)\lambda$$

8. MICHELSON'S INTERFEROMETER

Here, the amplitude of the light beam from a source is divided in two parts of equal intensities by partial reflection and transmission. These beams then brought together to produce interference fringes.

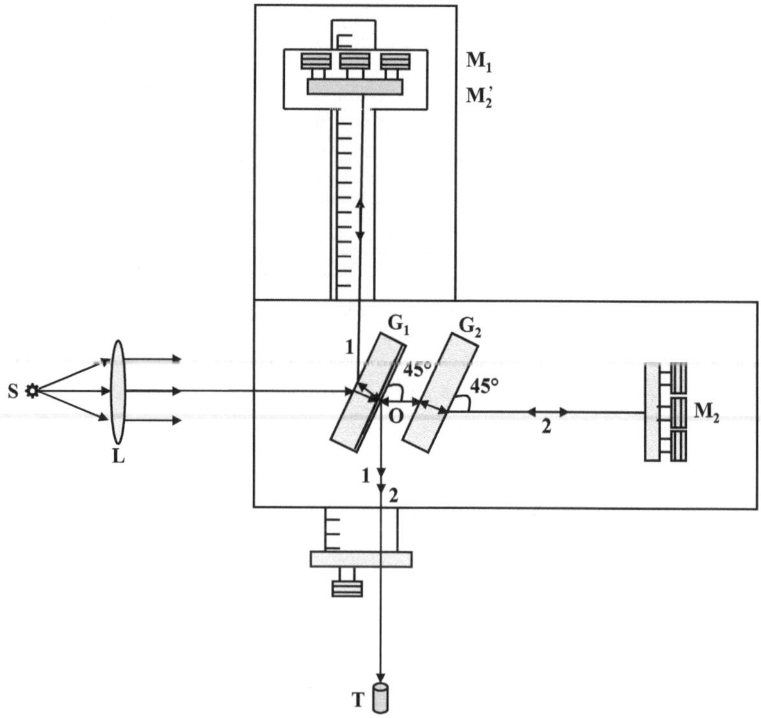

Fig. 1.22 *Construction of Michelson Interferometer*

Construction

As shown in Fig. 1.22 the apparatus consists of two optically plane, highly polished plane mirrors M_1 and M_2, which are at right angles to each other. G_1 and G_2 are two optically flat glass plates placed parallel to each other. G_1 and G_2 are also inclined at an angle of 45° with M_1 and M_2. The face of G_1 towards G_2 is semi-silvered. M_1 is mounted on a carriage whose movement is controlled by a very fine micrometer screw (up to 10^{-5} cm). M_1 and M_2 are provided with three leveling screws at their back. With these screws mirrors can be tilted about horizontal and vertical axes so that they can be exactly perpendicular to each other. The telescope T receives the reflected light from M_1 and M_2.

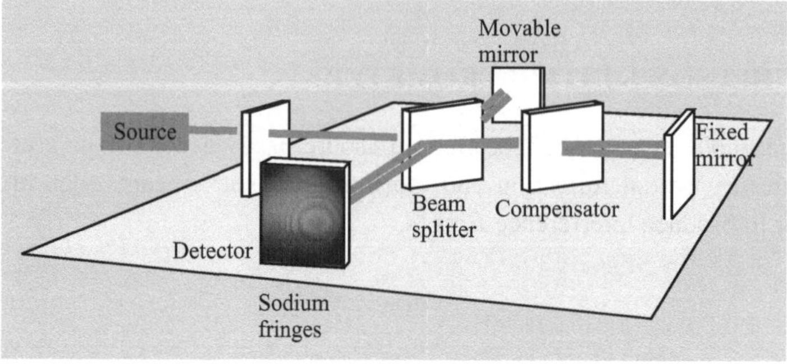

Fig. 1.23 *Experimental set up for Michelson Interferometer*

Working

Light from a monochromatic source S after being made parallel by a collimating lens L falls on G_1. It is then divided into two parts. First part ray 1 travels towards M_1 and second part ray 2 travels towards M_2. They then reflected back along their original paths meet again at G_1 and enter a short focus telescope T. These two rays produce interference fringes in the field of view of T. A desired path difference can be introduced between the two reflected rays by moving M_1.

Ray 1 passes through G_1 twice whereas 2 does not even once. So optical paths traveled by OM_1 and OM_2 are not equal. Hence a glass plate G_2 is introduced (same material and thickness) in path of 2 to compensate this difference. So G_2 is known as compensating plate. This is shown in Fig. 1.23.

Looking in the direction of M_1 from eye we observe M_1 and a virtual image M_2 of M_2 formed in G_1. Thus the two interfering beams come by reflection from M_1 and the other which is reflected from M_2 as if it had been reflected from M_2. Hence Michelson Interferometer is optically equivalent to an air film between M_1 and M_2. The interference fringes here may be straight, circular, parabolic, etc., depending upon path difference and the angle between M_1 and M_2.

I. Circular Fringes Condition

Here, M_2 is perpendicular to M_1, or, M_1 is parallel to M_2. Thus M_2M_1, or, $M_1 \| M_2$.

Here an air film of constant thickness is enclosed between the mirrors M_1 and M_2, or M_1 and M_2. The air film gives reflected beam to interfere. The path difference depends upon

(*a*) separation between M_1 and M_2,

(*b*) angle subtended on the eye,

(*c*) inclination between M_1 and M_2.

For maxima, $2\mu t \cos r + \lambda/2 = n\lambda$

Here, we find that each fringe is a locus of constant r. Since the loci of constant r are concentric circles, circular fringes are formed at infinity. Hence they are observed through a telescope focused at infinity.

II. Localized Fringes Condition

Here, M_2, not exactly perpendicular to M_1, or, not exactly M_2 not exactly parallel to M_1.

Here a wedge shaped air film of variable thickness is enclosed between them. The shape of the fringes depends on thickness of film and the angle of incidence. Their shapes are shown in Fig. 1.24. These fringes have curved shape towards decreasing thickness of air film enclosed between M_1 and M_2. If M_1 and M_2 images cross each other as shown in the middle diagram of Fig. 1.24, the pattern comes out in the form of straight lines.

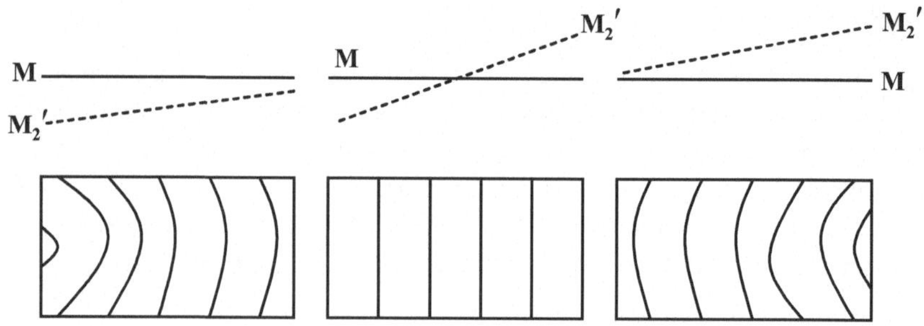

Fig. 1.24 *Localized fringes of different shapes for different M_1, M_2 positions*

These fringes are visible up to path difference comparatively much smaller that that in case of circular fringes.

III. Localized White Light Fringes

Using white light source coloured and curved localized fringes are obtained. The fringes of zero thickness are perfectly dark and straight. Other fringes are coloured due to overlapping of various colours.

IV. Applications

(*a*) **Determination of wavelength of monochromatic light :** We first set the instrument (Interferometer) for circular fringes. If M_1 and M_2 are at equal distance from the glass plate *G*, the field of view is completely dark.

For normal incidence ($r = 0°$, $\cos r = 1$), we have

$$2\mu t + \lambda/2 = n\lambda$$

Now M_1 is moved $\lambda/2$ away from M_2' then an additional path difference λ is introduced and hence, $(n + 1)^{\text{th}}$ bright spot appears at the centre of the field. Let *N* be the number of fringes that cross the centre of field when M_1 is moved away from initial position x_1 to final x_2, then we have

$$N\left(\frac{\lambda}{2}\right) = x_2 - x_1 \quad \Rightarrow \quad \lambda = \frac{2(x_2 - x_1)}{N}$$

Hence, is measured and also *N* is counted, using these λ is finally calculated.

(*b*) **Determination of difference in wavelengths of light :** Here, a source have two different wavelengths λ_1 and λ_2, lying very close to each other. So the instrument is now adjusted for circular fringes and λ_1 and λ_2 form their separate, but, practically coincident fringe patterns. Now, as M_1 is moved slowly the two patterns separate slowly and when thickness of air film is such that the dark fringe of λ_1 falls on bright fringe of λ_2 the result is maximum indistinctness. Now M_1 is further moved by a distance *x* to have next indistinct position. Here, if *n* fringes of λ_1 appear at the centre, $(n + 1)$ fringes λ_2 should appear at the centre.

Hence, for λ_1, $x = \frac{n\lambda_1}{2} \Rightarrow n = \frac{2x}{\lambda_1}$

and, for λ_2, $x = \frac{(n+1)\lambda_2}{2} \Rightarrow n+1 = \frac{2x}{\lambda_2}$

Then, $\frac{2x}{\lambda_2} - \frac{2x}{\lambda_1} = (n+1) - n$

$$\Rightarrow \quad \lambda_1 - \lambda_2 = \frac{\lambda_1\lambda_2}{2x} = \frac{\lambda_{av}^2}{2x}$$

Hence by knowing the value of x, we can get the value of $(\lambda_1 - \lambda_2)$.

(*c*) **Determination of thickness *t* of thin plate :** We first set the instrument for localized fringes of white light. The crosswire are set at central fringe. Now the plate is inserted into the path of one of interfering rays. So the path is increased by $2(\mu - 1)t$ and fringes shift their position by x. M_1 is now moved by x till the central fringe coincides with cross wire.

So, $2x = 2(\mu - 1)t \Rightarrow t = \frac{x}{\mu - 1}$

Thus using its thickness t can be evaluated.

2

Diffraction

When light falls on obstacles or small apertures whose size d is comparable with the wavelength of incident light, λ, there is a departure from straight lines propagation. There is a bending of light round the corners of obstacles and light enters in the region of geometrical shadow. This bending of light is called the Diffraction.

A plane wave cannot bend if $\lambda << d$, bending starts when $\lambda \approx d$. According to Huygens's theory, every point of a light source is a source of secondary wavelets spreading out radially as shown in Fig. 2.1. The diffraction is the result of the interference of these secondary wavelets at the point of observation. The diffraction pattern consists of interference maxima and minima and is observed on a screen. We can easily say that diffraction of light is basically the consequences of superposition of light waves from coherent sources on the same wavefront after the wavefront is changed by the small obstacle.

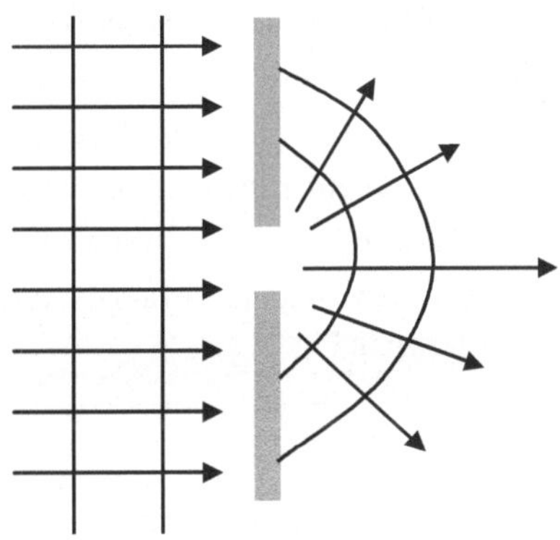

Fig. 2.1 *A plane wavefront incident on an obstacle*

Diffraction phenomenon is divided into two categories.

1(A) FRESNEL'S DIFFRACTION

In this diffraction:

- Source and screen are placed at finite distances from the aperture of obstacles having sharp edges.
- No lenses are used for making rays parallel or convergent.
- Incident wavefronts are either spherical or cylindrical.
- The distances are important, not the angular inclination.
- This class of diffraction is studied only with approximation.
- Examples are diffraction through circular aperture, by circular obstacle, by straight edge, etc.

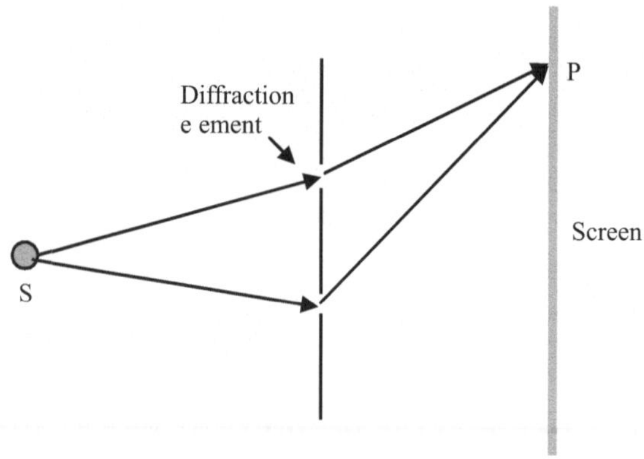

Fig. 2.2 *Fresnel's class of diffraction*

The experimental arrangement for Fresnel's diffraction is shown in Fig. 2.2

1 (B) FRAUNHOFER'S DIFFRACTION

In this diffraction:

- Source, screen and telescope are placed at infinity or near infinity.
- Lenses are used.
- Wavefront incident is plane.
- Here, angular inclinations are important and not the distances.
- This can be studied with greater accuracy.
- The experimental arrangement for Fraunhofer double slit diffraction is shown in Fig. 2.3

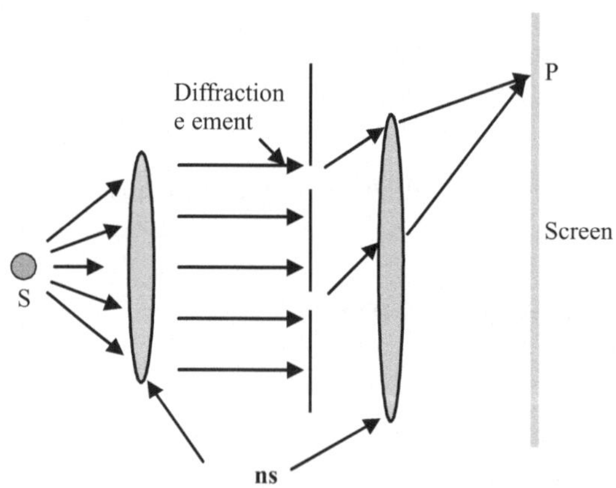

Fig. 2.3 *Fraunhofer double slit diffraction*

- Examples are diffraction through single slit, double slit, transmission grating, etc.

2. COMPOSITION OF SIMPLE HARMONIC MOTIONS OF EQUAL AMPLITUDE AND PERIOD AND PHASES INCREASING IN ARITHMETIC PROGRESSION

Let us consider a secondary wavefront, which consists of n nos. of vibration affecting a particle simultaneously each having amplitude of vibration a. δ be the common phase difference between successive vibrations as shown in Fig.4. We have to find the resultant amplitude R and the resultant phase change φ of the particle. A polygon (having amplitudes of each side equal to a) is constructed in Fig. 2.4. The closing side gives R and the angle made by it with the first vibration is φ.

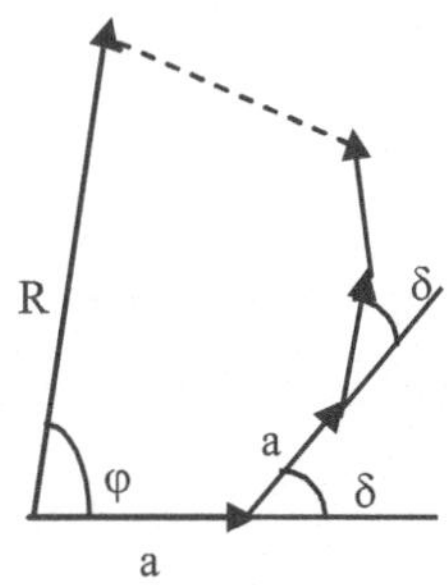

Fig. 2.4 *A polygon with same amplitude and common phase difference*

Now, resolving the amplitudes for their horizontal components, we get

$$R \cos \phi = a + a \cos \delta + a \cos 2\delta + \ldots + a \cos (n-1)\,\delta$$

$$\Rightarrow 2 \sin\left(\frac{\delta}{2}\right) R \cos \phi = a\left[2\sin\left(\frac{\delta}{2}\right) + 2\sin\left(\frac{\delta}{2}\right)\cos\delta + 2\sin\left(\frac{\delta}{2}\right)\cos 2\delta + \ldots + 2\sin\left(\frac{\delta}{2}\right)\cos(n-1)\,\delta\right]$$

$$= a\left[2\sin\left(\frac{\delta}{2}\right) + \left\{\sin\left(\frac{3\delta}{2}\right) - \sin\left(\frac{\delta}{2}\right)\right\} + \left\{\sin\left(\frac{5\delta}{2}\right) - \sin\left(\frac{3\delta}{2}\right)\right\} + \ldots + \left\{\sin\left(n-\frac{1}{2}\right)\delta - \sin\left(n-\frac{3}{2}\right)\delta\right\}\right.$$

(2cos A sin B = sin (A+B) – sin (A–B))

$$= a\left[\sin\left(\frac{\delta}{2}\right) + \sin\left(n-\frac{1}{2}\right)\delta\right] = 2a \sin\left[\frac{1}{2} + n - \frac{1}{2}\right]\frac{\delta}{2}\,.\cos\left[n - \frac{1}{2} - \frac{1}{2}\right]\frac{\delta}{2} = 2a \sin\frac{n\delta}{2}\,.\cos\,[n-1]\,\frac{\delta}{2}$$

$$\left(\sin A + \sin B = 2\sin\left(\frac{A+B}{2}\right)\cos\left(\frac{B-A}{2}\right)\right)$$

$$\Rightarrow \quad R\cos\phi = a\frac{\sin\frac{n\delta}{2}}{\sin\frac{\delta}{2}}\,.\cos\,[n-1]\,\frac{\delta}{2} \qquad \ldots(2.1)$$

Similarly, resolving the amplitudes for their vertical components, we get

$$R \sin \phi = 0 + a \sin \delta + a \sin 2\delta + K + a \sin (n-1)\delta$$

Finally as earlier,

$$R \sin \phi = a \frac{\sin \frac{n\delta}{2}}{\sin \frac{\delta}{2}} . \sin [n-1] \frac{\delta}{2} \qquad ...(2.2)$$

Using eqns.(2.1) and (2.2), we have

$$R^2 = R^2 \sin^2 \phi + R^2 \cos^2 \phi$$

$$= \left[a \frac{\sin \frac{n\delta}{2}}{\sin \frac{\delta}{2}} . \sin [n-1] \frac{\delta}{2} \right]^2 + \left[a \frac{\sin \frac{n\delta}{2}}{\sin \frac{\delta}{2}} . \cos [n-1] \frac{\delta}{2} \right]^2$$

$$= a^2 \left[\frac{\sin^2 \frac{n\delta}{2}}{\sin^2 \frac{\delta}{2}} \right]$$

$$\Rightarrow \qquad R = a \left[\frac{\sin \frac{n\delta}{2}}{\sin \frac{\delta}{2}} \right] \qquad ...(2.3)$$

This expression of eqn.(2.3) gives the resultant amplitude of n nos. of SHM having common phase difference and same amplitude between two successive vibrations.

3. FRAUNHOFER'S DIFFRACTION THROUGH A SINGLE SLIT

Here, the incident plane wavefront is obtained by using a convex lens in such a way that the source lies in the focal plane of the lens. The diffracted light from the slit is again focused on the screen with the help of another convex lens. In the Fig. 2.5 *AB* is a section of narrow slit of width *e* perpendicular to the plane of paper. A plane wavefront *WW'* of monochromatic light of wavelength λ propagating normally to the slit is incident on it.

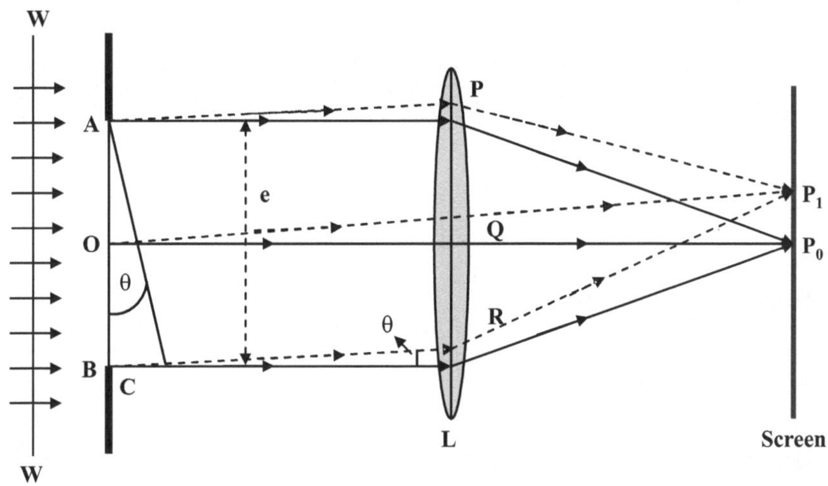

Fig. 2.5 *Diffraction from a single slit*

The diffracted light is focused by a convex lens L on a screen. Now according to the Huygens's principle every point on the slit is a source of secondary wavefronts spreading out in all directions. The secondary wavelets travel normally to the slit, *i.e.*, along OP_0. P_0 is a bright central image. The secondary wavelets traveling at an angle θ with the normal are focused at P_1. P_1 is of different intensity depending upon the path difference between the secondary waves originating from the corresponding points of the wavefront.

To determine the intensity at P_1 we draw perpendicular AC on BR. Hence, path difference between wavelets from A and B in the direction of θ is

$$BC = AB \sin\theta = e \sin\theta.$$

So phase difference is given by $\dfrac{2\pi}{\lambda} \times e \sin\theta$.

Now, we consider that the width of slit is divided into n equal parts and a is the amplitude of the wave from each part (as width of the each individual part is same).

Hence, the phase difference between any two consecutive waves is

$$= \frac{1}{n} \times \text{Total phase difference} = \frac{2\pi e}{n\lambda} \sin\theta .$$

Let, $$\delta = \frac{2\pi e}{n\lambda} \sin\theta .$$

Then using eqn.(2.3) resultant amplitude R at a point P is given by

$$R = a\frac{\sin(n\delta/2)}{\sin(\delta/2)}$$

$$\therefore \qquad R = a\frac{\sin\left(\dfrac{\pi e \sin\theta}{\lambda}\right)}{\sin\left(\dfrac{\pi e\sin\theta}{n\lambda}\right)} \qquad \left[\frac{nd}{2} = \frac{n}{2}\times\frac{2\pi e\sin\theta}{n\lambda} = \frac{\pi e\sin\theta}{\lambda}\right]$$

$$\Rightarrow \qquad R = a\frac{\sin\alpha}{\sin(\alpha/n)}$$

Here, $\alpha = \dfrac{\pi e \sin\theta}{\lambda}$.

As $\dfrac{\alpha}{n}$ is very small, $\sin\left(\dfrac{\alpha}{n}\right) \approx \dfrac{\alpha}{n}$

Hence, $$R = na\frac{\sin\alpha}{\alpha} = A\frac{\sin\alpha}{\alpha}; \qquad \text{...(2.4)}$$

(when $n \to \infty$, $a \to 0$, amplitude $= A = na$)

So, resultant intensity is given by

$$I = R^2 = A^2\frac{\sin^2\alpha}{\alpha^2} \qquad \text{...(2.5)}$$

Now, from eqn.(2.4),

$$R = A\frac{\sin\alpha}{\alpha} = \frac{A}{\alpha}\left[\alpha - \frac{\alpha^3}{3!} + \frac{\alpha^5}{5!} - K\right]$$

$$= A\left[1 - \frac{\alpha^2}{3!} + \frac{\alpha^4}{5!} - K\right] \qquad \text{...(2.6)}$$

I. Principal Maxima

This condition is achieved only when R will be maximum , *i.e.*, A.

This is possible only when $\dfrac{\sin\alpha}{\alpha} = 1$, *i.e.*, $\alpha \to 0$.

As from eqn.(2.4),

$$\alpha = \frac{\pi e}{\lambda}\sin\theta = 0 \quad \Rightarrow \quad \sin\theta = 0 \quad \Rightarrow \quad \theta = 0$$

and, $$R = A\frac{\sin\alpha}{\alpha} = A.$$

Hence, $I \propto A^2$.

The principal maximum is formed by those secondary wavelets which travel normally to the slit.

II. Minima

R will be minimum when only $\sin\alpha = 0$ and $\alpha \neq 0$

As for $\alpha = 0$, we are again going back to the maximum condition. So this value is omitted.

Hence the only possible values are $\alpha = \pm\pi, \pm 2\pi, \pm 3\pi, \ldots$

$$\therefore \quad \frac{\pi e \sin\theta}{\lambda} = \pm m\pi,$$

$$\Rightarrow \quad e\sin\theta = \pm m\lambda \quad \ldots(2.7), \quad \text{where } m \neq 0 \text{ and } m = 1, 2, 3, \ldots$$

This is the minima equation for a single slit.

III. Secondary Maxima

There are secondary maxima between two equally spaced minima. The pattern can be obtained by differentiating I with respect to α and then equating it to zero.

Hence, $$\frac{d}{d\alpha}(I) = \frac{d}{d\alpha}\left[A^2\left(\frac{\sin\alpha}{\alpha}\right)^2\right] = 0$$

$$\Rightarrow \quad A^2 \times 2\,\frac{\sin\alpha}{\alpha} \times \frac{\alpha\cos\alpha - \sin\alpha}{\alpha^2} = 0$$

$$\Rightarrow \quad \sin\alpha\,(\alpha\cos\alpha - \sin\alpha) = 0 \qquad [\text{as } \alpha \text{ and } A^2 \neq 0]$$

$$\therefore \quad \sin\alpha = 0 \quad \text{or} \quad \alpha\cos\alpha - \sin\alpha = 0$$

But, $\sin\alpha = 0$ is not possible as it is the condition for the minima.

So, $\alpha\cos\alpha - \sin\alpha = 0$

$$\Rightarrow \quad \tan\alpha = \alpha \qquad \ldots(2.8)$$

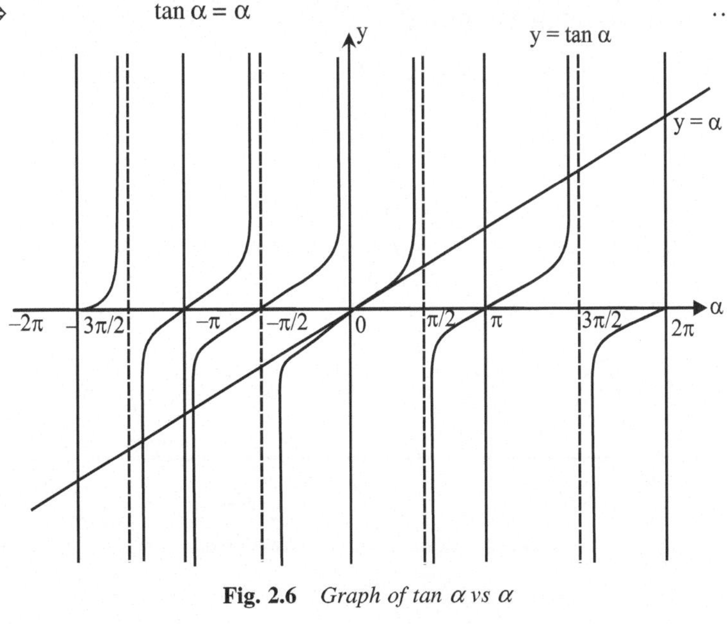

Fig. 2.6 *Graph of tan α vs α*

Eqn.(2.8) is the condition for secondary maxima. The values of α satisfying $\tan\alpha = \alpha$ are obtained by the points of intersection of the graphs between the equations $y = \alpha$ and $y = \tan\alpha$ curves as shown in Fig. 2.6.

The points of intersections are $\alpha \approx 0, \pm\frac{3\pi}{2}, \pm\frac{5\pi}{2}, K$

So, $\alpha = 0$ gives the principal maxima and others are the secondary maximas.

IV. Width of Central Maxima

The direction of first minima is given by

$$e\sin\theta = \pm\lambda$$

$$\Rightarrow \qquad \sin\theta = \pm\lambda/e \qquad \text{...(2.9)}$$

If the the focal length of the lens between the slit and the screen is f and the linear half-width of the central maximum is y, then we have

$$\sin\theta = \frac{y}{f}$$

$$\Rightarrow \qquad y = f\sin\theta \qquad \text{...(2.10)}$$

Using eqns.(2.9) and (2.10), we have

$$y = \frac{f\lambda}{e}$$

Thus, the width of the central maxima is

$$2y = \frac{2f\lambda}{e}$$

V. Intensity Distribution Graph

The graph is shown in Fig.2.7 along with the positions of maximas and minima as:

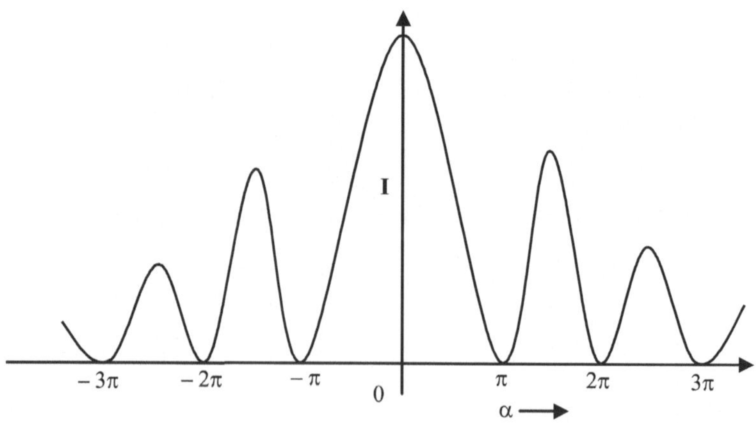

Fig. 7. *Single slit intensity distribution*

Principal maxima

$$\alpha = 0$$

Secondary maxima

$$\alpha = \pm\frac{3\pi}{2}, \pm\frac{5\pi}{2}, K$$

Minima

$$\alpha = \pm\pi, \pm 2\pi, \ldots$$

The typical diffraction pattern for a single slit with a helium-neon laser is shown in Fig. 2.8.

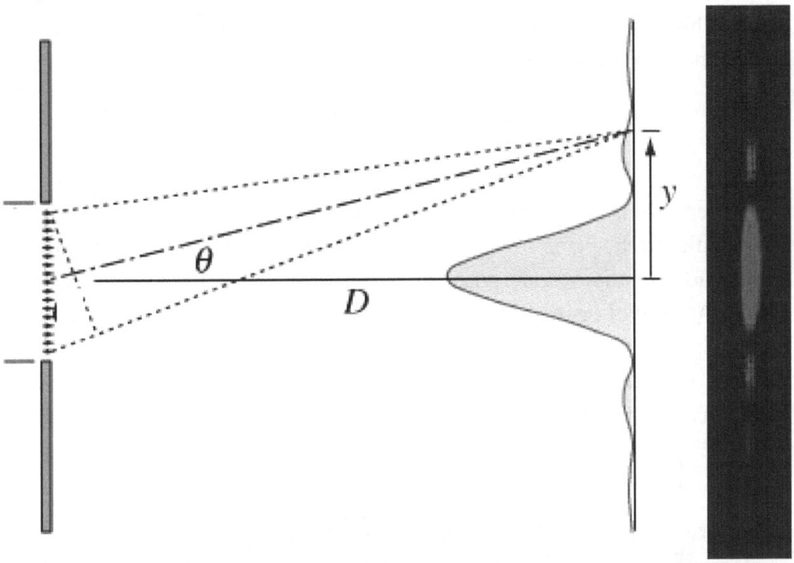

Fig. 2.8 *Typical single slit diffraction pattern using He-Ne laser*

Key Points

(*i*) The width of the principal maxima is proportional to λ.

(*ii*) The width of the central maximum is inversely proportional to the width of the slit.

(*iii*) The width of the principal maxima is twice that of any other maxima.

(*iv*) The ratio of the intensity of the principal to any order of secondary maxima is $\left(1 : \frac{4}{(n\pi)^2}\right)$, where $n = 3, 5, 7, \ldots$

4. PLANE TRANSMISSION DIFFRACTION GRATING

Grating is an arrangement consisting of large number of parallel slits of same width and separated by equal opaque spaces having their widths comparable to the wavelength of incident light.

Plane Transmission Diffraction Gratings are constructed by ruling equidistant parallel lines on a transparent material, such as, glass, with a fine point. The ruled lines are opaque to light, while space between any two lines acts as a slit. The first grating was made by Fraunhofer around the year 1800. Then Rowland produced grating by ruling fine lines (nearly upto 8000 lines per cm) on the surface of glass with a fine sharp diamond point. However, for an improved grating the ruling process is complicated and expensive. Commonly we use replica grating for its cost-effectiveness. The technique of preparing replica grating starts with the pasting of a thin layer of colloidal solution on the surface of a ruled grating. The liquid in colloid evaporates and thin film is thus stripped off from the grating surface. This film retains the impressions of the original grating and is fixed between the two glass plates. Hence it can now serve the purpose of a plane grating.

Theory

Fig. 2.9 represents a portion of plane transmission diffraction grating placed perpendicular to the plane of paper. Here, e and d are the width of the individual slit and width of the opaque portion respectively. $(e + d)$ is known as grating element. A parallel beam of light of wavelength λ is incident normally on the grating. The secondary wavelets traveling in the same direction of incident light will focus at P_0 on the screen XY as the screen is placed at the focal plane of the convex lens L. P_0 will be a central maximum.

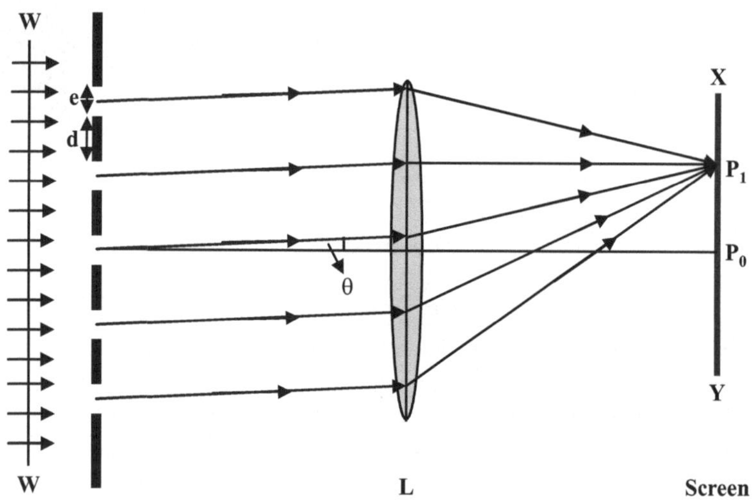

Fig. 2.9 *Diffraction from a Plane Grating*

Let us consider the secondary waves traveling in a direction inclined at an angle θ with the direction of incident light. These waves will reach P_1 on passing through convex lens L in different phases. So, on both sides of central maximum dark and bright bands are obtained. The intensity at P_1 may be considered by applying theory of Fraunhofer's diffraction at a single slit. The wavelets proceeding from all the points in a slit along the direction of θ are equivalent to a single wave of amplitude $\left[\frac{A \sin \alpha}{\alpha}\right]$ starting from the middle point of the slit.

Here, $\alpha = \frac{\pi e \sin \theta}{\lambda}$.

For N slits, we have N numbers of diffracted waves. The path difference between two consecutive slits is $(e + d) \sin \theta$. So phase difference δ is given by

$$\delta = \frac{2\pi}{\lambda} \times (e + d) \sin \theta = 2\beta$$

where, $$\beta = \frac{\pi}{\lambda} \times (e + d) \sin \theta .$$

Now, using eqn.(2.3) the resultant amplitude is given by

$$R = a \frac{\sin (n\delta/2)}{\sin (\delta/2)}$$

Here, $$a = \left[\frac{A \sin \alpha}{\alpha}\right], n = N \text{ and } \delta = 2\beta.$$

So, $$R = \left[\frac{A \sin \alpha}{\alpha}\right] \times \frac{\sin\left(\frac{N.2\beta}{2}\right)}{\sin \beta} = \left[\frac{A \sin \alpha}{\alpha}\right] \times \frac{\sin (N\beta)}{\sin \beta} \qquad \ldots(2.11)$$

Hence, $$I = R^2 = \left[\frac{A \sin \alpha}{\alpha}\right]^2 \times \left[\frac{\sin (N\beta)}{\sin \beta}\right]^2 \qquad \ldots(2.12)$$

In eqn.(2.12) first term on right hand side, $\left[\frac{A \sin \alpha}{\alpha}\right]^2$, gives the distribution of intensity of a single slit, and, the second term, $\left[\frac{\sin (N\beta)}{\sin \beta}\right]^2$, gives the distribution of intensity of the combined effect of all the slits.

I. Principal Maxima

This condition is achieved only when $\left[\frac{\sin (N\beta)}{\sin \beta}\right]^2$ is maximum.

This is possible only when

$$\sin \beta = 0 \Rightarrow \beta = \pm n\pi, \qquad \text{where } n = 0, 1, 2, \ldots$$

and at the same time, $\sin N\beta = 0$,

so, that $\left[\dfrac{\sin(N\beta)}{\sin\beta}\right]$ becomes indeterminate.

Hence, we have to apply L'Hospital's rule of differentiation.

On applying the same, we get

$$\underset{\beta \to \pm n\pi}{Lt} \frac{\sin N\beta}{\sin\beta} = \underset{\beta \to \pm n\pi}{Lt} \frac{\dfrac{d}{d\beta}(\sin N\beta)}{\dfrac{d}{d\beta}(\sin\beta)}$$

$$= \underset{\beta \to \pm n\pi}{Lt} \frac{N\cos N\beta}{\cos\beta} = N(\pm 1) = \pm N$$

So, $\underset{\beta \to \pm n\pi}{Lt}\left[\dfrac{\sin N\beta}{\sin\beta}\right]^2 = N^2$.

The resultant intensity is

$$I = N^2\left[\frac{A\sin\alpha}{\alpha}\right]^2.$$

The principal maximas are more intense.

Hence, for maxima : $\beta = \pm n\pi, \quad \Rightarrow \quad \beta = \frac{\pi}{\lambda}\times(e+d)\sin\theta = \pm n\pi,$

or, $\quad (e+d)\sin\theta = \pm n\lambda$ ...(2.13)

This is the grating equation.

For zero order maximum, $n = 0$, and $n = 1, 2, 3, ..$ for 1st, 2nd, 3rd, ...order maximum.

II. Minima

When $\sin N\beta = 0$, $\sin\beta \neq 0$. $\Rightarrow N\beta = \pm m\pi$

Here, m has all integral values except, 0, N, $2N$, ..., nN, for which it again yields the condition of principal maxima.

$$\therefore N.\frac{\pi}{\lambda}.(e+d)\sin\theta = \pm m\pi$$

$\Rightarrow \quad N(e+d)\sin\theta = \pm m\lambda$ ...(2.14)

Hence, they are adjacent to principal maxima.

III. Secondary Maxima

As there are (N-1) minima between two adjacent principal maxima, there must be (N-2) other secondary maximas between two consecutive principal maximas.

Those secondary maximas can be obtained by differentiating eqn.(2.12) with respect to β.

So, $$\frac{dI}{d\beta} = 0$$

$$\Rightarrow \left[\frac{A\sin\alpha}{\alpha}\right]^2 \times 2 \times \left[\frac{\sin(N\beta)}{\sin\beta}\right] \times \left[\frac{N\cos N\beta \sin\beta - \sin N\beta \cos\beta}{\sin^2\beta}\right] = 0$$

$$\Rightarrow N\cos N\beta \sin\beta - \sin N\beta \cos\beta = 0$$

$$\Rightarrow \qquad N\tan\beta = \tan N\beta \qquad \ldots(2.15)$$

This is the condition of secondary maxima.

The roots of this equation other than those which corresponds to principal maxima condition $\beta = \pm n\pi$ give the positions of secondary maxima.

IV. Intensity of the Secondary Maxima

For the determination of the intensity we have to evaluate $\frac{\sin^2 N\beta}{\sin^2\beta}$ from the equation $N\tan\beta = \tan N\beta$. For that we have drawn the triangle as shown in Fig. 2.10 with respective sides in such a way that $\tan N\beta = \frac{N\tan\beta}{1}$.

From the Fig. 2.10, we have

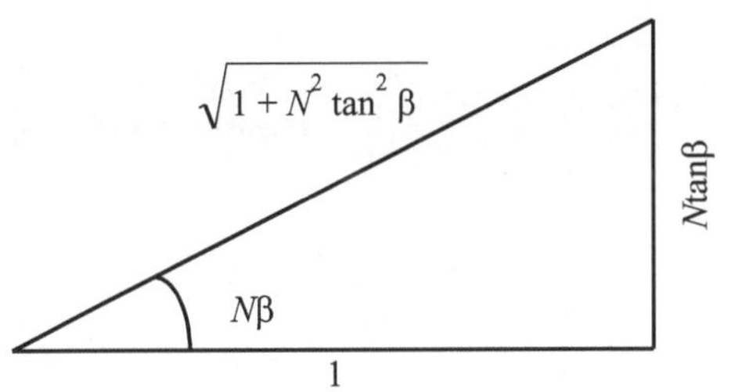

Fig. 2.10 *Geometrical figure to obtain* 'sin $N\beta$'

$$\sin N\beta = \frac{N\tan\beta}{\sqrt{1+N^2\tan^2\beta}}$$

$$\Rightarrow \sin^2 N\beta = \frac{N^2\tan^2\beta}{1+N^2\tan^2\beta}$$

$$\Rightarrow \frac{\sin^2 N\beta}{\sin^2\beta} = \frac{N^2\tan^2\beta}{(1+N^2\tan^2\beta)\sin^2\beta} = \frac{N^2}{\cos^2\beta + N^2\sin^2\beta} = \frac{N^2}{1+(N^2-1)\sin^2\beta}$$

Hence, the intensity of the secondary maxima is given by

$$I_{SM} = A^2 \frac{\sin^2 \alpha}{\alpha^2} \frac{N^2}{1 + \left(N^2 - 1\right)\sin^2 \beta} \quad \text{...(2.16)}$$

Hence, the ratio of the intensity of secondary maxima to the intensity of principal maxima is

$$\frac{I_{SM}}{I_{PM}} = \frac{1}{1 + \left(N^2 - 1\right)\sin^2 \beta} \quad \text{...(2.17)}$$

It is clear from eqn.(2.17) that for increasing value of N, the secondary maxima weakens. In actual case these intensities are so weak that they normally invisible. Fig. 2.11 depicts how in the resultant intensity of a grating depends on the combination of individual contribution and the slits combination as a whole.

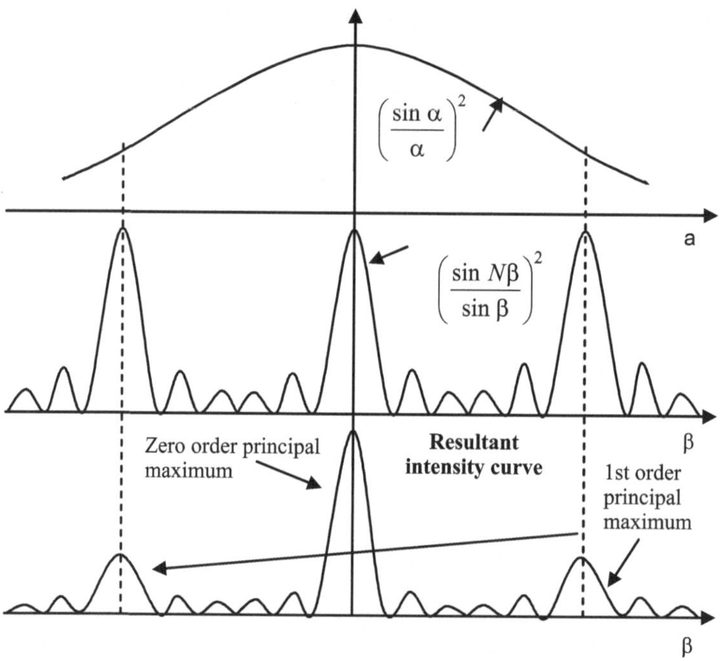

Fig. 2.11 *Intensity distribution pattern of a plane diffraction grating*

V. Width of the Principal Maxima

The width of principal maxima of any order is defined as the angular separation between first two minima on either side of the principal maxima.

The nth order principal maxima grating equation is obtained from eqn.(2.13) as

$$(e + d) \sin \theta_n = n\lambda$$

Thus, according to the definition of the width of principal maxima if $d\theta_n$ is the angular half width of nth principle maxima, then, the direction of minima is given by

$$N(e+d)\sin(\theta_n \pm d\theta_n) = m\lambda$$

$$\Rightarrow \quad N(e+d)\sin(\theta_n \pm d\theta_n) = (nN \pm 1)\lambda$$

$$\Rightarrow \quad N(e+d)\sin\theta_n \cos(d\theta_n) \pm N(e+d)\cos\theta_n \sin(d\theta_n) = nN\lambda \pm \lambda$$

As $(d\theta_n)$ is very small

So, $\cos(d\theta_n) = 1$, $\sin(d\theta_n) = (d\theta_n)$

$$N(e+d)\sin\theta_n \pm N(e+d)\cos\theta_n \,.\, (d\theta_n) = nN\lambda \pm \lambda$$

Using grating maxima eqn (2.13)

$$N \times n\lambda \pm N(e+d)\cos\theta_n \,.\, (d\theta_n) = nN\lambda \pm \lambda$$

$$\Rightarrow \quad N(e+d)\cos\theta_n \,.\, (d\theta_n) = \lambda$$

$$\Rightarrow \quad d\theta_n = \frac{\lambda}{N(e+d)\cos\theta_n}$$

Thus, as N is increasing $(d\theta_n)$ decreases.

5. APPLICATIONS

I. To determine the wavelengths of prominent lines of white light source

Theory

A diffraction grating has large no. of slits of equal width and separation as shown in Fig. 2.12. Let there are N parallel slits each of width e separated by opaque space d. So the grating element is $(e + d)$. Let a collimated beam is incident on the grating normally as shown in Fig. 2.12. The rays are diffracted by various slits at the same angle θ at point P on the screen/eye piece as shown in Fig. 2.9. The resultant intensity at P due to all the rays is obtained from eqn.2.12 as

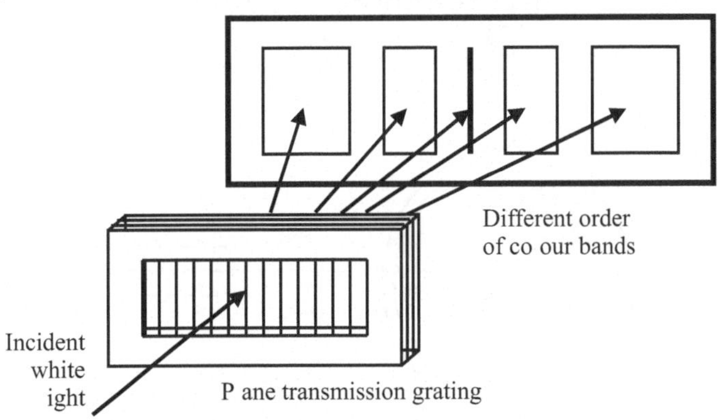

Fig. 2.12 *Spectra formation by Plane transmission grating*

$$I = \left[\frac{A\sin\alpha}{\alpha}\right]^2 \cdot \left[\frac{\sin N\beta}{\sin\beta}\right]^2$$

The intensity will be maximum when,

$$(e+d)\sin\theta = \pm n\lambda$$

Here, n is the order of the diffraction pattern.

Thus the wavelength of a spectral line can be calculated by

$$\lambda = \frac{(e+d)\sin\theta}{n} \qquad \ldots(2.18)$$

Spectrometer Design

The apparatus involved in measurement of θ is Spectrometer. The spectrometer consists of a collimator, a prism table and telescope as in above Fig. 2.13. The collimator consists of two hollow concentric metal tubes, one (the smaller) sliding under the other. The free end of the longer tube carries an achromatic lens and that of the smaller tube has a slit adjustable width. The slit is adjusted in the focal plane of lens in order to obtain a parallel beam. The collimator also has two screws for adjusting the inclination of the axis of the collimator. The prism table is a circular table, which moves over a graduated circular scale carried by the telescope. The relative positions of the prism table and the telescope can be read over these two vernier scales. There are three leveling screws at the lower surface of the prism table and a tangent screw for slow motion. The telescope too possesses two fine crosswire. Like prism table, the hollow concentric tubes like collimator. The free end of the tube facing prism table has an achromatic objective lens, and the other free end carries a eye piece along with a telescope can be clamped to the main body of the spectrometer. In clamped condition the telescope can be given fine motion using tangent screw.

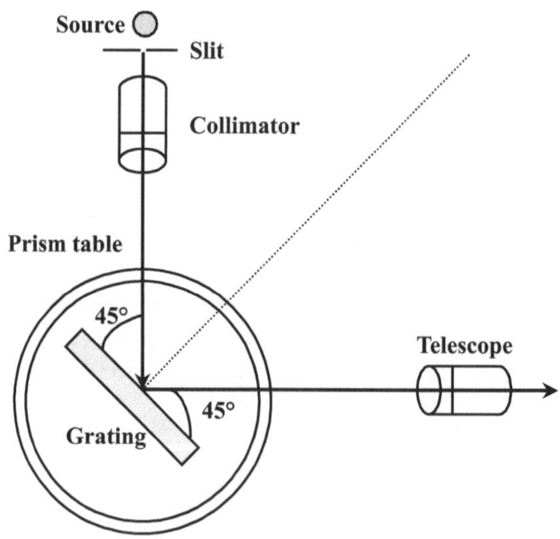

Fig. 2.13 *Experimental arrangement using Spectrometer*

Experimental Steps

(*i*) The main body of the spectrometer including prism table is leveled the screws provided at the base and with the help of a spirit level.

(*ii*) The slit screw is adjusted to get a sharp and vertical slit.

(*iii*) The grating is mounted over the prism table vertically.

(*iv*) For setting the grating for normal incidence first the direct image of the slit is seen through the telescope. The readings of the two verniers are noted. The telescope is then rotated on either side by 90° and clamped. The prism table is rotated so that the reflected image of the slit is seen on the crosswire through the telescope.

(*v*) Again the readings of the two verniers are taken and the prism table is turned from this position by 45° or 135° such that the face of the grating normal to the incident beams. Now the set up is ready for observations.

(*vi*) The telescope is rotated to the left of the direct image to get the first order spectrum. (Direct image is the zeroth order). The telescope is clamped. The different spectral lines are adjusted colourwise one by one on the crosswire and the readings of verniers V_1 and V_2 are noted for each colour.

(*vii*) Now the telescope is rotated towards the right side of the direct image. Again 1st order spectrum of same colours are obtained and readings of verniers V_1 and V_2 for each colour are noted as shown in Fig. 2.14.

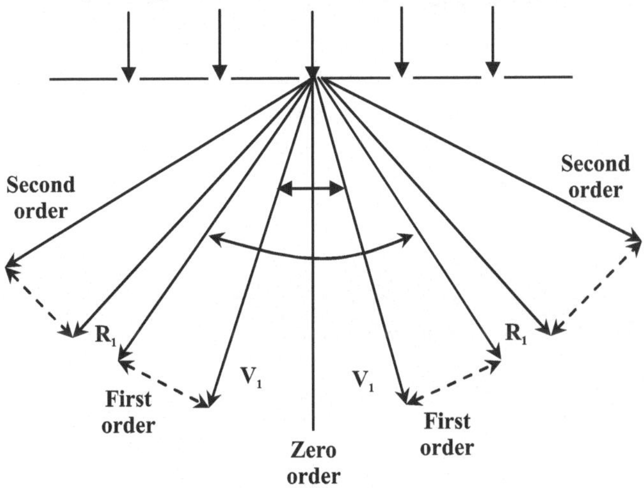

Fig. 2.14 *Position of spectra of different orders*

(*viii*) The difference of readings of same verniers for various colours for 1st orders calculated, half of the angle so obtained is the angle of diffraction for that particular colour and particular order.

(*ix*) The various wavelengths are now determined using

$$\lambda = \frac{(e+d)\sin\theta}{n}.$$

6. ABSENT SPECTRA WITH A GRATING

It is observed that in grating spectra some orders are absent, while others are present. This happens, when for a given angle of diffraction, θ, the path difference between the two diffracted rays from the two extreme ends of one slit becomes equal to an integral multiple of λ for certain order of the spectrum. Let the path difference be λ. Now considering a slit composing of two halves, the path difference between the secondary waves from the corresponding points in the two halves will be λ/2. Now they will cancel each other's effect resulting zero intensity.

Mathematically:

We have grating equation for principal maxima from eqn.(2.13) as

$$(e + d) \sin \theta = n\lambda \qquad \ldots(2.19)$$

In the same direction of single slit minima is satisfied of mth order, then, we get from eqn.(2.7) as

$$e \sin \theta = m\lambda \qquad \ldots(2.20)$$

Here, $n = 0, 1, 2, \ldots$ and $m = 1, 2, \ldots$

If both the eqns.(2.19) and (2.20) are satisfied simultaneously, then the maxima of nth order will be missing in the grating spectrum,

$$\frac{(e+d)\sin\theta}{e \sin\theta} = \frac{n\lambda}{m\lambda}$$

$$\Rightarrow \qquad \frac{e+d}{e} = \frac{n}{m}.$$

This is the condition for Absent Spectra.

Example

When the 2nd order has to be suppressed, then, $n = 2$ and $m = 1$.

So, $\frac{e+d}{e} = \frac{2}{1} \Rightarrow e = d$. It implies that when the ruling width equals the slit width, the second order spectrum will be suppressed.

7. MAXIMUM NUMBER OF ORDERS AVAILABLE WITH A GRATING

This is the maximum number of orders available by a grating for a fixed wavelength of incident light. The equation for principal maxima is

$$(e + d) \sin \theta = n\lambda \Rightarrow n = \frac{(e+d)}{\lambda} \sin\theta.$$

For a maximum value of θ as 90°, $n = n_{max}$

and so $$n_{max} = \frac{(e+d)\sin 90°}{\lambda} = \frac{e+d}{\lambda}.$$

Example

When grating element, $(e + d) < 2\lambda$, then, $n_{max} < \frac{2\lambda}{\lambda} \Rightarrow n_{max} < 2$, we will see in this condition only one order of spectra.

8. DISPERSIVE POWER OF GRATING

It is defined as the rate of variation of angle of diffraction with wavelength of incident light.

Mathematically, dispersive power (DP) = $\frac{d\theta}{d\lambda}$

Derivation

The nth order principal maxima for wavelength λ is given by $(e + d) \sin \theta = n\lambda$.

Taking derivative of both sides with respect to λ, we have

$$\frac{d}{d\lambda}\left[(e+d)\sin\theta\right] = \frac{d}{d\lambda}(n\lambda)$$

$$\Rightarrow \quad (e+d)\cos\theta\frac{d\theta}{d\lambda} = n$$

$$\Rightarrow \quad \frac{d\theta}{d\lambda} = \frac{n}{(e+d)\cos\theta}$$

So, $$DP = \frac{d\theta}{d\lambda} = \frac{n}{(e+d)\cos\theta}$$

Thus, DP is

(*i*) proportional to order of the spectrum, n,

(*ii*) inversely proportional to grating element, $(e + d)$, or, directly proportional to the number of rulings in the grating, and

(*iii*) inversely proportional to the cosine of angle of diffraction, θ.

9. RESOLVING POWER

An optical instrument should be able to resolve two point objects if the corresponding diffraction patterns are distinguishable from each other.

The ability of the optical instrument to produce their separate patterns is known as the resolving power (*RP*) of the instrument. It is defined as the ratio of the wavelength of any spectral line (λ) to the difference in wavelengths between this line and a neighboring spectral line (λ + dλ) such that the two spectral lines can be just resolved.

Mathematically, $RP = \frac{\lambda}{d\lambda}$, where, $d\lambda$ and λ are the smallest difference in two wavelengths just resolvable and wavelength of either of them or mean of them respectively.

Rayliegh's Criterian of Resolution

According to it the two objects of same intensity are resolvable by an optical instrument when the central maximum in the diffraction pattern of one falls over the first minimum in the diffraction pattern of the other and *vice versa.*

This is explained in the Fig. 2.15 (*a*) and 2.15 (*b*).

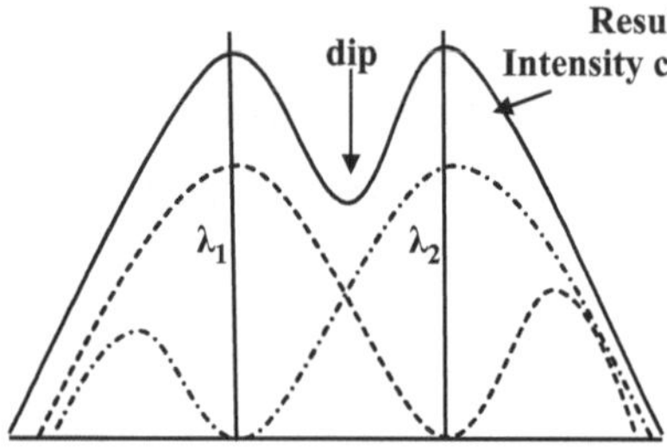

Fig. 2.15 (*a*) *Resultant intensity pattern when λ_1 and λ_2 are just resolved*

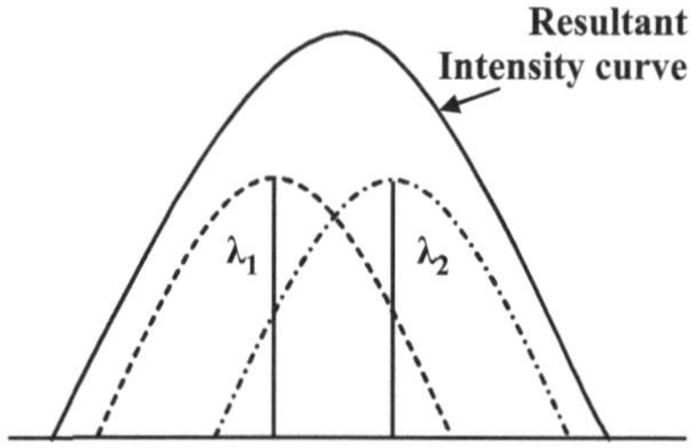

Fig. 2.15 (*b*) *Resultant intensity pattern when λ_1 and λ_2 are not resolved*

The dip is observed only when the wavelengths are just resolved.

(*a*) **Resolving Power of Grating :** The *RP* of a plane diffraction grating is defined as the capacity of it to form separate diffraction maxima of two wavelengths which are very close to each other.

Theory

In Fig.2.16, *AB* represent the surface of a plane diffraction grating with grating element $(e + d)$ and N be the total number of slits. Let a beam of light having wavelengths λ and (λ + dλ) is incident normally on the grating. *XY* is the field of view of telescope. P_1 is the nth primary maximum of a spectral line of λ at an angle of diffraction θ_n and P_2 is the nth primary maximum of a spectral line of $(\lambda + d\lambda)$ at an angle of diffraction $(\theta_n + d\theta_n)$. So by Rayleigh's criterion, λ and $(\lambda + d\lambda)$ will be just resolved only when nth order of primary maximum of $(\lambda + d\lambda)$ in a direction $(\theta_n + d\theta_n)$ falls over next minima of λ in the same direction of $(\theta_n + d\theta_n)$.

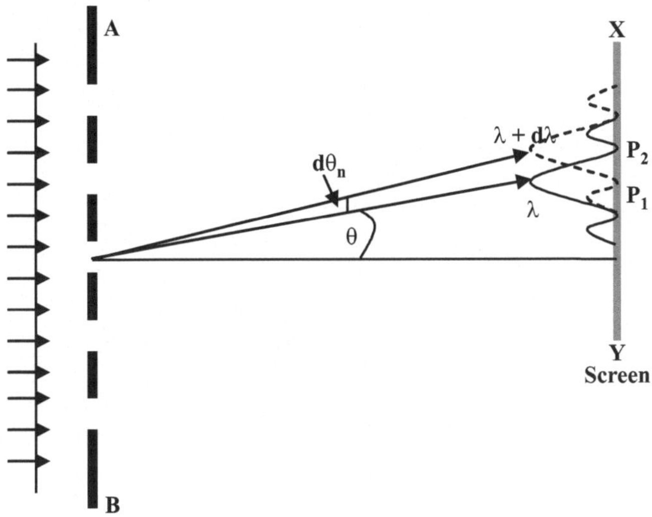

Fig. 2.16 *Intensity distribution of diffracted beam for wavelengths λ and $(\lambda + d\lambda)$*

Now, principal maximum of λ along θ_n is

$$(e + d) \sin \theta_n = n\lambda \qquad \text{...(2.22)}$$

The equation of minima of a grating from eqn.(2.14), is given by

$$N(e + d) \sin \theta = m\lambda \qquad \text{...(2.23)}$$

where, m is an integer, except $m = 0, N, 2N, 3N, \ldots\ldots$

Thus, the 1st minimum adjacent to nth principal maximum along $(\theta_n + d\theta_n)$ can be obtained by putting $m = nN + 1$. So 1st minimum along $(\theta_n + d\theta_n)$ is

$$N(e + d) \sin (\theta_n + d\theta_n) = (nN + 1)\lambda \qquad \text{...(2.24)}$$

The principal maximum of $(\lambda + d\lambda)$ along $(\theta_n + d\theta_n)$ is

$$(e + d) \sin (\theta_n + d\theta_n) = n(\lambda + d\lambda) \qquad \text{...(2.25)}$$

Using eqn.(2.25) into eqn.(2.24), we have

$$N.n(\lambda + d\lambda) = (nN + 1)\lambda$$

$$\Rightarrow \quad nN\lambda + nN.d\lambda = nN\lambda + \lambda$$

$$\Rightarrow \quad RP = \frac{\lambda}{d\lambda} = nN \qquad \text{...(2.26)}$$

Also, $RP = \dfrac{\lambda}{d\lambda} = nN = N(e + d)\sin \theta_n \times \dfrac{n}{(e + d)\cos \theta_n}$.

Hence, $RP =$ Total aperture $\times$ DP

(*b*) **Difference between dispersive and resolving powers of Grating**

(*i*) Let us consider the case of two spectral lines of wavelengths λ_1 and λ_2 of different maxima width as shown in Fig. 2.17(*a*) and 2.17(*b*). In both the figures dispersive power for both of them are same, but resolving power is different. Fig. 2.17(*b*) shows fully resolved spectral lines, while Fig. 2.17(*a*) has overlapping region and is not in resolved position. In both cases nos. of lines per cm is same, but N is large for Fig. 2.17 (*b*). Thus, resolving power depends on the width of the slit.

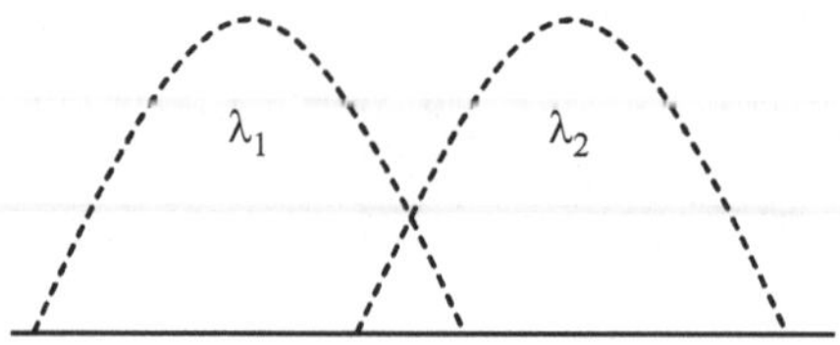

Fig. 2.17 (*a*) *Intensity patterns of two overlapped wavelengths λ_1 and λ_2*

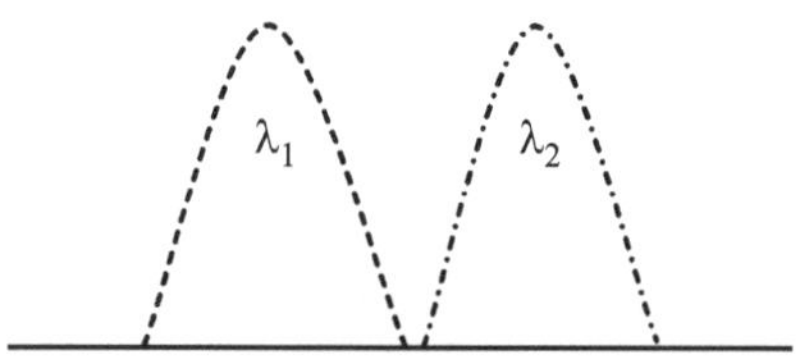

Fig. 2.17 (*b*) *Intensity patterns of two full resolved wavelengths λ_1 and λ_2*

(*ii*) Dispersive power gives the idea about angular separation between two spectral lines and resolving power tells the limit of just resolution of two close objects.

(*iii*) $DP = \dfrac{d\theta}{d\lambda} = \dfrac{n}{(e+d)\cos\theta}$, and $RP = \dfrac{\lambda}{d\lambda} = nN$

(*iv*) With an increase of N, resolving power increases, however, dispersive power remains unchanged.

(*v*) $DP \propto \dfrac{1}{e+d}$, whereas resolving power remains unchanged.

3

Polarization

Explanation of interference and diffraction phenomenon leads to the wave nature of light. But it does not tell about the character of the light wave whether it is longitudinal or transverse. Transverse nature of light was postulated by Thomas Young in 1817 to explain the absence of interference between light waves vibrating in mutually perpendicular planes.

The natural light is unpolarised which can be transformed into different types of polarisation using simple optical devices. Light is an electromagnetic wave consisting of electric and magnetic field vibrating perpendicular to each other and to the direction of propagation. Since electric vector is more effective therefore most of the time we refer to the plane of vibration constituted by electric vector and direction of propagation in describing the polarisation. Polarised light has many applications in the engineering industry.

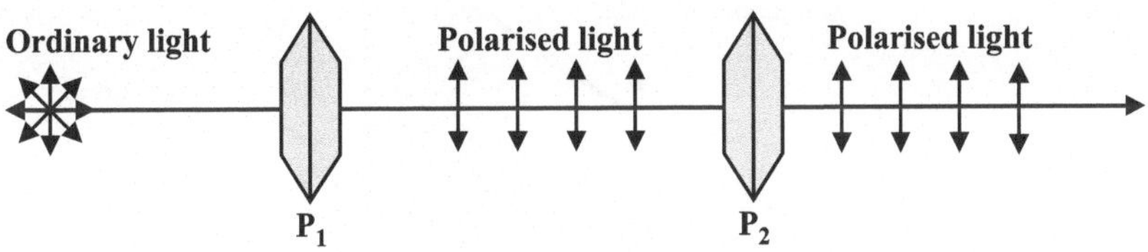

Fig. 3.1 (*a*) *Maximum intensity position of polarised light*

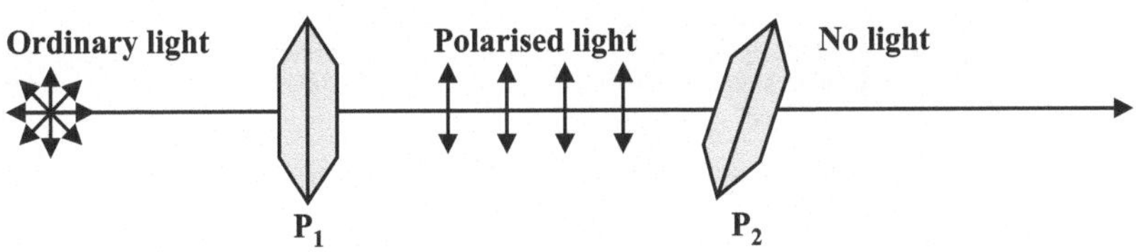

Fig. 3.1 (*b*) *Minimum intensity position of polarised light*

One of the most fascinating application of polarization of light is in liquid crystal display devices like wrist watches, calculator, TV screen, video displays etc.

Polarisation of light phenomenon confirms about the transverse nature of light. When light is passed through a pair of crystal plates (P_1 and P_2) called 'Tourmaline' (occurring in nature) with their planes at right angles to the direction of light, the intensity is maximum as shown in Fig. 3.1(*a*). However, when the plane of P_2 is rotated through 90° the intensity is minimum. This is shown in the Fig. 3.1(*b*). So, by this way we have restricted the vibrations in one particular direction only so that in other non-parallel orientation of the crystal, light is not allowed to pass through. This shows that ordinary light is transverse in nature where the vibrations are in perpendicular direction to the direction of propagation of wave.

The phenomenon of confinement of electric vector vibration in certain direction about the direction of propagation of light is called as polarisation of light. Thus this phenomenon may be defined as asymmetrical light vector vibration about the direction of propagation of light. It means light having the property of one sidedness is called polarised light.

1. POLARISATION FUNDAMENTALS

I. Plane of Vibration

The plane in which the electric vector vibrations of polarised light are confined is known as Plane of Vibration. It contains the electric vector vibration along with direction of propagation.

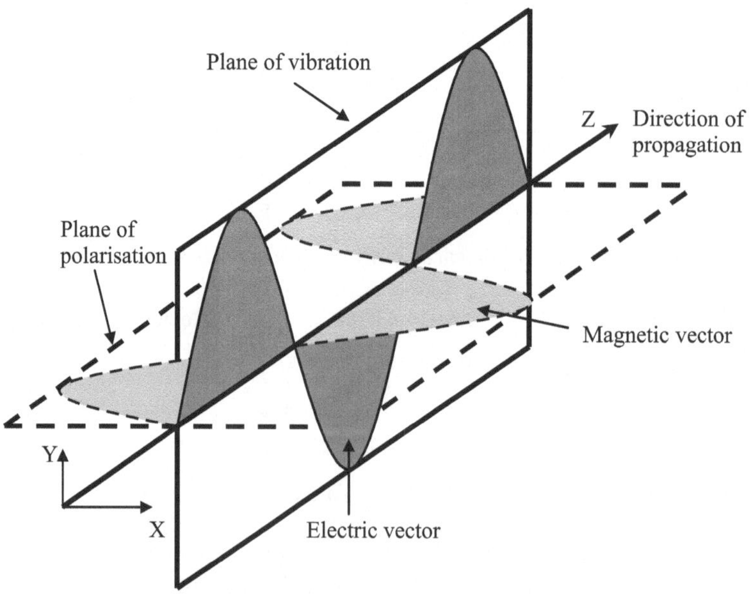

Fig. 3.2 *Light wave propagation with plane of vibration and plane of polarisation*

II. Plane of Polarization

A plane passing through the direction of propagation and perpendicular to the plane of vibration is known as Plane of Polarisation. It contains the direction of propagation of light and magnetic field vector. Fig. 3.2 represents the plane of vibrations and plane of polarisation associated with the propagation of light. In the figure wave is propagating along Z-axis, electric vectors and magnetic vectors are along Y-axis and X-axis respectively. The schematic representation of plane of polarisation and plane of vibration is shown in Fig. 3.3.

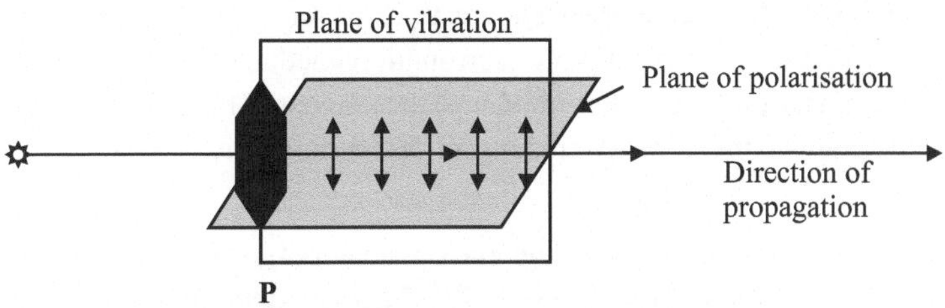

Fig. 3.3 *Schematic representation of plane of vibration and plane of polarisation*

III. Schematic Representation of Unpolarised and Polarised Light

In an unpolarized light, vibrations take place along all possible directions perpendicular to the direction of propagation of light. Schematically it is represented by star as shown in Fig. 3.4 (*a*). While in polarized light vibrations are

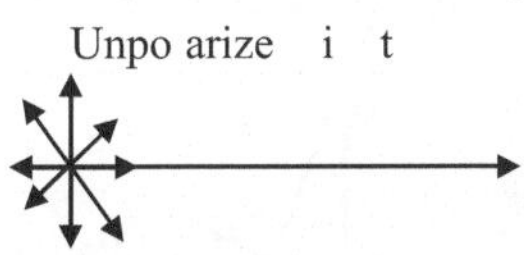

Fig. 3.4 (*a*) *Unpolarised light*

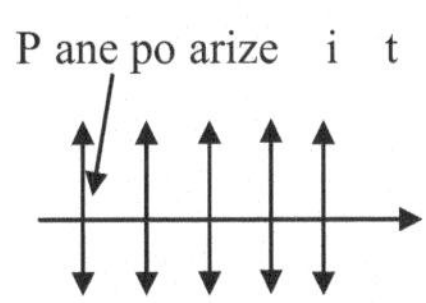

Fig. 3.4 (*b*) *Plane polarised light*

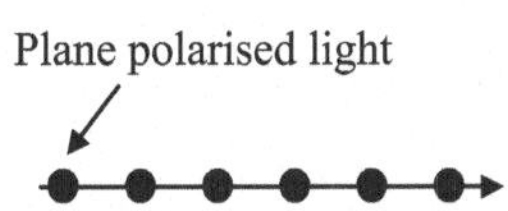

Fig. 3.4 (*c*) *Plane polarised light*

confined in certain direction therefore it is represented by arrow as in Fig. 3.4(*b*). If vibrations are perpendicular to the plane of paper, then the polarised light is represented by dot as in Fig. 3.4 (*c*).

IV. Brewster's Law

In 1812, Sir David Brewster observed that at a particular incident angle of unpolarized light on a transparent medium, the reflected light becomes completely polarised. This incident angle is known as polarising angle and depends upon the refractive indices of the media. The dependency is given by

$$\tan i_p = \frac{\mu_2}{\mu_1}$$

Where i_p is called as the polarising angle, μ_2 is the refractive index of reflecting surface and μ_1 is the refractive index of surrounding media. It means that if the ray is entering from air into the matter (medium) the tangent of the polarising angle is numerically equal to the refractive index μ of the medium, *i.e.*,

$$\mu = \tan (i_p)$$

Polarisation of unpolarized light by glass plate at polarising angle is shown in Fig. 3.5. Polarising angle for air-glass is 57°. This law determines that when the incident light is at polarising angle on medium of refractive index μ,the reflected and refracted light are mutually perpendicular to each other.

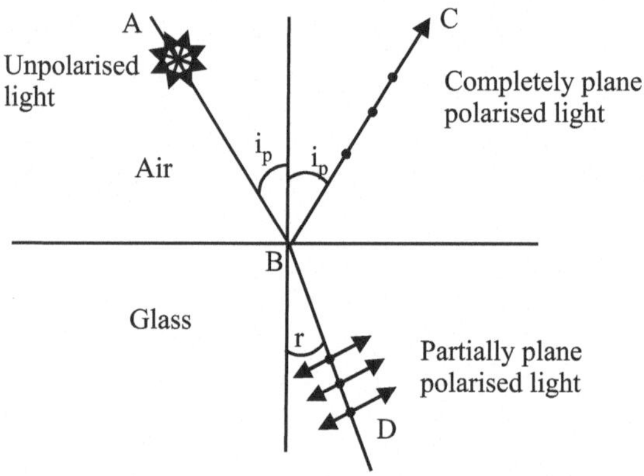

Fig. 3.5 *Production of plane polarised light from glass plate*

Mathematically it can be derived as follows. Let us consider unpolarised light *AB* is incident on a glass plate at polarising angle i_p. Let *BC* be the reflected light and *BD* be the refracted light with refraction angle *r*.

According to Brewster's law, we have

$$\mu = \tan i_p = \frac{\sin i_p}{\cos i_p} \quad \text{...(3.1)}$$

From Snell's law of refraction,

$$\mu = \frac{\sin i_p}{\sin r} \quad \text{...(3.2)}$$

From eqns. (3.1) and (3.2), we have

$$\frac{\sin i_p}{\sin r} = \frac{\sin i_p}{\cos i_p}$$

$$\Rightarrow \qquad \cos i_p = \sin r$$

$$\Rightarrow \qquad \sin\left(\frac{\pi}{2} - i_p\right) = \sin r$$

$$\Rightarrow \qquad \frac{\pi}{2} - i_p = r \Rightarrow r + i_p = \frac{\pi}{2} \qquad \text{...(3.3)}$$

If θ is the angle between refracted and reflected light, then we get

$$r + i_p + \theta = 180° \qquad \text{...(3.4)}$$

Using eqns. (3.3) and (3.4),

$$\theta = 90°$$

Hence at polarising angle, the angle between refracted light and reflected light is 90°.

A specific application of Brewster's angle is in the design of helium – neon gas laser tube to avoid reflecting losses between the mirrors.

V. Malus Law

In 1809, E.L Malus discovered a law for intensity of emerging light intensity from analyzer. This relation is also known as Malus's cosine square law.

According to this law, the intensity of emerging light from analyzer is proportional to the square of cosine of the angle between the plane of vibration of the incident polarised light wave to the analyzer and the principal plane of analyzer. Therefore, if I is the intensity of emerging light from analyzer and θ is the angle, then according to the law

$$I \propto E_1^2 \cos^2 \theta$$

$$\Rightarrow \qquad I = I_o \cos^2 \theta$$

Where E_1 is the amplitude of electric field vector in certain direction of incident polarised light wave to the analyzer.

Thus intensity of light through analyzer is maximum if transmission axes of the polariser and the analyzer are parallel *i.e.,* when θ = 0° or θ = 180°.

If the unpolarised light is incident on the polariser, the intensity of emergent light becomes half of its initial value *i.e.,*

as $$I = I_o (\cos^2 \theta)$$

$$\Rightarrow \qquad I = \frac{I_o}{2}$$

Here I_o is intensity of unpolarized light and I is the intensity of light after passing through polarizer.

VI. Anisotropic Medium

To polarize the incident light, the optical device must be anisotropic in nature. In anisotropic medium the properties vary in different directions. Examples are crystals of Quartz, Tourmaline, calcite, etc.

2. DOUBLY REFRACTIVE CRYSTAL

When an ordinary beam of light (unpolarized) is passed through a certain crystal, the refracted light splits up into two refracted rays, this phenomenon is known as double refraction and the material is known as doubly refracting crystal.

There are two types of doubly refracting crystals.

I. Uniaxial

There is only one direction along which two refracted rays travel with same velocity. Such crystal are of two types: positive and negative. In positive uniaxial double refracting crystal, the velocity of ordinary ray (O-ray) is greater than that of the extra-ordinary ray (E-ray), as the angle of refraction for ordinary ray r_1 is greater than angle of refraction r_2 for extra ordinary ray.

Example are quartz, iron oxide.

In negative uniaxial doubly refractive crystal $r_2 > r_1$, so the velocity of E-ray is greater than the velocity of O-ray.

Examples are tourmaline, calcite, sapphire, apatite, ruby and emerakl.

II. Biaxial

There is another class of doubly refracting crystals in which refracted rays have same velocity in two directions. Thus, these crystals have two optic axis. Therefore, these crystal are called as Biaxial double refracting crystal. This class was discovered by Brewster.

Examples of this class are topaz, aragonite, mica, borax and selenite.

In this chapter, we are going to discuss only uniaxial double refracting crystal due to its large number of application in the fabrication of popular optical devices like Nicol's prism.

III. Construction of a Doubly Refracting Crystal

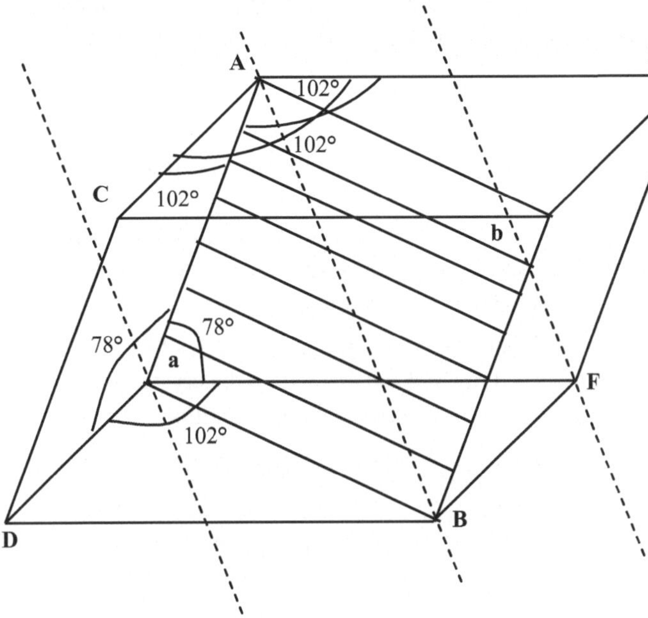

Fig. 3.6 A doubly refracting uniaxial crystal

Calcite ($CaCO_3$) crystal is a uniaxial crystal (natural) with six faces of rhombohedron on cleavage. These six faces are parallelograms. At each corner of the faces two angles are acute angle of 78° and one of obtuse angle 102° except two points *A* & *B*, here the three angles of the faces are obtuse of 102° each. These corners are known as Blunt Corners as shown in Fig. 3.6.

A line equally inclined to the three edges meeting at one of the blunt corners *A* or *B* or any other line parallel to its direction is the optic axis.

IV. Optic Axis of a Crystal

For a uniaxial crystal the direction along which the velocity of the E-ray is equal to the velocity of the O-ray, is called the Optic Axis.

Notes: (*i*) Optic axis is a direction, not a particular line.

(*ii*) The crystal is symmetric about it.

(*iii*) The velocities of O & E-rays are same along optic axis.

V. Principal Section/Plane

A plane containing the optic axis and perpendicular to the faces of the crystal is called the Principal Plane or Section for that face. As shown in Fig. 3.7 is the principal plane for the top and bottom faces of a calcite crystal.

For every point there are three principal planes, one each for each pair of opposite faces. One typical principle plane of a crystal is shown in Fig. 3.7. It is a parallelogram of angles 109° and 71°.

VI. Double Refraction Phenomenon

If an object is viewed through the doubly refracting crystal, two images of the object are observed. One image correspond to O-ray and other to E-ray. This phenomenon was first observed by Dr. E.Batholinus in 1669. When an ordinary beam of light (unpolarized) is passed through a calcite crystal, the refracted light splits up into two refracted rays as shown in Fig. 3.7.

Let a ray *SM* be incident on calcite crystal at an angle of incidence *i*. This ray splits up into two refracted ray *MN* and *MP* within the crystal. Among them one ray *MP* obeys ordinary laws of refraction and has vibrations perperdicular to the principal section. This is known as Ordinary (O) ray *i.e.,* O-ray. Since calcite crystal is doubly refracting negative uniaxial crystal therefore here the velocity of E-ray is faster than O-ray. So ray *MN* will have higher refraction angle r_2 as compared to refraction angle r_1 for ray *MP*. Ray *MN* does not obey ordinary laws of refraction and has vibrations in the principal section. This is known as Extra-ordinary (*E*) ray. Both these rays are plane polarized. Thus the incident ray forms two images corresponding to O-ray and E-ray.

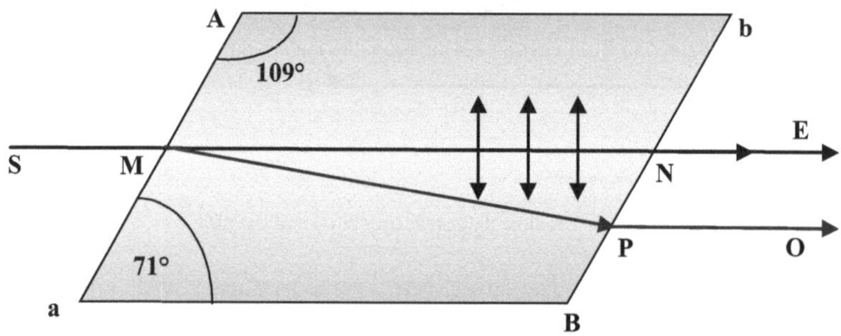

Fig. 3.7 *Double refraction through principal section*

When an ink dot is marked in a white paper and this ink dot is seen through calcite crystal, placed over it. Two images of an ink dot is observed. Now if the crystal is rotated about vertical axis and the images are seen vertically above the crystal, it is observed that one image correspond to O-ray remains fixed while other image correspond to E-ray is rotated. This fixed image is known as ordinary image and other image is known as extraordinary image. Both O-ray image and E-ray image can be identified by this experiment.

3. DIFFERENCE BETWEEN POSITIVE AND NEGATIVE CRYSTALS

<table>
<tr><th>Negative</th><th>Positive</th></tr>
<tr><td>(i) O-ray wavefront lies inside the E-ray wave front as shown in Fig. 3.8 (a).</td><td>(i) E-ray wavefront lies inside the O-ray wavefront as shown in Fig. 3.8 (b).</td></tr>
<tr><td>
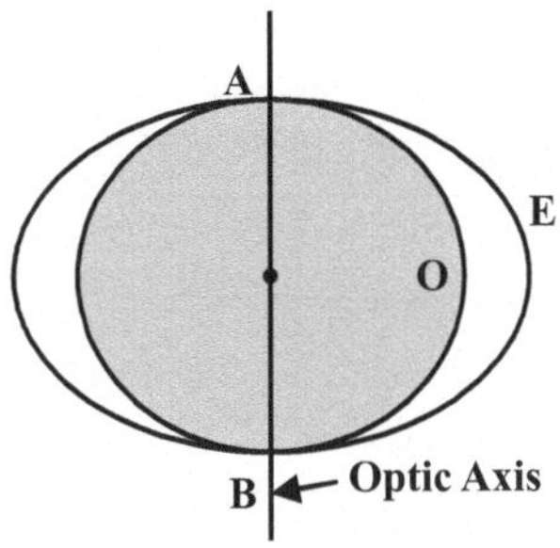

Fig. 3.8 (a) O-ray and E-ray wave-front for a negative crystal</td><td>
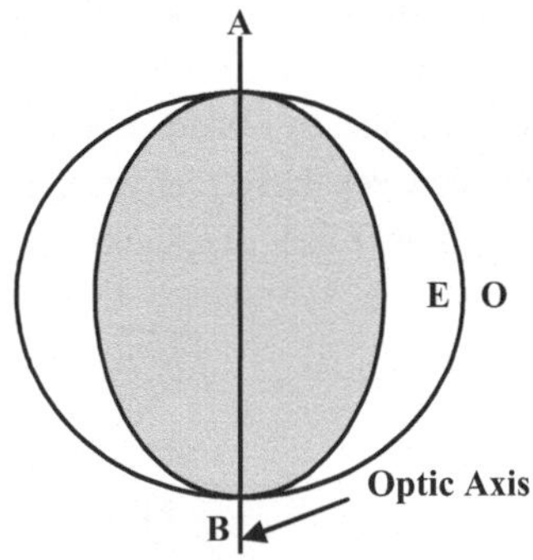

Fig. 3.8 (b) O-ray and E-ray wavefront for a positive crystal</td></tr>
<tr><td>(ii) Velocity of O-ray is constant in all direction.</td><td>(ii) Velocity of O-ray is constant in all direction.</td></tr>
<tr><td>(iii) Velocity of E-ray is minimum along optic axis & maximum in a direction perpendicular to the direction of the optic axis.</td><td>(iii) Velocity of O-ray is minimum along optic axis & maximum in a direction perpendicular to the direction of the optic axis.</td></tr>
<tr><td>(iv) Refractive index of O-ray is greater than E-ray i.e., $\mu_E < \mu_o$</td><td>(iv) Refractive index of E-ray is greater than O-ray $\mu_o < \mu_E$</td></tr>
<tr><td>(v) Velocity of E-ray is greater than O-ray i.e., $v_E > v_o$</td><td>(v) Velocity of O-ray is greater than E-ray i.e., $v_o > v_E$</td></tr>
<tr><td>(vi) The birefringence of these crystals, $\Delta\mu = \mu_e - \mu_o$, is negative.</td><td>(vi) The birefringence of these crystals, $\Delta\mu = \mu_e - \mu_o$, is positive.</td></tr>
<tr><td>(vii) Examples: Calcite, Tourmaline.</td><td>(vi) Examples: Quartz, Zicrom.</td></tr>
</table>

4. NICOL'S PRISM

This device was invented by William Nicole in 1828. It is a device used to roduce a plane polarized light.

Principle

In this device Nicole eliminated the ordinary ray from the emergent ray beams in calcite crystal by total internal reflection at a thin film of Canada balsam separating two pieces of calcite crystal. This device is known as Nicol's Prism. Thus, only extraordinary ray, which is plane polarized, is transmitted through the Nicole prism made by calcite crystal.

Construction

A calcite crystal, with length three times as its width, is taken. The end faces are grounded such that the angles in the principal section become 68° and 112° instead of 71° and 109° earlier. This is done to enhance the field of view. The crystal is cut in two pieces from a blunt corner in a direction perpendicular to the plane of principal section as well as the end faces *PR* and *QS* as shown in Fig. 3.9 (*a*) and 3.9 (*b*). The two cut surfaces are grounded and polished to get optical flatness, and then, cemented together by Canada Balsam. Layer. The refractive index of Canada balsam (μ_{cb}) lies between the refractive index for ordinary (μ_o) and extra-ordinary (μ_e) rays for calcite. For D-lines of sodium, $\mu_o = 1.658$, $\mu_{cb} = 1.55$ and $\mu_e = 1.486$.

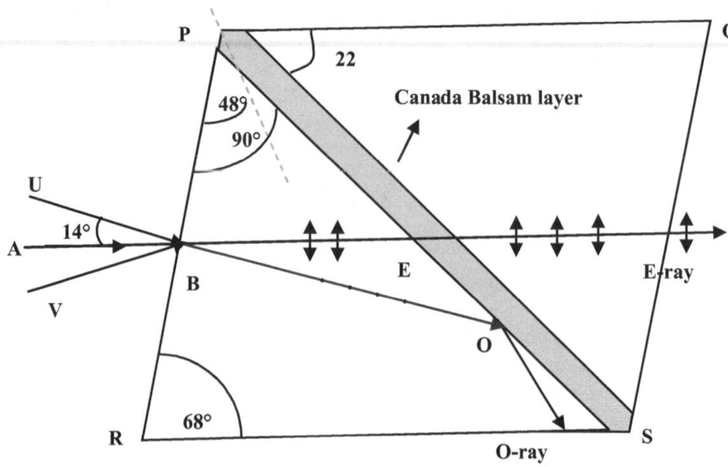

Fig. 3.9 (*a*) *Principal plane of Nicol prism*

Action

When the ray *AB* enters face *PR* in direction parallel to the long side, it is doubly refracted into ordinary *BO* and extraordinary *BE* plane polarized beam.

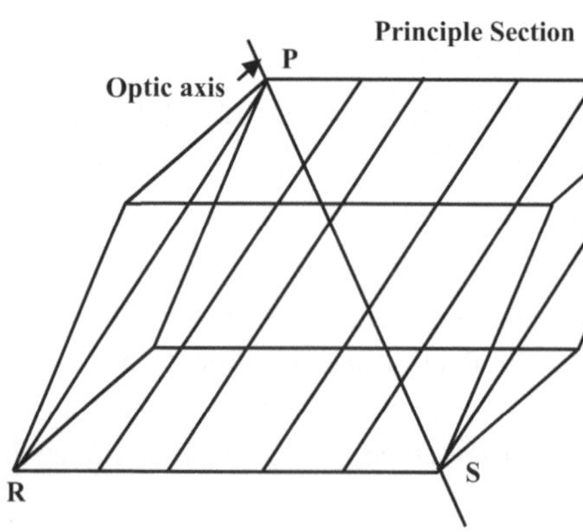

Fig. 3.9 (*b*) *A doubly refracting crystal with principal plane*

Now, $\mu_o > \mu_{cb}$, so Canada balsam acts as a rarer medium for ordinary (*O*) ray, and, $\mu_e < \mu_{cb.}$ so Canada balsam acts as a denser medium for extra-ordinary (*E*) ray. Also the dimensions of crystal are so chosen that the angle of incidence of '*O*' ray at Canada balsam surface becomes greater than the corresponding critical angle (*C*) of 69° at Canada balsam layer.

According to Snell's law, at critical angle

$$\sin C = \frac{1}{\mu}$$

$$= \frac{1}{\mu_{O-ray|w.r.t.canadabalsam}}$$

$$= 1\bigg/\frac{1.658}{1.55} \quad \Rightarrow \quad C = 69°$$

So, then O ray is completely reflected and is finally absorbed by the tube containing Nicol's Prism. The E ray, however, is not totally reflected as $\mu_e < \mu_{cb}$ and transmitted with no great loss of intensity. It is slightly displaced, but come out parallel to its original direction. Thus, only E ray is transmitted through Nicol prism and we get single polarized light.

Limitations

Nicol prism acts as a polarizing device effectively only with a slightly convergent or slightly divergent beam of light. For O-ray to get internal reflection from Canada balsam layer limit is nearly 14°. This is because that if the angle of incidence $\angle VBA$ with face *PR* is greater than 14°, the angle of incidence at Canada balsam layer for O-ray becomes less than critical angle therefore no internal reflection will take place for O-ray and both plane polarized light corresponding to E-ray and O-ray will be transmitted through Nicole prism. Hence, Nicole prism will not serve its purpose to produce single plane polarized light when $\angle VBA$ becomes greater than 14°.

Similarly, *E* ray has also a limitation; otherwise, total internal reflection of E ray will occur. This limit is also ≈ 14°. This is because that if the angle of incidence of incident beam $\angle UBA$ is greater than 14° with face *PR*, *E* ray have more refraction and more parallel to optic axis and $\mu_o \approx \mu_e$, this leads to elimination of E-ray by total internal reflection. Therefore, there won't be emergence of any plane polarized light out of the prism. Here, also prism will not serve its purpose to produce a plane polarized light.

I. Application of Nicol Prism

Nicol Prism as an Analyzer

Nicole prism can be used both as a polarizer and an anlyzer. When two Nicol prisms are arranged coaxially as shown in Fig. 3.10 (*a*) and (*b*), first prism P opposite to ordinary light is called as polarizer. This produces a plane polarized light which have vibrations in the direction parallel to the principle section of polarizer. Second prism *A*, which analyses the polarized light, is known as Analyzer. Such a combination of two Nicol prism is called as Polariscope.

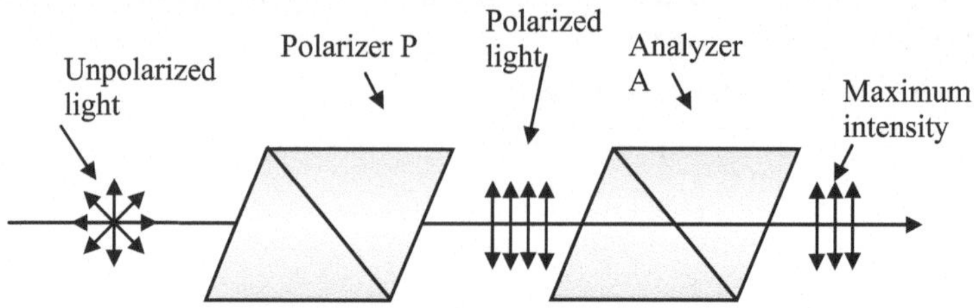

Fig. 3.10 (*a*) *Parallel Nicol prism position*

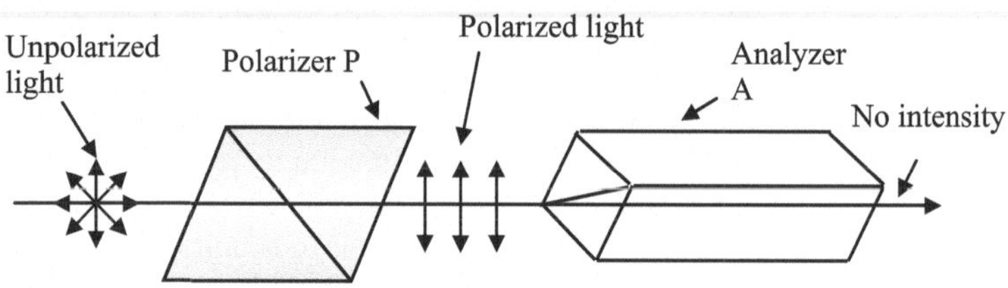

Fig. 3.10 (*b*) *Crossed Nicol prism position*

When the principle section of two Nicol prisms are parallel to each other as shown in Fig. 3.10 (*a*), the emergent E-ray vibrations are parallel to both the principle section of polarizer and analyzer. Therefore, E-ray beam from Nicol prism *P* is freely transmitted through Nicol prism *A*. The intensity of emergent light in this setting is maximum. Same intensity is also achieved when the angle between principle section of both P and A prism is 180°. Both situations belong to parallel Nicol prism position.

When the analyzer is rotated through 90° from the position of parallel Nicols, the principle section of two Nicols are mutually perpendiculars to each other so no light is transmitted by this setting of two Nicols. This position is shown in Fig. 3.10 (*b*) and is known as crossed Nicol prism position. In this position emergent E-ray vibrations from Nicol prism *P* is perpendicular to the principle section of Nicole prism *A*. Therefore, it starts to act as ordinary ray in Nicol prism *A*. Hence it is totally internally reflected and no light comes out of Nicol prism *A*. The intensity of emergent light gradually decrease from parallel Nicol prism position to crossed Nicol prism position.

5. THEORY OF PRODUCTION OF PLANE, CIRCULARLY AND ELLIPTICALLY POLARIZED LIGHTS

According to Maxwell's electromagnetic theory of light, a light wave consists of electric and magnetic vectors vibrating perpendicular to each other and as well as perpendicular to the direction of propagation of light. The electric vector is generally responsible for optical effect.

- For plane polarized light the orientation of light vector (electric vector) remains fixed, while its magnitude changes during vibration.
- When two plane polarized waves are allowed to superimpose in such a way that both electric vector vibration are of equal amplitude differ in phase by 90°, then the resultant vector rotates about the direction of propagation of light. The magnitude of the resultant light vector remains constant, but its orientation varies continuously, the light is then said to be circularly polarized light. If the resultant electric vector rotation takes place in clockwise direction, the light is said to be right circularly polarized light. If the rotation is in anticlockwise direction then the light is said to be left circularly polarized light.
- When two plane polarized waves are allowed to superimpose in such a way that both electric vector vibration are of different amplitude and also differ in phase by 90°, then the resultant electric vector start to rotate about the direction of propagation of light. In this condition both magnitude and orientation of resultant electric vector vary continuously, the light is then said to be elliptically polarized light. If the rotation of resultant electric vector takes place in clockwise direction, then the light is said to be right elliptically polarized light. If the rotation occurs anticlockwise, then it is said to left elliptically polarized light.

ANALYTICAL TREATMENT

I. The electric vectors of waves are vibrating in same direction with variable phase difference

Let us consider the case of two linearly polarized light vibrating in mutually perpendicular direction with the same frequency and travelling in the same direction, say Z-direction.

The two electric vectors are

$$X_1 = a \sin \omega t \text{ and } X_2 = b \sin (\omega t + \delta)$$

along x-axis and y - axis respectively.

Here a, b are the amplitudes of waves and ä be the phase difference between them.

Then, the resultant wave becomes

$$\begin{aligned} X &= X_1 + X_2 \\ &= a \sin \omega t + b \sin (\omega t + \delta) \\ &= a \sin \omega t + b \sin \omega t \cos \delta + b \sin \delta \cos \omega t \\ &= \sin \omega t\,(a + b \cos \delta) + b \cos \omega t \sin \delta \\ &= A \cos \phi \sin \omega t + A \sin \phi \cos \omega t \\ &= A \sin (\omega t + \phi) \end{aligned} \quad ...(3.5)$$

Here, $A \cos \phi = a + b \cos \delta$

And, $A \sin \phi = b \sin \delta$

Equation (3.5) represents the resultant of two waves. If the phase difference between two waves varies at random, then the two waves will be incoherent and the direction of resultant vector X varies randomly.

When $X_1 = X_2$, the resultant vector will give unpolarized light.

If $X_1 \neq X_2$, then the resultant electric vector will give partially polarized light.

II. The electric vectors of waves are vibrating perpendicular to each other with constant phase difference

Now let us consider the case when the phase difference between two waves remains constant. In this case if the electric vector vibration are along x-axis and y axis then the resultant electric vector will describe a certain closed curve in x-y plane. Here, the two electric vectors are along X-axis and Y-axis.

So along X-axis electric vector magnitude is given by

$$X = a \sin (\omega t + \delta) \quad \text{...(3.6)}$$

Along Y-axis electric vector magnitude of wave is given by

$$Y = b \sin \omega t \quad \text{...(3.7)}$$

From eqn. (3.7), $\sin \omega t = Y/b$

$$\Rightarrow \qquad \cos \omega t = \sqrt{1 - (Y/b)^2}$$

Hence, using above equation, eqn. (3.6) becomes

$$\frac{X}{a} = \sin \omega t \cos \delta + \cos \omega t \sin \delta$$

$$= \left(\frac{Y}{b}\right) \cos \delta + \sqrt{1 - \left(\frac{Y}{b}\right)^2} . \sin \delta$$

$$\Rightarrow \left(\frac{X}{a} - \frac{Y}{b} . \cos \delta\right) = \sqrt{1 - \left(\frac{Y}{b}\right)^2} . \sin \delta$$

$$\Rightarrow \left(\frac{X}{a} - \frac{Y}{b} . \cos \delta\right)^2 = \left[1 - \left(\frac{Y}{b}\right)^2\right] \sin^2 \delta$$

$$\Rightarrow \frac{X^2}{a^2} + \frac{Y^2}{b^2} \cos^2 \delta - 2\frac{XY}{ab} \cos \delta = \sin^2 \delta - \frac{Y^2}{b^2} \sin^2 \delta$$

$$\Rightarrow \frac{X^2}{a^2} + \frac{Y^2}{b^2} - 2\frac{XY}{ab} \cos \delta = \sin^2 \delta \quad \text{...(3.8)}$$

Equation (3.8) is a general equation of an ellipse.

Special Cases

I. Plane Polarized Light

When $\delta = 0 \text{ or } n\pi$

$\Rightarrow \sin\delta = 0, \cos\delta = 1$

Hence, using eqn. (3.3), we have,

$$\Rightarrow \frac{X^2}{a^2} + \frac{Y^2}{b^2} - 2\frac{XY}{ab} = 0$$

$$\Rightarrow Y = \pm \frac{b}{a} X \qquad \text{...(3.9)}$$

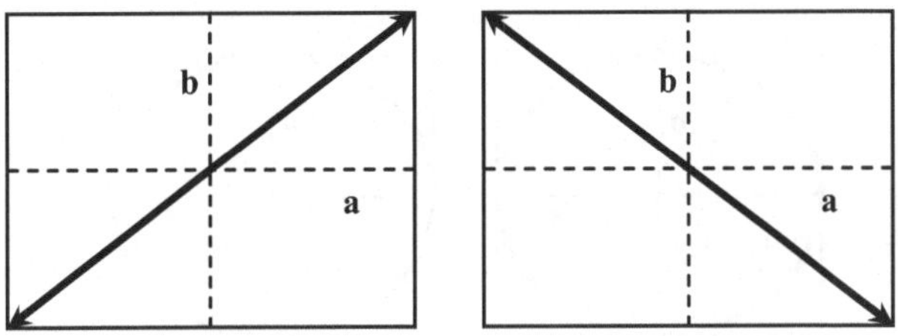

Fig. 3.11 (*a*) **Fig. 3.11 (*b*)**

Fig. 3.11 *Resultant electric vectors of plane polarized light*

Equation (3.9) is a straight line as shown in Fig. 3.11 (*a*). Hence the emergent light will be linearly plane polarized, vibrating along one direction. In both case of phase difference the resultant vector vibration will be as shown in the Fig. 3.11, where Fig. 3.11 (*a*) shown positive slope while Fig. 3.11 (*b*) shows negative slope.

II. Elliptically Polarized Light

When $\delta = \left(\frac{2n+1}{2}\right)\pi$

Here, $n = 0, 1, 2, 3, \ldots$

$\Rightarrow \sin\delta = \pm 1, \cos\delta = 0$

Hence, using eqn. (3.8), we have

$$\Rightarrow \frac{X^2}{a^2} + \frac{Y^2}{b^2} = 0 \qquad \text{...(3.10)}$$

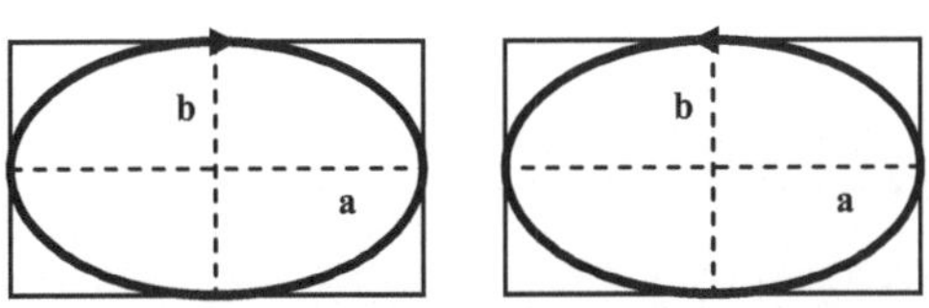

Fig. 3.12 (*a*) **Fig. 3.12 (*b*)**

Fig. 3.12 *Elliptical rotation of resultant electric vector*

This is the equation of an ellipse. Hence the emergent light will be elliptically polarized.

The resultant electric vector rotation about direction of propagation of light in elliptically polarized light is shown in Fig. 3.12 (*a*) and Fig. 3.12 (*b*)

III. Circularly Polarized Light

In eqn. (3.10), when $a = b$

And the thickness of plate is such that

$$\delta = \left(\frac{2n+1}{2}\right)\pi \text{, where } n = 0, 1, 2, 3$$

So we have,

$$X^2 + Y^2 = a^2$$

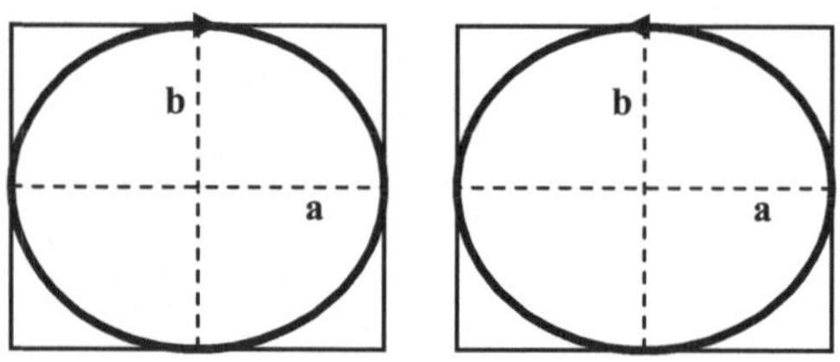

Fig. 3.13 (*a*) **Fig. 3.13 (*b*)**

Fig. 3.13 *Circular rotation of resultant electric vector*

This is an equation of a circle of radius a. Hence the emergent light is circularly polarized. The resultant electric vector rotation about direction of propagation in circularly polarized light is shown in Fig. 3.13 (*a*) and Fig. 3.13 (*b*).

6. QUARTER WAVE PLATE

We consider a calcite plate whose refracting faces cut parallel to the optic axis. When a plane polarized light falls normally on a thin plate of uniaxial crystal (here calcite) cut parallel to its optic axis, the light splits up into *O* and *E* rays. They travel the same path but with different velocities. This is shown in Fig. 3.14 for a plate of thickness *t*. The velocity of *E* ray is greater than that of *O* ray. Hence, a path difference is introduced between them. If the thickness of the plate is such that it introduces a phase difference of $\pi/2$ radian or path difference of $\lambda/4$, then the plate is called 'Quarter Wave Plate'.

Theory

Path difference between the two rays is

$$\mu_0 t - \mu_e t = t(\mu_0 - \mu_e)$$

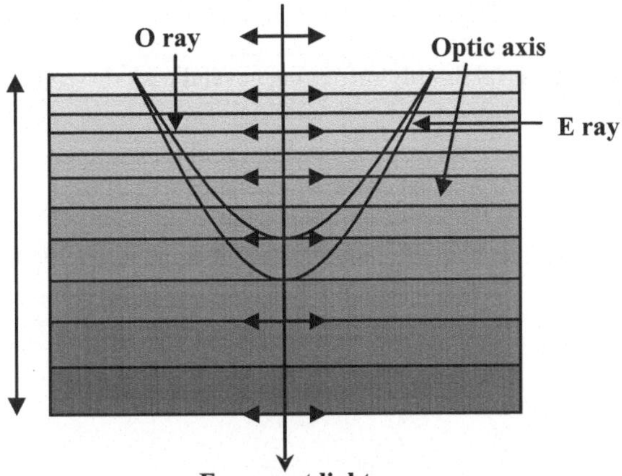

Fig. 3.14 *O-ray and E-ray propagation through retardation plate*

For a quarter wave plate, we have

$$t(\mu_0 - \mu_e) = \frac{\lambda}{4}$$

$$\Rightarrow \qquad t = \frac{\lambda}{4(\mu_0 - \mu_e)}$$

For positive crystal like quartz

$$t(\mu_e - \mu_o) = \frac{\lambda}{4}$$

$$\Rightarrow \qquad t = \frac{\lambda}{4(\mu_e - \mu_o)}$$

This quarter plate is used to produce elliptically and circularly polarized light. When the incident polarized light vibrations make an angle 45° with the optic axis of quarter wave plate then it will produce circularly polarized light, otherwise this plate produces elliptically polarized light.

7. HALF WAVE PLATE

In this case doubly refracting uniaxial crystal faces are cut parallel to the optic axis. When a plane polarized light falls normally on a thin plate of uniaxial crystal (here calcite) cut parallel to its optic axis, the light splits up into O-and E-rays. They travel the same path but with different velocities. This is shown in Fig. 3.14 for calcite crystal for a plate of thickness t. The velocity of E-ray is greater than that of O-ray. Hence, a path difference is introduced between them. Here, the thickness of the plate is such that it introduces a phase difference of π radian or path difference of $\pi/2$, therefore, the plate is called 'Half Wave Plate'.

Theory

In case of negative uniaxial doubly refracting crystal path difference between the two rays is

$$\mu_0 t - \mu_e t = t(\mu_0 - \mu_e)$$

For half wave plate this path difference should be equal to $\frac{\lambda}{2}$, therefore, we have

$$t(\mu_0 - \mu_e) = \frac{\lambda}{2}$$

$$\Rightarrow \qquad t = \frac{\lambda}{2(\mu_0 - \mu_e)}$$

For positive crystal like quartz, we have

$$t(\mu_e - \mu_o) = \frac{\lambda}{2}$$

$$\Rightarrow \qquad t = \frac{\lambda}{2(\mu_e - \mu_o)}$$

This retardation plate is used to produce linearly polarized light. But in this case when a polarized light is made to fall on this plate, the emergent light is linearly polarized light having vibrations making angle 2θ with original direction of vibration. Here θ is the angle of incident light vibration with the optic axis of plate.

8. OPTICAL ACTIVITY

When a plane polarized light passing through some materials like quartz, sugar crystals, sugar solutions, etc., its plane of vibration is slowly rotated about the direction of propagation. In the emergent light plane of vibration is different from the incident light.

The property of rotating the plane of vibration of plane polarized light about its direction of travel by some crystal is known as Optical Activity and the substances are known as optically active substance. The optical activity of the optically active substance is shown in Fig. 3.15.

There are two types of optically active substances.

I. Dextro-rotatory Crystal

These are the substance which rotates the plane of vibration of incident polarized light in clockwise direction about the direction of propagation of light. The examples of these substances are cane sugar, quartz, etc.

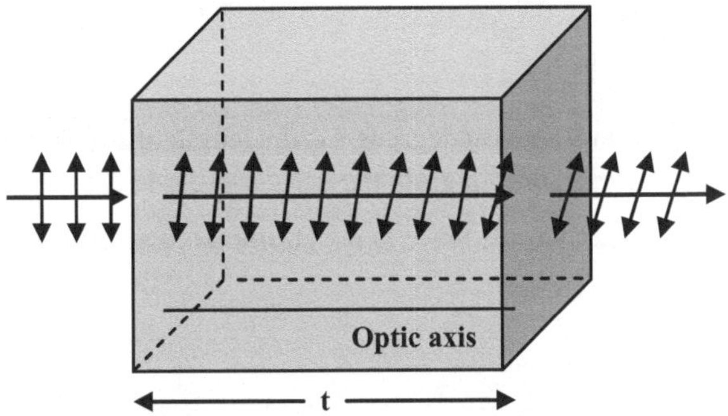

Fig. 3.15 *Optical rotation of plane of polarization*

II. Laevo—rotatory Crystal

These are the substance which rotates the plane of vibration of incident polarized light in anti-clockwise direction about the direction of propagation of light. The examples of these substances are fruit sugar, some crystals of quartz.

In 1815, Biot observed that the amount of rotation of plane of vibrations of incident polarized light is dependent on thickness and concentration of the substance used. It also depends upon the wavelength of light used and the temperature of the substance. According to Biot's law,

(*i*) Amount of rotation, θ, is proportional to the distance of travel of light into the material, t, *i.e.*, $\theta \propto t$.

(*ii*) For a liquid solutions or vapours, θ for a fixed t is proportional to the concentration (C) of the solutions or vapours, *i.e.*, $\theta \propto C$.

(*iii*) Rotation varies inversely as the square of λ of light used,

i.e., $$\theta \propto \frac{1}{\lambda^2}.$$

(*iv*) Total rotation produced by n number of optically active substances is the algebric sum of rotations produced individually, *i.e.*,

$$\theta = \theta_1 + \theta_2 + \theta_3 + ... = \sum_{i=1}^{n} \theta_i$$

Thus, phenomenon determines percentage of optically active materials present in the solution. The measurement of optical rotation determines the sugar present in a diabetic patient. It has got wide applications in sugar industry also.

9. SPECIFIC ROTATION

The specific rotation of a substance, S, at a particular temperature and for a given wavelength of light is defined as the rotation of plane of vibration of polarized light produced by one decimeter length of its solution with the concentration of one gram per *cc*, *i.e.*,

$$S = \frac{\theta}{l \times C}$$

where, θ is the angle of rotation in degrees, l is the length of solution in decimeter and C is solution concentration in gram per *cc*.

If length of optical active substance is in centimeter then specific rotation S is given by

$$S = \frac{10\,\theta}{l \times C}$$

In case of solid substances the specific rotation is defined as amount of rotation produced by one decimeter length of the substances, *i.e.*,

$$S = \frac{\theta}{l}$$

10. LORENTZ HALF SHADE POLARIMETER

Polarimeters are the instruments designed to measure the angle of rotation of the substance. By measuring rotation the specific rotation of solution can be determined. Saccharimeters are the polarimeters designed to measure the concentration of the sugar solution.

Construction

The schematic diagram of the Laurent's half shade polarimeter is shown in Fig. 3.16. A is a monochromatic source of light, L is a convex lens to make the rays parallel, N_1 and N_2 are two Nicol prisms, one acting as an analyzer and the other as polarizer. N_2 is capable of rotation about a common axis of N_1 and N_2. The rotation of N_2 can be measured in a circular scale CS. Light after passing through the polarizer becomes plane polarized with its vibration in the principal plane of N_1. Now this light is passing through a half shade device HS and then through glass tube T containing optically active solution. The concentration of this solution has to be measured. Usually T is a hollow glass tube having a large diameter at middle so that no air bubbles come across the path of light when solution is inside. The emergent light on passing through analyzer N_2 is viewed through the telescope T.

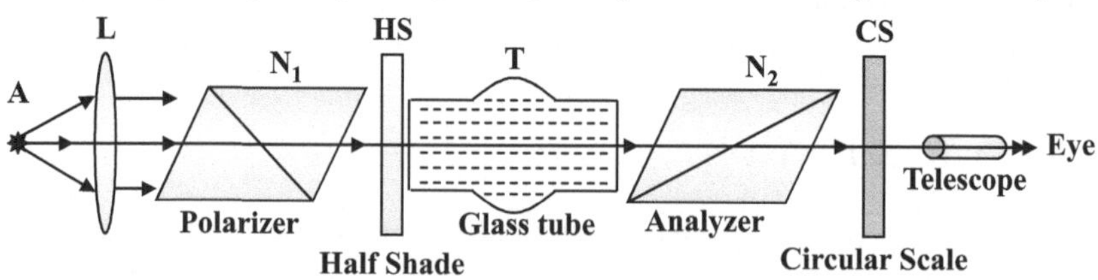

Fig. 3.16 *Schematic diagram of Laurent's half-shade polarimeter*

Action of Half Shade

The Laurent's half shade device consists of a semi-circular half wave plate *ABC* of quartz and a semi circular glass plate *ADC* as shown in Fig. 3.17 (*a*) and 3.17 (*b*).

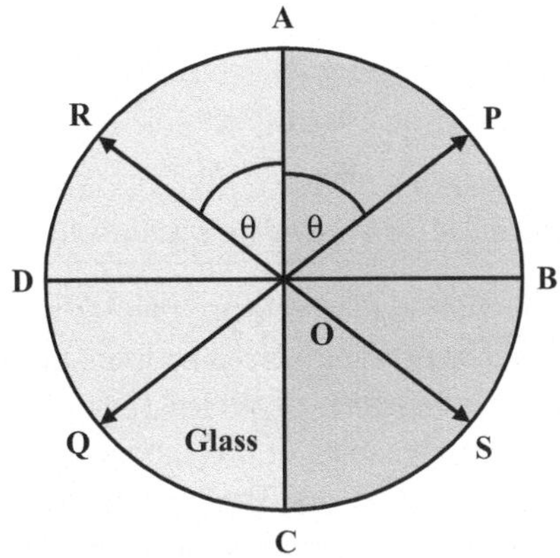

Fig. 3.17 (*a*) *Rotation of plane of vibration in half shade device*

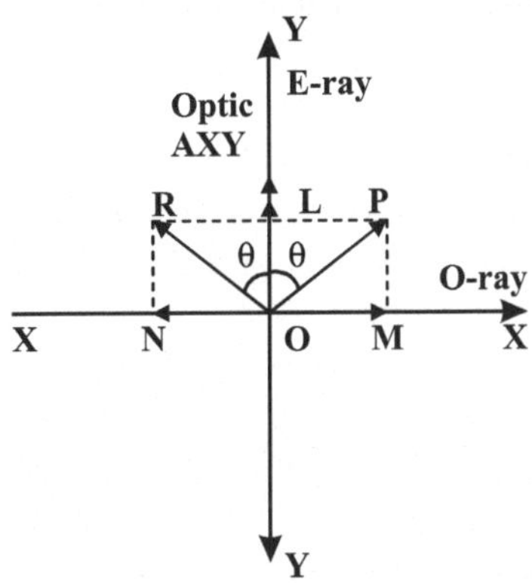

Fig. 3.17 (*b*) *Direction of plane of vibration in half shade device*

The half wave plate introduces a phase difference of 180° between O-ray and E-ray passing through it. The two plates are cemented along *AC*. The thickness of glass plate is such that it absorbs the same amount of light as the quartz plate. Let the plane of vibration of the plane polarized light incident normally on the half shade plate is along *PQ*. *PQ* makes an angle θ with *AC* as in Fig. 3.17 (*a*). The vibrations emerge from glass plate along *PQ*. Inside quartz plate light is divided into two components, one, *O* ray along *XX*', and the other, E-ray, parallel to optic axis, *YY*' as shown in Fig. 3.17 (*b*). The two components travel along same direction but with different speeds. As quartz is a positive uniaxial doubly refracting crystal,

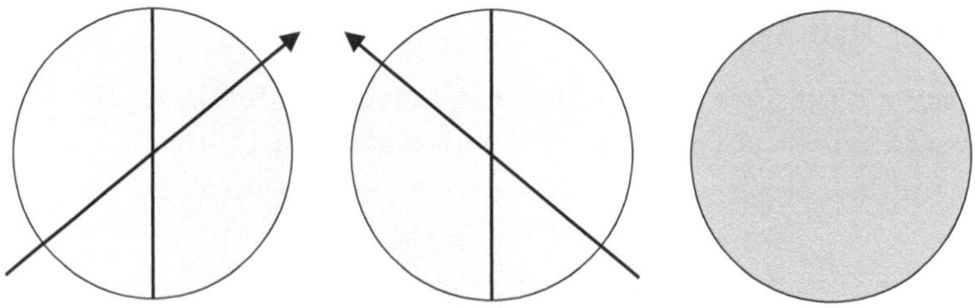

Fig. 3.18 (*a*) Fig. 3.18 (*b*) Fig. 3.18 (*c*)

Fig. 3.18 *Various field of view for different positions of the analyzer*

E ray component moves with higher velocity than *O* ray component and on emergence a phase difference of π introduced between them. Due to this phase difference the direction of *O* components reversed. If initial direction is *OM*, then, it finally becomes *ON*. So, the resultant of *E* component *OL* and *ON* will be *OR* now, which, makes an angle θ with Y-axis. The vibration of beam of emergent light out of quartz will be along *RS*. If the principal plane of N_2 is along *PQ*, the light pass unobstructed through glass plate *ADC*, but partly obstructed by *ABC*, so the glass half will appear brighter than the quartz half as shown in Fig. 3.18 (*a*). Other way, if the principal plane of N_2 is along *RS*, reverse situation occurs, so then quartz half will appear brighter than the glass half as shown in Fig. 3.18 (*b*). However, if the principal plane of N_2 is parallel to *AC* (Y axis), it is equally inclined both for *O* and *E* rays. The field of view becomes equally illuminated as shown in Fig. 3.18 (*c*). Thus, half shade serves the purpose of dividing the field of view in two halves. When there is a rotation of N_2 from the position of equal brightness or illumination a marked change in the intensity of two halves is observed.

Application

Determination of specific rotation of sugar

For the determination of the specific rotation of sugar first of all glass tube is filled with water in a polarimeter and then measure the equal brightness position in circular scale by rotating Nicol prism N_2. Now, we add sugar to the water solution with known concentration. Since sugar is an optically active substance, therefore, there will be rotation of plane of vibration of incident plane polarized light by the substance. Again measure the position of equal brightness position in circular scale by rotating Nicol prism N_2. The difference between two angular position with sugar and without sugar measure the optical rotation of sugar solution for known concentration. The length and concentration of glass tube containing optical substance is known to us. So, we can calculate the specific rotation of sugar solution by using the following relation

$$S = \frac{\theta}{l \times C}$$

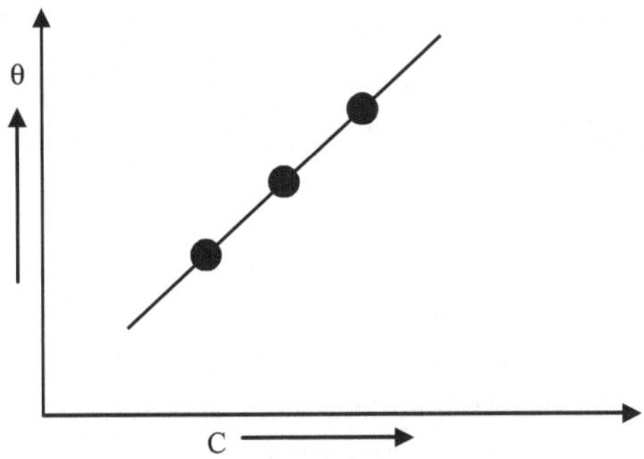

Fig. 3.19 *Graph indicating variation of rotation of plane of polarization with the concentration of the solution*

where, θ is the angle of rotation in degrees, l is the length of solution in decimeter and *C* is solution concentration in gram per *cc*. A graph between optical rotation and concentration of sugar solution is appear to be a straight line as shown in Fig. 3.19.

11. BI-QUARTZ POLARIMETER

This polarimeter is more accurate and sensitive than the former to find the angle of rotation θ. The experimental parts of Bi-quartz Polarimeter are the same as in section 10.I except the half shade device is replaced by bi-quartz plate and monochromatic light is replaced by white light.

Action

It consists of two semi-circular plates of quartz (one left handed, another right handed) each of thickness 3.75 mm. Both are cut perpendicular to the optic axis and joined along diameter *PQ* as shown in Fig. 3.20.

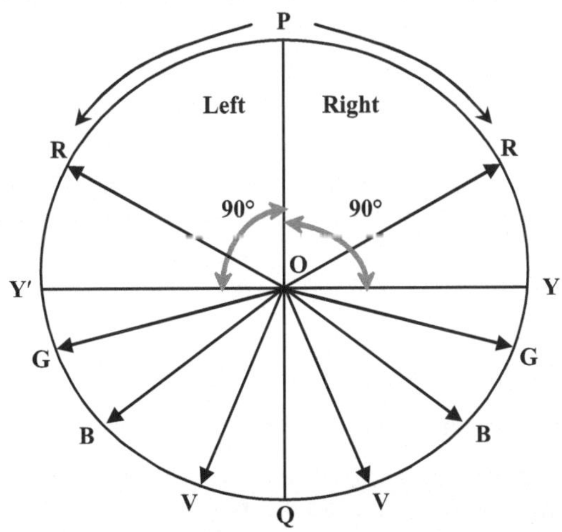

Fig. 3.20 *Rotation of plane of polarization of polarized light by Biquartz polarimeter*

The thickness of each plate rotates the plane of polarization for yellow light by 90°. When white light rendered plane polarized with a polarizer, travels through a bi-quartz normally, the phenomenon of rotatory dispersion occurs in both bi-quartz plates. Because light is along the optic axis. The planes of vibration of different colours are rotated through different angles. For yellow, it is 90° along YOY'. If the plane of N_2 is parallel to *POQ*, yellow light will not be transmitted through N_2 and the appearance of two halves in similar. A grayish violet tint is there, this is sensitive tint or tint of the passage. When N_2 is rotated to one side of this position, one half appears blue and other appears red. If N_2 is rotated in opposite direction, colours become interchanged. The position of sensitive tint is very sensitive and is used for accurate determination of optical rotation.

Application

Determination of specific rotation of sugar

For the determination of specific rotation of sugar solution (optical substance) by this polarimeter, we follow the same procedure as in the case of Laurent's half shade polarimeter given in section 10.III(*a*). In this case, a grayish violet position, known as sensitive tint, is measured during experiment. It is more accurate method to determine the specific rotation of optical active substance.

12. PHOTOELASTICITY

This branch of science deals with the analysis of stress distribution in structural and mechanical components with the help of polarized light.

Some transparent substances like glass, celluloid, bakelite, etc., are not normally doubly refracting, but they show the above phenomenon only under stress. When specimens of those substances are placed between two crossed Nicols they do not change the field of view. However, when stress is applied to those, the emergent light, in general, is elliptically polarized and hence partly transmitted the analyzing Nicol. Thus, field of view is changed. When these certain materials are subjected to external force there occur distortion in molecular arrangement of the material and the body becomes optically anisotropic (having different properties in different direction).This is known as induced birefringence and is determined by $\Delta\mu$, which is difference of refractive indices of extraordinary and ordinary rays. Thus, we have

$$\Delta\mu = \mu_e - \mu_o$$

This anisotropy in material can be induced by the action of

(*a*) mechanical strain,

(*b*) an electric field, or

(*c*) a magnetic field.

Anisotropy caused by mechanical strain can be measured by following relation

$$\Delta\mu = k\sigma$$

Here k is proportionality constant and σ is the stress applied to the specimen.

So, when these certain specimen is kept in crossed Nicol prism position under no stress, there is no change in field of view, *i.e.,* field of view is dark. As these materials are subjected to external force, colour contours are observed. The dark contours indicate the absence of strain in those areas. While colour contours indicate the deformed areas of material. In this way the contours enable us to study the distribution of the stresses in the material. This method of stress analysis is known as photo-elastic analysis and widely used in civil and mechanical practices. The mechanically induced birefringence is used to study stresses in girders, beams etc.

Uses

(*i*) For testing glass used for making optical instruments.

(*ii*) Measurement of strains in the buildings. The model is usually made of Xylonite.

(*iii*) The stress distribution is complex engineering structures, *i.e.,* bridges, girders, gear teeth, etc., can be analyzed by the photo-elastic studies of their transparent models. These models are subjected to loads and stresses are investigated when placed between crossed Nicols. This helps to determine the stresses to be expected in actual practice.

4

Laser

LASER stands for Light Amplification by Stimulated Emission of Radiation. It is the outgrowth of MASER (Microwave Amplification by Stimulated Emission of Radiation) first developed by Gorden & Townes, and, Basov & Prokhorov simultaneously in 1954. Although the principle of LASER is first predicted by Einstein way back in 1916, experimentally laser was first developed by Mainman in 1960. In 1961 gas filled Helium-Neon laser was developed by Javan. It 1964 Nobel Prize was awarded to Townes, Basov & Prokhorov in this field. Later on, with the development of semiconductor technology it was possible to develop semiconductor lasers. These made them possible to use laser for almost all practical applications, specially, in communication area.

LASER is the only artificially developed coherent source of light. It is regarded as a perfect source of coherent light source for optical communication. However, coherence property can be rechecked only after a certain time and after a certain distance traveled by the beam.

1. COHERENCE PROPERTIES OF LASER

The coherence properties of light between two sources maintain a constant phase difference between them. However, we have to see to what extent the radiations from different atoms of a given source may be coherent with respect to space and time. This gives rise to two criterion of coherence, firstly, up to how much distance traveled by the waves, this coherence property is maintained, it is known as Spatial Coherence, and, secondly, up to how much time this property of coherence will be maintained, it is known as temporal Coherence.

I. Temporal Coherence

A coherent source always produce perfect sinusoidal wave for all values of time. However, transition from a higher state to a lower state producing light emission is within 10^{-10} sec. So after that phase changes abruptly. Hence the average time interval for which the field, *i.e.*, light wave remains sinusoidal is known as

Coherence Time (τ) or, time for temporal coherence. The distance L for which the field remains sinusoidal is called Coherence Length and is given by $L = \tau c$. Here, c is the velocity of light. The coherence length can easily be determined by using Michelson Interferometer experiment. We can get interference fringes only when the two coherent sources are present. So within coherence length we can get sharp fringes otherwise not. For sodium light, $L \approx 3$ cm, for Neon (6328 Å), but for Cadmium (6438 Å), $L \approx 30$ cm.

So for N*a*, $$\tau = \frac{L}{c} = \frac{3}{3 \times 10^{10}} = 10^{-10} \text{ sec},$$

but for Cadmium, $$\tau = \frac{L}{c} = 10^{-9} \text{ sec}.$$

Here the constant phase difference property is maintained for a certain time span. So this is a longitudinal coherence. This depends on bandwidth and so the monochromaticity gives a measure of temporal coherence.

II. Spatial Coherence

The spatial coherence is the phase relationship between the waves of propagation along its path at different points in space. This gives the spatially uniform distribution of wave in space in a direction perpendicular to the direction of propagation. Hence it is lateral or transverse coherence. Conventional light sources have few mm to few cm of coherent length. For example, green line of mercury has nearly 1.1 cm, sodium light have ≈ 3 cm, and orange light of krypton have 21 cm. However, for a laser beam, it has a minimum few hundreds of km coherent length. The more coherent the light is the less divergent the beam is. Hence divergence property is an outcome of spatial coherence.

2. SPONTANEOUS AND STIMULATED RADIATION

Light is absorbed or emitted by atoms, molecules, ions during their transition from one energy state to another. When a particle is excited by taking (absorbing) energy externally, it reaches the excited state, E_2, from the ground state, E_1, as shown in Fig. 1. The numbers of particles in excited states is lesser than in non-excited state. The life time, τ, or the stay of the particle at excited state is very small, *e.g.*, for H-atom, $\tau \approx 10^{-8}$ sec.

Some excited states in which life time is more than 10^{-8} sec. ($\tau \approx 10^{-5}$ sec) are known as Metastable States.

Important points

(*i*) Transitions are possible within the states only allowed by the Selection Rules.

(*ii*) Transitions from one energy state to other can be non-radiative, *i.e.*, dissipation will be in form of heat.

The probability of transitions from higher energy state to the lower one with emission of radiation is made up of two factors, one is a constant factor and the other is a variable one.

I. Constant Factor : Spontaneous Emission

As in Fig. 4.1 the particle passes from higher energy state, E_2, to the ground state, E_1, with a spontaneous emission of photon with energy, $h\nu$. Similarly, spontaneous emission (emission by its own) from state 2 to 1 emitting a photon of frequency í is given by $\nu = \dfrac{E_2 - E_1}{h}$.

The phenomenon of spontaneous emission is shown in Fig. 1 stepwise.

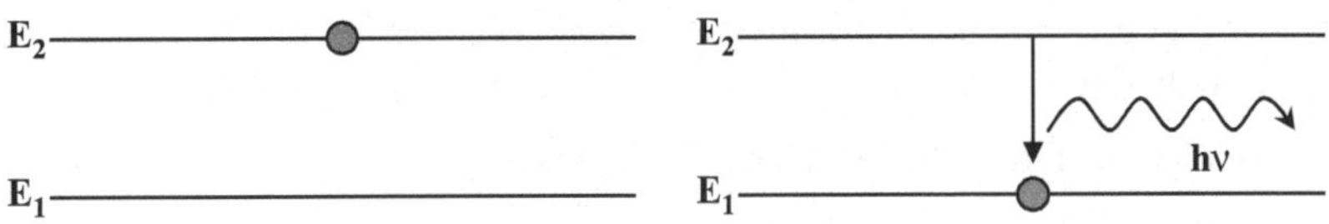

Fig. 4.1 *(a) Initial condition* **Fig. 4.1** *(b) Final condition*

Fig. 4.1 *Spontaneous emission*

The emission is random. For large number of atoms, radiation has random direction and a random phase. The waves coincide neither in same wavelength nor in phase. Thus the radiation is incoherent and has a broad spectrum. These beams have no constancy in phase between them. So there is no multiplication of the individual waves and monochromaticity is not at all maintained.

II. Variable Factor: Stimulated Emission

For spontaneous emission, the atom in the excited state has to wait. However, if artificially we can push the atoms from its excited state to downwards, it leads to stimulated emission predicted first by Einstein in 1916. A photon of energy exactly equal to the transition energy gap difference gives the stimuli and the atom in the excited state drop back to the lower energy state emitting same energy as of incident photon. They will also be in same polarization state. This externally controlled process has photon multiplication feature also. One photon induces an atom to emit second photon and while traveling, these two unidirectional photons deexcite another two atoms in their path produces four and this process of photon multiplication continues in an avalanche manner. These confirm the possibility to obtain large intensity of light by their constructive interference. This factor depends on the presence of the radiation of same frequency. It implies that a particle can pass from excited to ground state emitting a photon not only simultaneously but also when forced to it, under the effect of another external quantum. The phenomenon of stimulated emission is shown in Fig. 4.2 stepwise.

If an atom is in excited state then an incident photon of correct energy may cause the atom to jump to lower energy state, emitting an additional photon of the same frequency.

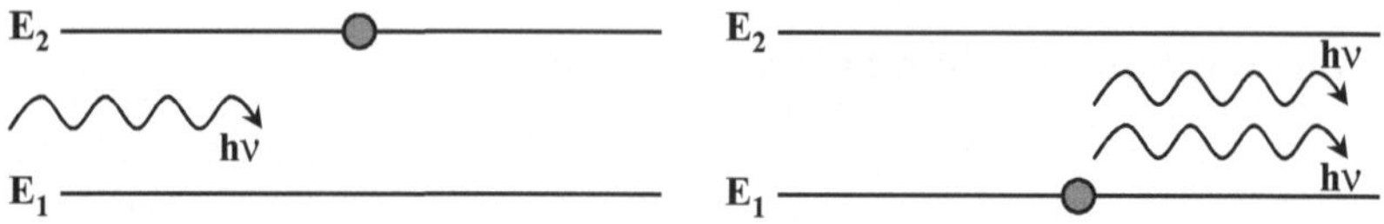

Fig. 4.2 (*a*) *Initial condition* **Fig. 4.2** (*b*) *Final condition*

Fig. 4.2 *Stimulated emission*

Thus, two photons of same frequency are present in radiation. This phenomenon is called stimulated emission of radiation. These two photons are coherent and travel in the same direction. This is coherent with the stimulating incident radiation having same frequency and phase as the incident one.

The radiation of the particle which is in excited state, stimulates the emission of a similar radiation by the particle, by transition to lower energy state.

3. PRINCIPLE OF LASER ACTION

LASER action consists of following necessary conditions or parameters.

I. Active Medium

- This is the medium where light is getting amplified. This may be solid, liquid or gas.
- This is the medium where only a fraction of atoms of a particular species takes part in stimulated emission in LASER action. These atoms are also known as Active Centers. The remaining material acts as Host. For example, in helium – neon (He-Ne) laser, neon is the active center and helium is the host.

II. Population Inversion

Emission of high intensity of radiation by stimulated emission in laser requires a large amount of excited atoms in the upper levels. Normally the higher energy levels have less occupancy than the lower levels. However, for laser emission we need to have the opposite. This population inversion is an artificial situation required to achieve stimulated radiation. The number of active atoms occupying an energy state is called population of that state.

We consider N_1 and N_2 are the populations of the lower E_1 and upper E_2 energy levels respectively. In thermal equilibrium these populations are fixed by the Boltzmann factor. From Boltzmann's statistics,

$$N_1 \propto e^{-E_1/kT} \text{ and } N_2 \propto e^{-E_2/kT}.$$

Hence, $$\frac{N_1}{N_2} = e^{-(E_2 - E_1)/kT} \quad \ldots(4.1)$$

The negative sign in the exponent implies that $N_2 \langle\langle N_1$ at equilibrium. Here more atoms are in lower energy level E_1. This is the normal state. Atoms at E_1 will occasionally absorb radiation and make transition to E_2. Atoms in E_2 will occasionally emit radiation and make transition to the lower level. This is spontaneous emission. Laser operation requires obtaining stimulated emission almost exclusively. There must be more atoms in the upper level than in the lower level in order to achieve stimulated emission exclusively. Therefore, a non-equilibrium state is to be produced in which the population of the upper energy level exceeds to a large extent the population of the lower energy level. When this situation occurs, the population distribution between the levels E_1 and E_2 is said to be inverted and the medium is said to have gone into the state of Population Inversion. Extending eqn.(4.1) in this case, we observe that N_2 can exceed N_1 only if the temperature is negative. Therefore, the state of population inversion is referred to as a negative temperature state. The name only suggests that this state is a non-equilibrium state. The population inversion is actually achieved at normal temperature.

(*a*) **Factors affecting population inversion:** There are three phenomena that can prevent population inversion. First one prevents lower laser level from undergoing its normal decay process in a gas laser. This is known as Radiation Trapping. The second, applicable to gas or plasma lasers, tends to reduce the population of the upper laser level and move it to the lower laser level via collisions. This is the Electron Collisional Thermalisation. The third is the absorption within the gain medium by various processes present within the medium that are not associated with upper and lower lasing levels.

(*b*) **Negative absorption:** Intensity of light decreases exponentially as it propagates through a medium as $I = I_0 e^{-\alpha L}$.

Here I_0 is the initial intensity; L is the length traveled by light in the medium and α is the absorption coefficient. So, the amplification of light may be described by the above equation but with negative α. In this point of view a medium in the state of population inversion is called a medium with a negative absorption coefficient and the light amplification is referred to as negative absorption.

- Here the numbers of particles (N_2) in higher energy state E_2 becomes more than the same (N_1) in lower energy state E_1 unlike normal situation. Thus the name 'Population Inversion' arises.

- This is possible only when temperature is negative under normal case. So it is also referred as Negative Temperature State.

However, this is totally hypothetical as we can not go below 0°K and it is obtained generally at room temperature. Generally light intensity decreases while propagating into a medium, so a medium in a state of Population Inversion is also called a medium of Negative Temperature Coefficient.

(*c*) **Pumping**

- Here the numbers of particles (N_2) in higher energy state E_2 becomes more.
- Sustaining the condition of population inversion requires the atoms have to be supplied continuously to excited state by any means.
- This process of supplying energy to the medium to transfer it into the state of population inversion is known as Pumping.

There are several methods of pumping processes.

(*i*) Optical pumping using light source (in Solid State Laser, *e.g.*, ruby).

(*ii*) Electrical pumping using gas discharge (in Gas Laser, *e.g.*, He-Ne).

(*iii*) Direct pumping using electrical energy which is ultimately converted into light energy (semiconductor laser).

(*iv*) Chemical pumping using energy from chemical reactions (carbon dioxide laser).

Although it seems to have two energy levels to attain laser action, however, this is not possible. So, three or four level pumping schemes are often used and four level pumping shows maximum lasing efficiency. The base level is called the lower lasing level and the upper level as upper lasing level. The topmost level is known as pumping level. So, for the four level lasers we have two upper levels. In three level schemes laser transition takes place between upper and lower lasing level. However, the laser transition is in between two upper levels only in four level laser. The general energy level diagrams of three and four level lasers are shown in Fig. 4.3.

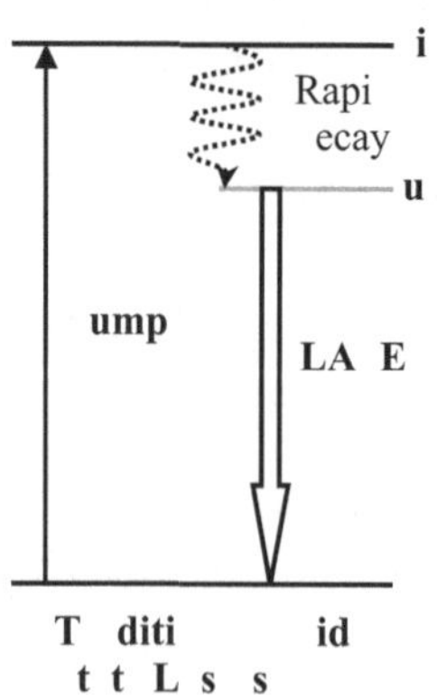

Fig. 4.3 (*a*) *Three level laser*

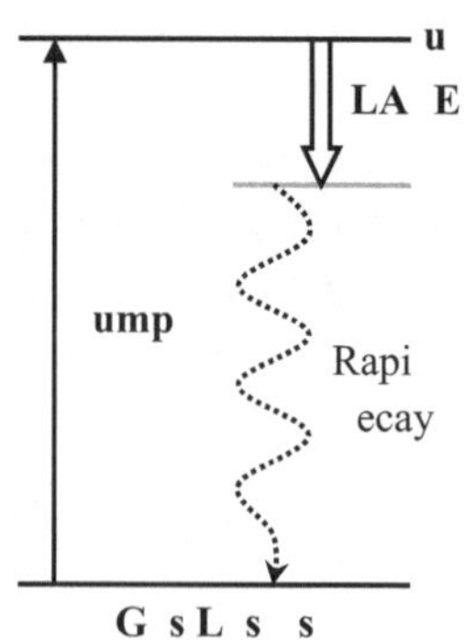

Fig. 4.3 (*b*) *Three level laser*

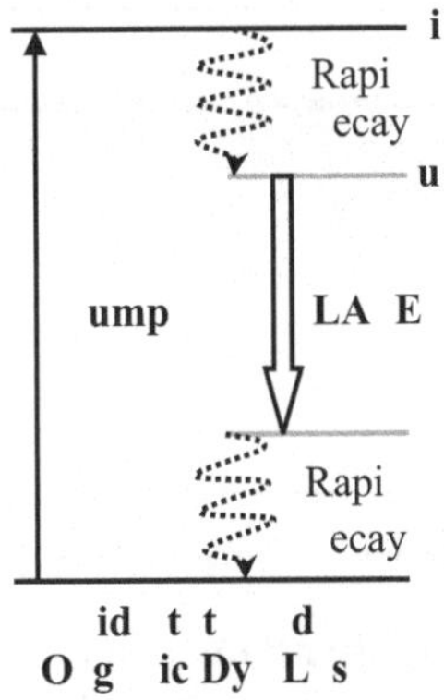

Fig. 4.3 *(c) Four level laser*

Fig. 4.3 *Laser transitions in various energy level schemes*

(d) **Optical resonator**

- For the continuation of stimulated emission and the increase of light intensity in a cumulative way a positive feedback of light into the medium is a must.
- Spontaneous photons are emitted by atoms independently at random which produce spontaneous emissions in different directions. So the final result is incoherent light.
- An intense beam of light can be available only when the optical gains of the system equals the optical losses.

All these three requirements are met with an Optical Cavity Resonator, which converts the active medium into a Light generator.

An optical resonator consists of two opposing plane parallel mirrors with the active material placed in between as shown in Fig. 4.4. One mirror is semi-transparent and the other one is totally reflecting. The mirrors are set normal to the optic axis of the materials. Now, when in an active medium pumping is taking place to achieve population inversion state. Spontaneous photons are emitted in initial stage. In order to have a coherent light output, we should have to select photons of specific directions and others are treated as reject one and these stimulated photons made to pass through the medium a number of times. So, here the mirrors constituting the resonator cause the directional selectivity. Photons emitted in any direction other than parallel to the optic axis (as clear from Fig. 4.4), will pass out from the sides of the resonator and are lost. Photons emitted parallel to the axis are augmented. On reaching semi-transparent mirror some are transmitted out and rest reflected and they de-excite more atoms and build up their strength. At fully reflecting mirror major portion will be reflected. At each reflection at the front mirror, the beam is partially transmitted through it and partially fed back into medium. Thus, the photon density of the medium enhances. As the oscillations build up to enough intensity it emerges out through the front mirror as a highly collimated intense beam of LASER.

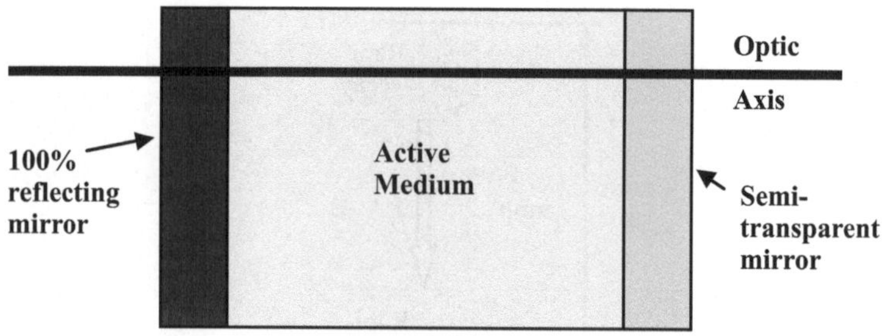

Fig. 4.4 *Optical resonator*

(*i*) **Mechanism of light amplification:** Light can be amplified by an active medium taken into the state of population inversion. A photon spontaneously emitted by an atom acts as the input and starts stimulated emission. To sustain it and to increase the light intensity in a cumulative way, a positive feedback of light into the medium is must. *Secondly*, spontaneous photons are emitted by atoms independently in various directions which produce stimulated emissions in different directions. The resultant effect is a production of incoherent light and this way it squanders away advantages inherent in stimulated emission. *Thirdly*, an intense light beam can be available only when the optical gain of the system exceeds the optical losses. *Finally*, a large incident radiation energy density is required to keep the stimulated emissions dominating. All these requirements are met with the help of an optical cavity resonator which converts active medium into a light generator.

(*ii*) **Action :** Fig. 4.5 depicts the successive stages of the action of an optical resonator. Initially the active centers in the medium are in ground state as in Fig. 4.5 (*a*). Through suitable pumping mechanism the material is taken to a state of population inversion (Fig. 4.5 (*b*)). Spontaneous photons are emitted in the initial stage in every possible directions as are the stimulated photons (Fig. 4.5 (*c*)). To generate a coherent light output it is imperative that photons with specific directions are selected while others are rejected. Also to attain the maximum possible amplification of light the stimulated photons are to be made to pass through the medium a number of times. The mirrors constituting the resonator cause the directional selectivity. Photons emitted in any direction other than parallel to the optic axis will pass out of the sides of the resonator and are lost. Photons emitted parallel to the axis are augmented. On reaching the semitransparent mirror some of the photons are transmitted out and part of them will be reflected back.

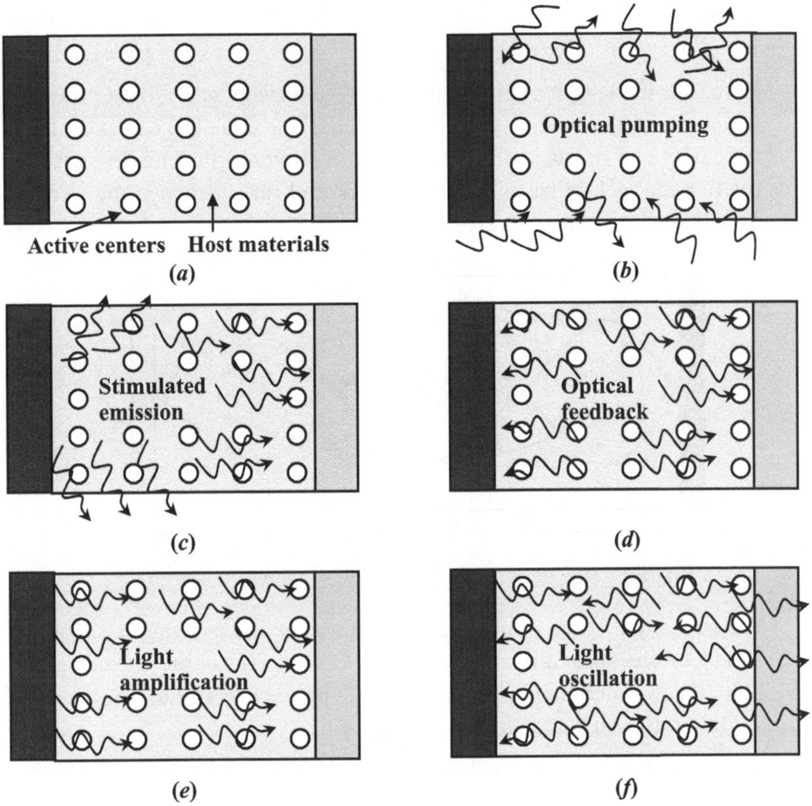

Fig. 4.5 *Stepwise laser action in an optical resonator*

While propagating in the opposite direction they de-excite more and more atoms and build up their strength (Fig. 4.5 (*d*)). At the 100% reflecting mirror most of the photons will get reflected. The amplified beam will move along the same path as the starting photon (Fig. 4.5 (*e*)) and undergoes multiple reflections at the mirrors and gain in strength. At each reflection at the mirrors the beam is partially transmitted through it and partially fed back into the medium. Laser beam oscillation begins when the amount of amplified light becomes equal to the amount of light lost through sides of the resonator, through the mirrors and through the absorption by the medium. As the oscillations build up to enough intensity it emerges through the front build up to enough intensity it emerges through the front mirror as a highly collimated intense beam (Fig. 4.5 (*f*)). However, the wave properties of light require an additional condition to be fulfilled. The waves propagating within the cavity resonator should take on a standing wave pattern, so that a continuous increase in the wave amplitude occurs.

The condition for forming a standing wave within the resonator is that the optical path traveled by a wave between the consecutive reflections should be of an integral multiple of the wavelength.

Thus, $2L = m\lambda_m$

$\Rightarrow$ $L = (m\lambda_m)/2$ ...(4.2) $[m = 1, 2, 3, \ldots]$

Here, L is the length of the cavity; λ_m is the wavelength of light produced within the material. Only wavelengths satisfying eqn.(4.2) can exist inside the cavity in a steady state. Waves of other wavelengths interfere destructively with each other as they pass back and forth between the mirrors. Thus, they attenuate quickly.

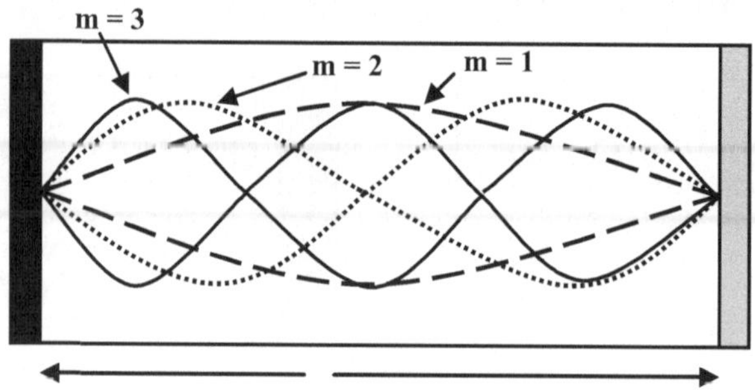

Fig. 4.6 *Standing wave formation in an optical resonator*

We now say that the optical cavity is resonant at wavelengths satisfying eqn.(4.2). Because of its relative length as compared to the wavelength of light, the resonator may simultaneously support several standing waves of slightly different wavelengths as in Fig. 4.6.

They are called longitudinal modes. Each mode has a distinct frequency and is given by

$$\nu_m = \frac{c}{\lambda_m} = \frac{c}{2L/m}$$

$$\Rightarrow \quad \nu_m = \frac{mc}{2L} \qquad \ldots(4.2)$$

Here, m is called the mode number, ν_m are the frequencies which are sustained in the cavity and hence the emerging beam is restricted to these frequencies only. Out of this broad range, the resonator selects and amplifies only a certain narrow bands and if desired only one such band. For this reason only the monochromaticity in Laser beam is obtained.

(*iii*) **Losses inside the cavity:** In resonant cavity (passive) the light irradiance gradually declines with travel distance owing to various losses in the medium inside the cavity or escapes through the side surface and is given by $I(z) = I_0 e^{-\alpha z}$. Here α is the loss factor.

(*a*) **Radiant Losses:** A proportion of the field energy leaves the cavity through the output mirror as Laser radiation. This loss is the radiant loss.

The loss factor due to this, α_{rad} is given by $\alpha_{rad} = \dfrac{\ln(1/R)}{2L}$

Here, L is the resonator length and R is the reflectivity of the output mirror.

(*b*) **Diffraction Losses :** These are an important class of losses caused by the fact that the aperture (lateral dimensions) of the resonator's mirror, as well as other components are of limited dimension which are fit for diffraction on the mirror edge. This entails a loss of a proportion of the flux. The collimated light beam of mirror size can be well aligned with the mirror, yet owing to the wave properties of light a portion of beam energy will walk off the mirror edge, as it were, and leave the cavity. These losses are called Diffraction Losses. Diffraction losses increase as the mirror diameter a increases and the wavelength and resonator length increase.

4. RUBY LASER

Ruby is sapphire (Al_2O_3) crystal doped with 0.05% (by weight) of chromium (Cr^{3+}) ions. At this concentration there are 10^{25} (approx.) Cr^{3+} ions per cubic meter. Cr^{3+} ions are the active centers and aluminium, oxygen atoms are inert ions act as the host. Due to the presence of Cr^{3+}, the transparent Al_2O_3 crystal shows a pink or reddish colour. It operates in a pulsed mode with laser wavelength of 6943 Å. The laser is flash lamp operated with pulse durations ranging from a fraction of millisecond to a few milliseconds.

Structure

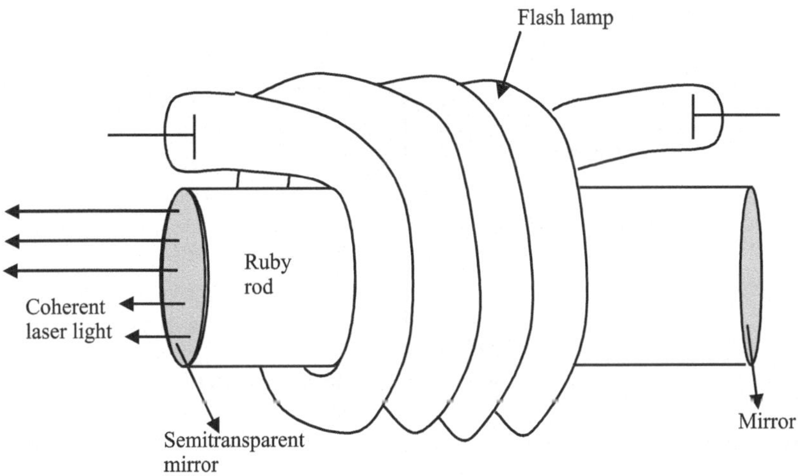

Fig. 4.7 *Schematic diagram of Ruby laser*

The schematic of ruby laser is shown in Fig. 4.7. Ruby rod is of diameter up to 25 mm and length up to 20 cm. Its ends are grounded and polished so that the end faces are parallel and also perpendicular to the axis of the rod. One face is silvered to form a mirror of 100% reflectivity and the other end is made semitransparent. The silvered faces constitute the mirrors of Fabry Prot resonator. The rod is surrounded by a helical photographic Xenon flash lamp to produce white light flashes. Water is generally used as a coolant to flow inside the pumping cavity to remove excess heat from the ruby rod. Water is transparent in the pumping wavelength regions, so has little effect on reduction of pumping intensity.

Working

It is a three level laser. The energy levels of Cr^{3+} ions are shown in Fig. 4.8. E_1, E_2, E_3 and E_{316}' are the ground state, metastable state, first and second excited states respectively. E_3 and E_3' are highly unstable. Ions pumped from E_1 to E_3 by green light and E_1 to E_3' by blue light undergoes non-radiative transitions to the metastable state E_2 and accumulate there. After some time population inversion is achieved with respect to E_1 and finally laser emission happens through semitransparent mirror. After laser emission population inversion stops and so, lasing action stops till population inversion is again achieved. So, laser output here is pulsed only, not continuous. It consists of pulses of microsecond duration. In ruby laser, the efficiency is low and also a large excitation energy is required for pumping to achieve population inversion.

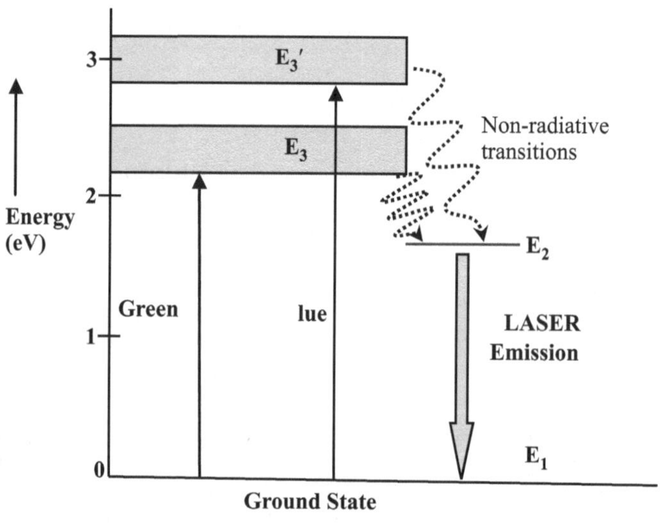

Fig. 4.8 *Energy level diagram of Ruby laser*

Use

For its high pulse energy it is used in holography. It is also used to remove tattoos and also skin etches.

5. HELIUM NEON LASER (HE-NE) (GAS LASER)

In ruby laser we are not getting a continuous laser beam. It is basically a pulsed laser. To overcome this difficulty gas laser was developed. Helium Neon Laser

(He-Ne) was the first successful gas laser. This laser is trouble free and has extremely long operating lifetime. They operate in a low pressure mixture of helium and neon gases and the laser transitions occur within the neutral atomic species. The most common wavelength is the 632.8 nm transition in the red portion of the spectrum. One particularly strong transition 3.39 μm competes with the 632.8 nm transition, having a common upper laser level, and therefore must be quenched by creating high loss in the laser cavity at that wavelength to allow only the 632.8 nm transition to lase. These lasers generally produce powers in the range of 0.5 – 50 mW in the red.

Construction

A laser cavity of length 0.3 mt would thus allow several longitudinal modes to lase simultaneously within the laser cavity. The complete laser assembly is typically 0.15 – 0.5 mt in length, with lateral dimensions of the order of several tens of mm with operating life time of 50,000 hours. It consists of a fused quartz tube as shown in Fig. 9 and is filled with a mixture of He and Ne gas (approximately 15% Ne). The gain region is an anode region where a high voltage electrode exists and an cathode region where a high voltage electrode is located as shown in Fig. 4.9.

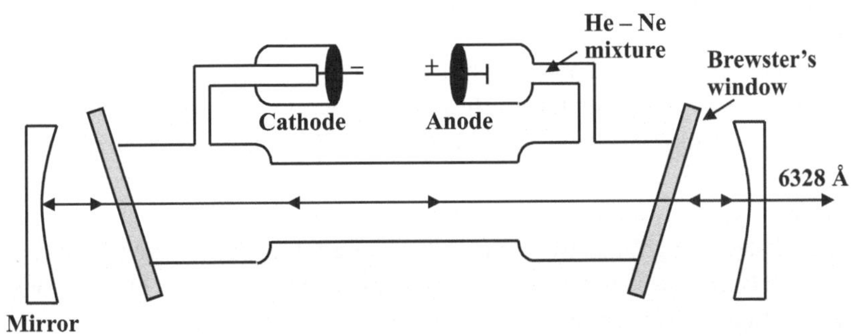

Fig. 4.9 *Schematic diagram of He-Ne laser*

Electrodes are connected to high voltage power supply to produce gas discharge. On the axis of tube two reflectors are fixed (which form Fabry Perot Resonator). The distance between the mirrors is adjusted such that it equals $m\lambda/2$ to support standing wave pattern. The optimum gas pressure is determined by the product of the pressure and the tube bore diameter. The optimum is 3.6 – 4.9 torr-mm with a He : Ne ratio of 5 : 1. thus, for a discharge bore diameter of 1.5 mm, the optimum total gas pressure would be 2.5 torr (1 torr is equal to the pressure of 1 mm column of mercury). The laser mirrors are located internally in the anode and cathode regions that comprise the optical cavity. If a polarized output is desired from the laser, a Brewster angle window is also inserted in the laser cavity. A higher starting voltage is used for a few microsecond or less to produce initial ionization within the discharge. The laser power is emitted from a 1% transmitting flat mirror at the anode end and the cathode has 99.9% reflecting spherical mirror whose curvature is determined by the desired cavity length. Actual construction of a He-Ne laser is shown in the Fig. 4.10.

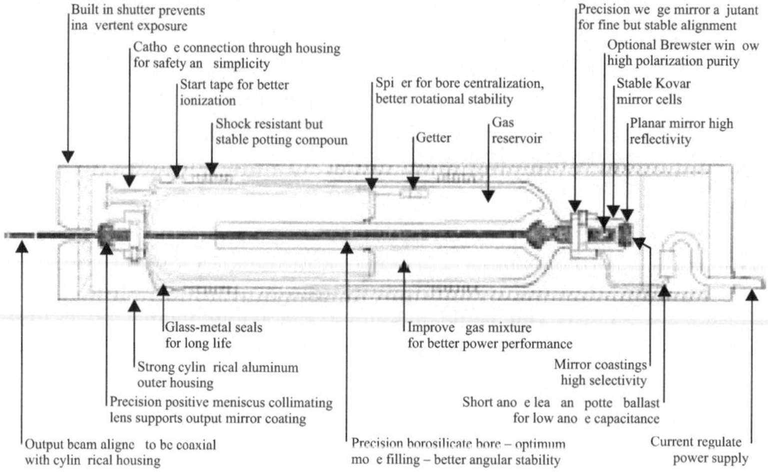

Fig. 4.10 *He-Ne laser with its various components*

Working

He-Ne laser uses four level pumping systems. With power the electric field ionizes some of the atoms in the mixture of He and Ne. Electrons and ions are accelerated between anode and cathode as shown in Fig. 4.11. The energetic electrons excite He-atoms through collisions from F_1 to F_2 and F_3. These are Metastable states. With more current through discharge tube more He-atoms accumulate in the excited states though radiative transitions are forbidden, the excited He-atoms can return to ground state by transferring their energy to Ne-atoms by collisions. Such an transfer (or transition) takes place when two colliding atoms have identical energy states. F_2 and F_3 levels nearly coincide with E_4 and E_6 levels of Ne respectively. So Ne-atoms acquire energy and go to the excited state on collisions with He-atoms, which return to F_1 after transferring the energy. This is the main pumping process of He-*N*e laser. Ne-atoms are active centre. The probability of energy transfer from He-atom to Ne-atom is more as He- and Ne-atoms are in 5 : 1 ratio. The E_4 and E_6 levels are Metastable states. So, as collisions are going on, at ordinary temperature, E_3 and E_5 levels are sparsely populated, and a state of population inversion is achieved between E_5 and $E_{6,}$ and, E_4 and E_3 levels. Consequently, three laser transitions occur.

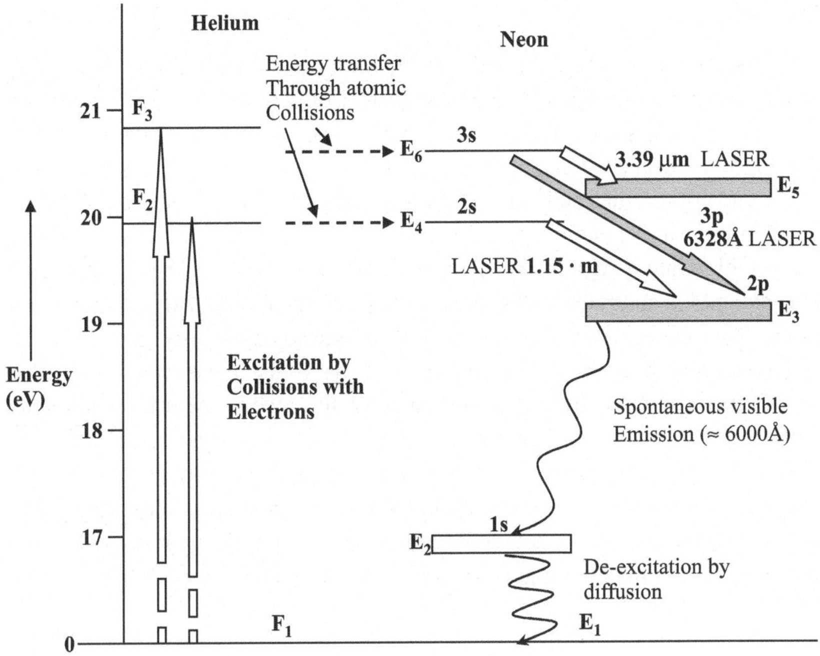

Fig. 4.11 *Energy level diagram of He-Ne laser*

(*i*) E_6 to E_3 : Main one with wavelength 6328Å,

(*ii*) E_4 to E_3 : wavelength 1.15 μm, and

(*iii*) E_6 to E_5 : wavelength 3.39 μm.

Special uses

(*i*) It provides very stable, single transverse mode reference beam necessary to identify optical properties of materials and surface roughness.

(*ii*) The well characterized beam is used as a writing source on photosensitive material to provide detailed pattern.

6. SEMICONDUCTOR LASER

With tremendous growth in semiconductor or solid state technology, it was possible to get laser action in the semiconductor devices. This is known as Semiconductor laser. They are small, efficient laser devices with dimensions less than a square millimeter. Their operating wavelengths are from 0.6 to 1.55 μm depending on the laser medium materials. Although most semiconductor lasers can be optically pumped, however, mostly they are directly electrically pumped due to convenience. The first semiconductor lasers were made of gallium arsenide material to emit laser of wavelength 0.8 μm.

Principle

It is basically a forward biased pn junction made with direct band gap semiconductors with two side walls roughened to help in achieving resonating condition. This is shown in Fig. 4.12. Along the other two side walls (made optically flat) through the depletion region the laser beam is coming out. Here both *p* and *n*-type materials are preferably heavily doped. When a current is passed through a *pn* junction, *p*-region being positively biased, holes are injected from *p*-region into *n*-region, and, electrons from *n* to *p* region. The electrons and holes recombine and release of energy takes place in or very near to the junction region. This energy is called the activation energy or energy gap. In some semiconductors like GaAs and others, the energy is released as light. The wavelength of emitted light depends on activation energy of crystal. Photons emitted at the moment of recombination of an electron with a hole, will stimulate recombination of other carriers. The result will be stimulated emission of radiation. If these radiations moving in the plane of the junction are made to move back and forth in the plane of the junction by reflection at opposite parallel sides and perpendicular to the plate of junction, radiation can be produced. When exciting currents are small, only a small part of carriers undergo recombination. This process is in random and incoherent. But when the current density is increased, the emission becomes more and more coherent, and radiation intensity increases enormously. Fig. 4.13 shows a typical output of a semiconductor laser versus impressed current flowing through the laser. At low impressed currents the junction region radiation emission by spontaneous process increases slowly. At the threshold current, I_{th} (typically few mA), where the gain exceeds the losses, laser action begins and the output increases dramatically.

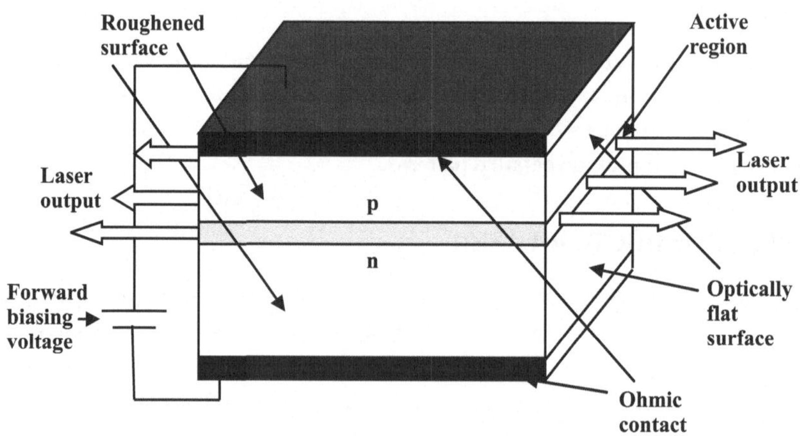

Fig. 4.12 *A typical semiconductor laser in action*

These lasers have (1) small size, (2). can easily be pakagable into the circuit, (3). higher efficiency, (4) more output power, and, (5) high stability of the output frequency, (6). very easily productionised. Commonly available semiconductor lasers are at room temperature: GaAs at 9000Å, GaAsP at 6500Å.

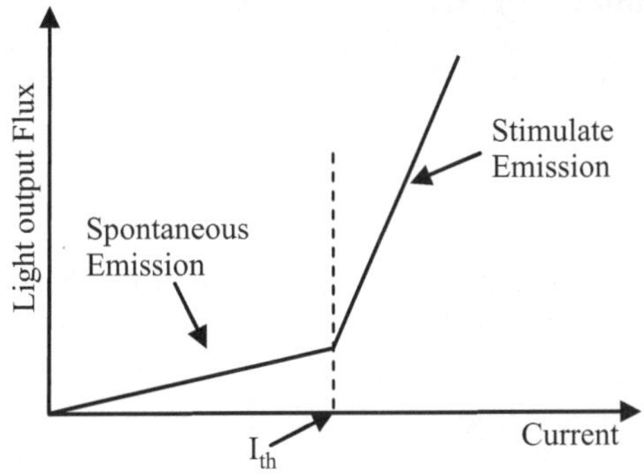

Fig. 4.13 *Lasing response with respect to forward current*

7. CHARACTERISTICS OF LASER BEAM

I. Coherence

The waves emitted by laser source will be in phase and are of same frequency. So light is coherent with coherent length ≈ few kilometers. (for ordinary light, coherent length ≈ few millimeters.)

II. Directionality

Lasers emit light in one direction unlike ordinary light.

III. Divergence

Ordinary light spreads out in the form of spherical wavefronts and is highly divergent. However, for laser, angular divergence is extremely small. The little divergence present is due to wave properties of light.

IV. Intensity

Intensity of ordinary light decreases rapidly with distances. In contrast, laser emits light in form of a narrow beam. As the energy concentrated in a very narrow region, its intensity is extremely high and does not fall with distance. [Laser light for 1 mW laser is 10,000 times brighter than sunlight at earth's surface.]

V. Monochromaticity

Laser light is highly monochromatic.

8. USES OF LASER

I. Microwave Communication

Due to narrow bandwidth more channels of message can be accommodated to be carried simultaneously by carrier waves.

II. Communication

Due to narrow angular spread it is used for communication between the earth and satellites.

III. Storage of Information

Due to narrow bandwidth more information can be stored in the computers.

IV. Industrial Applications

Blasting holes in Diamonds, making holes in Hard steels, etc.

V. Medical Applications

(*a*) Treatment of retina and other optical treatments.
(*b*) Surgery (without blood),
(*c*) Treatment of cancers,
(*d*) Stone removal from gall bladder, kidney, etc.

VI. Military Applications

(*a*) War weapon,
(*b*) Radar communication.

VII. Chemical Applications

(*a*) Investigation of structure of molecules,
(*b*) Laser Raman spectroscopy, etc.

5

Fibre Optics

Electronic communications use radio waves and microwaves to carry information through copper wires and co-axial cables. The information carrying capacity of these wires is highly restricted and does not suffice for the modern needs. The use of light waves, in place of radio waves or microwaves, enhances the number of signals tremendously that can be transmitted simultaneously. By 1960, it has been established that light could be guided by a glass fibre. The availability of laser, which is a coherent source of light wave, presented communication engineers with a suitable carrier wave capable of carrying enormously large amount of information compared with radio waves and microwaves. Although a dream of carrying millions of telephones audio or video channels through a single light beam is yet to be realized, the technology is slowly edging towards making this a dream into reality.

In 1970 with the fabrication of low loss glass fibres by Corning glass work in the United States made practical the use of optical fibre as a viable transmission medium in light wave communication systems. In a communication system, the optical fibre acts as the transmission channel carrying the light beam loaded with information. One of the key elements in the fibre optic revolution has been the dramatic improvement in the transmission characteristic of optical fibres. These include the attenuation of the light beam as well as the distortion in optical signals as they race through the optical fibre. In near future we hope that fibre optics technology will be as common as electric wires in household.

Optical fibres are glass or plastic conduits, as thin as human hair, designed to guide light waves along their length.

An optical fibre works on the principle of total internal reflection of light. When light enters one end of the fibre it undergoes successive total internal reflections from side walls and travels down the length of the fibre along a zigzag path as shown in Fig. 5.1. A small fraction of light may escape through side walls, but, a major portion emerges out from the other end of the fibre. Cross sectional view of

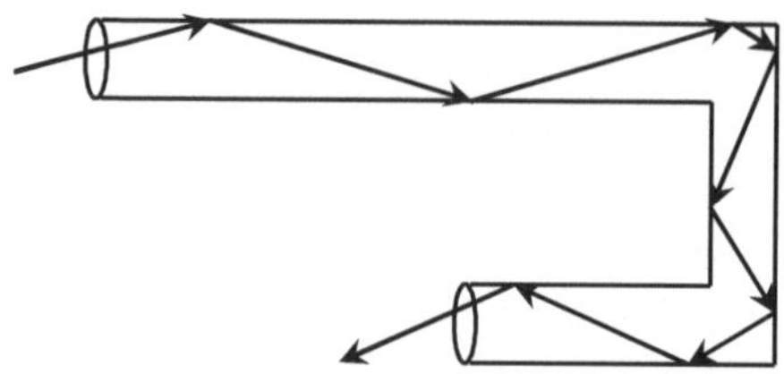

Fig. 5.1 *Propagation of light wave inside a fibre optic cable*

a multimode optical fibre is shown in Fig. 5.2. It has in general three co-axial regions. The innermost region is the light guided region, known as core. It is surrounded by another co-axial region known as cladding. The outermost layer is known as sheath or jacket. It is commonly made of plastic or polymer. The refractive index of cladding is always lower than that of core. The purpose of cladding is to make light concentrated (confined) to the core. The core acts like a continuous layer of two parallel mirrors. A signal is first encoded into a light beam and then the light launched into the core and striking the core to cladding surface

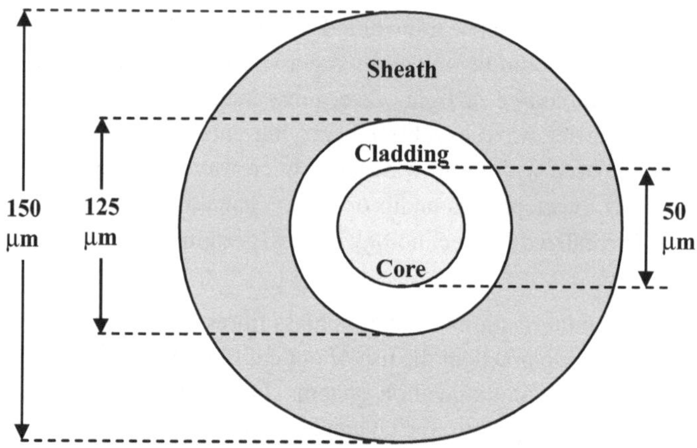

Fig. 5.2 *Schematic diagram of the cross-sectional view of the fibre optic cable*

interface at greater than critical angle will be reflected back into the core. Since the angles of incidence and reflection are equal, light will continue to rebound and propagate through the fibre. If a large number of fibres are packed together, light may leak out and interfere with the lights of the other. Cladding has also a major role for the prevention of this leakage. The sheath protects the cladding and core from abrasions, contamination and other harmful influence of moisture. Also it enhances mechanical strength of the fibre. Although a variety of optical fibres are available, the fibres in most use today are so called single mode fibres with core diameter of 10 micrometer and overall diameter of 125 μm and multimode fibre with a core diameter of 50 μm and an overall diameter of 125 μm.

1. LIGHT RAY PROPAGATION THEORY IN OPTICAL FIBRE

The transmission of light via dielectric waveguide structure was first proposed and investigated at the beginning of the twentieth century. However, a transparent dielectric rod typically of silica glass with a refractive index of around 1.5, surrounded by air, proved to be an impractical waveguide due to its unsupported structure and the excessive losses at any discontinuity of the glass-air interface. This problem led to a proposal later on that a waveguide consisting of a transparent core with a refractive index n_1, surrounded by a transparent cladding of slightly lower refractive index n_2. These proposal stimulated tremendous efforts in reducing the attenuation of light energy through waveguide by purification of the materials. The light gathering ability of optical fibre depends on some factors like core size, acceptance angle and numerical aperture etc. They are described as given below.

I. Numerical Aperture

We consider an optical fibre and light enters through one end (launching end) as in Fig. 5.3. Let n_1, n_2 and n_0 are the refractive indices of core, cladding and incident medium respectively. Here, $n_2 > n_1$. Now, a ray enters at A at an angle i refracted at an angle r. The angles here are measured with respect to the fibre axis. The refracted ray then strikes cladding in-surface at an angle φ, internally reflected with angle φ (when φ is more than critical angle φ_C).

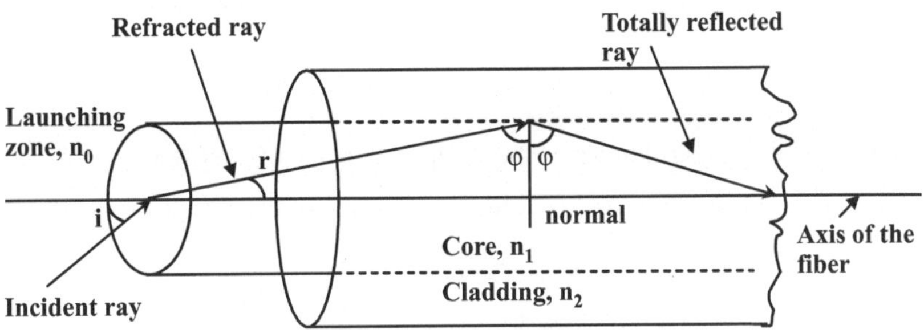

Fig. 5.3 *Propagation of light wave through a fibre optic cable*

So, $$\frac{\sin i}{\sin r} = \frac{n_1}{n_0} \quad \ldots(5.1)$$

Also, $$\sin r = \sin(90° - \varphi) = \cos\varphi \quad \ldots(5.2)\ [\text{as } \varphi + r = 90°]$$

Using equations (5.1) and (5.2), we have

$$\sin i = \frac{n_1}{n_0}\cos\varphi \quad \ldots(5.3)$$

For total internal reflection to occur, $\phi = \phi_C$ and $i = i_{max}$.

Hence, from equation (5.3)

$$\sin i_{max} = \frac{n_1}{n_0}\cos(\varphi_C) \qquad \ldots(5.4)$$

But, we know, for any refracting surface, from the condition of total internal reflection,

$$\sin \phi_C = \frac{n_2}{n_1} \qquad \ldots(5.5)$$

[at core and cladding interface]

Using equations (5.4) and (5.5)

$$\sin i_{max} = \frac{n_1}{n_0}\cos(\varphi_C)$$

$$= \frac{n_1}{n_0}\cdot\sqrt{1-\sin^2(\varphi_C)}$$

$$= \frac{n_1}{n_0}\cdot\sqrt{1-\left[\frac{n_2}{n_1}\right]^2}$$

$$\Rightarrow \qquad \sin i_{max} = \frac{\sqrt{n_1^2 - n_2^2}}{n_0} \qquad \ldots(5.6)$$

When the incident medium is air, then, $n_0 = 1$. So, from equation (5.6)

$$\sin i_{max} = \frac{\sqrt{n_1^2 - n_2^2}}{1} \Rightarrow \sin\theta_0 = \sqrt{n_1^2 - n_2^2}$$

Here, $\theta_0 = i_{max}$, and is known as the acceptance angle of the fibre.

The numerical aperture (*NA*) is defined as the sine of the acceptance angle.

So, $\quad NA = \sin\theta_0 = \sqrt{n_1^2 - n_2^2}$.

Generally, $\quad NA = 0.13 - 0.50$.

Normally for short distance communications, $NA = 0.1 - 0.3$.

However, it is observed that smaller NA is not an advantage because it makes it difficult to launch or push power into the fibre.

II. Acceptance Angle and Acceptance Cone

Having considered the propagation of light wave in an optical fibre through total internal reflection at the core-cladding interface, it is useful to enlarge upon the geometric approach with a reference to light ray entering the fibre. Since only rays with a sufficient incident angle at the core cladding interface are transmitted by the total internal reflection. This means not all rays entering the fibre will continue to be propagated down its length. In the equation (5.6), $i_{max} = \theta_0$ is known as acceptance angle.

Acceptance angle can be defined as the maximum angle that a light ray can have at core-air interface in launching end of fibre in such a way that it attains critical angle for total internal reflection at core cladding interface in the optical fibre. If the ray enter into the fibre core at angle grater than $i_{max} = \theta_0$, it will be transmitted to core cladding interface at an angle less than Φ_C and will not be totally internal reflected.

Acceptance cone can be defined as the cone formed by light ray at an acceptance angle about the fibre axis.

The light rays contained within the cone having a full angle, $2\theta_0$, are accepted and transmitted along the fibre as shown in Fig. 5.4. Therefore, this cone is known as Acceptance Cone.

The larger the diameter of cone, larger will be the acceptance angle and cone.

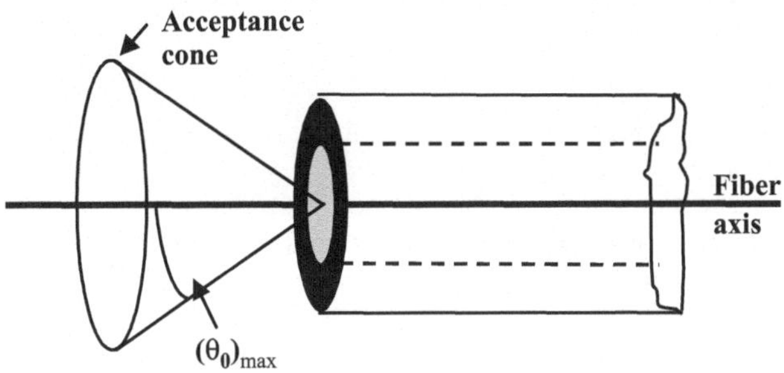

Fig. 5.4 *Acceptance cone of optical fibre*

III. Number of Modes and Cut-off Parameters

The number of modes supported by a fibre is determined by a parameter known as 'Cut-off Parameter' or 'Normalized Frequency of Cut-off' or simply '*V*-Number'. *V*-number is given by,

$$V = \frac{2\pi a}{\lambda} . \sqrt{n_1^2 - n_2^2} .$$

Here, a is the radius of the core and λ is the free space wavelength.

So, $$V = \frac{2\pi a}{\lambda} . \sqrt{n_1^2 - n_2^2} = \frac{2\pi a}{\lambda} . NA$$

However,

$$NA = \sqrt{n_1^2 - n_2^2}$$

$$= \sqrt{(n_1 - n_2)(n_1 + n_2)}$$

$$= \sqrt{\frac{(n_1 + n_2)}{2} . \frac{(n_1 - n_2)}{n_1} . 2n_1}$$

$$\Rightarrow \qquad NA \approx \sqrt{\frac{(n_1+n_1)}{2} \cdot \frac{(n_1-n_2)}{n_1} \cdot 2n_1}$$

$$\approx \sqrt{n_1 \cdot \frac{(n_1-n_2)}{n_1} \cdot 2n_1} = \sqrt{2n_1^2 \cdot \Delta} = n_1 \cdot \sqrt{2\Delta}$$

Hence,

$$V = \frac{2\pi a}{\lambda} \cdot NA = \frac{2\pi a}{\lambda} \cdot n_1 \cdot \sqrt{2\Delta}$$

V-number gives information about the important design variable for the fibre; namely the core radius a, the relative refractive index difference Δ and operating wavelength λ.

Total number of guided modes for multimode step index fibre M_s is related to V value for fibre and is given as

$$M_s = \frac{V^2}{2}$$

This determine as estimate of the number of guided modes propagating in a particular multimode step index fibre. Total number of guided modes for graded index fibre M_g is given as

$$M_g = \frac{V^2}{4}$$

IV. Index Profile

A plot of refractive index of the core and cladding layer with the distance from the core axis is known as Index Profile.

2. TYPES OF OPTICAL FIBRE

There are two types of optical fibre.

I. Single Mode Fibre (SMF)

It has a smaller core diameter and can support only one mode of propagation. Here refractive index of the core changes abruptly at the core – cladding interface, as shown in Fig. 5.5 (*a*). So this index profile is also known as 'Step Index (SI)' and the fibre is called as SI fibre.

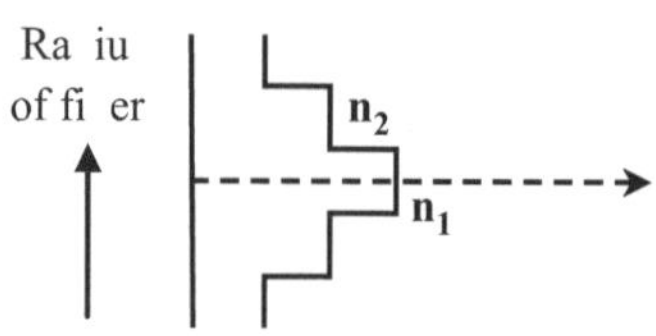

Fig. 5.5 (*a*) *Index profile of SMF fibre*

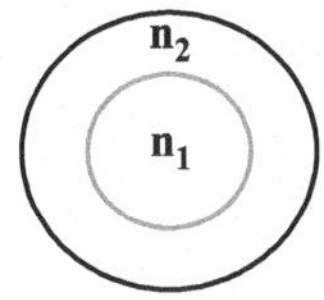

Fig. 5.5 *(b) Cross-sectional view of SMF fibre*

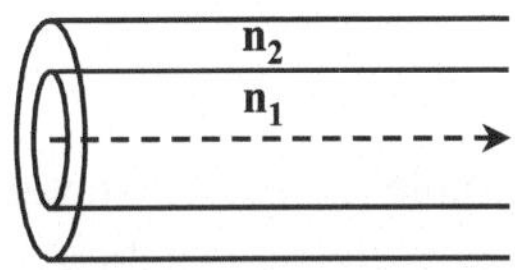

Fig. 5.5 *(c) Side view of SMF fibre*

A single mode step index fibre has a very thin core of refractive index (n_1) of higher value which is surrounded by a cladding of lower refractive index (n_2) as shown in Fig. 5.5 (*b*). Fig. 5.5 (*c*) shows the side view of it. A typical SMF has a core diameter of 4 - 10 μm. Light travels along a single path along the axis. A SMF has a very small value of 'Fractional Change of Refractive Index ', **Δ,** given by,

$$\Delta = \frac{n_1 - n_2}{n_1}.$$

Here, $\Delta \approx 0.002$. It has the distinct advantage of low dispersion but at the same time launching through the fibre is very difficult.

II. Multi Mode Fibre (MMF)

It has a larger core diameter and can support a number of modes of propagation. There are two types of MMF.

(*a*) **Step Index Type :** Multimode step index fibre allows the propagation of a finite number of guided modes along the channel. The number of guided modes is dependent upon the physical parameter (*i.e.*, relative refractive index difference (Δ),core radius) of the fibre and the wavelength of the transmitted light. Here, only core diameter is large (≈ 100 μm) as shown in Fig. 5.6.

So light follow zigzag paths inside the fibre. Many such zigzag paths are allowed in MMF. The *NA* value is larger (≈ 0.3) as core diameter is large. The rays entering at different angles travel different path and emerge out of the fibre at different time in different directions.

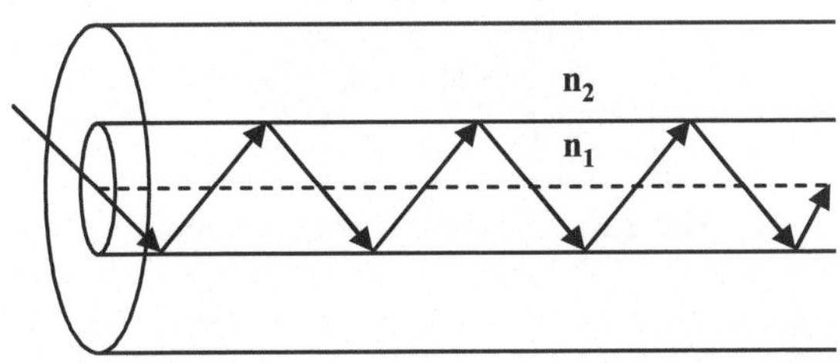

Fig. 5.6 *Propagation of light wave inside a Step Index MMF*

For, *SI* fibre maximum number of modes, N_m, supported is

$$N_m \cong \frac{1}{2} . V^2$$

Numerically, $V < 2.405$, $N_m \approx 1$ for SMF, and, $V > 2.405$, $N_m > 1$ for MMF. So, corresponding cut off frequency is

$$\frac{1}{\lambda_C} = \frac{1}{\lambda V / 2.405}$$

In these fibres considerable dispersion (broadening) may occur due to different group velocities of the propagating modes.

(*b*) **Graded Refractive Index Type (GRIN) :** Here also core diameter is large and core consists of concentric layers of different refractive Graded index fibre *i.e.*, it does not have a constant refractive index core. In these fibres, core refractive index n_1 decrease with radial distance from a maximum value at the fibre axis to a constant value $n_{2,}$ the refractive index of cladding as in Fig. 5.6 (*a*) and Fig. 5.6 (*b*). The transmission of light ray in a multimode graded index fibre is shown in Fig. 5.6 (*c*). It is observed that the light rays follow curved path through the fibre core. Using the concepts of geometric optics, the gradual decrease in refractive index from the centre of core creates many refraction of the rays with total internal reflection at different angles on the fibre axis because they are incident on a large number of high to low refractive index interfaces within a core.

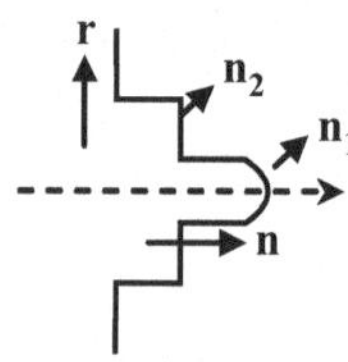

Fig. 5.6 (*a*) *Index profile of GRIN fibre*

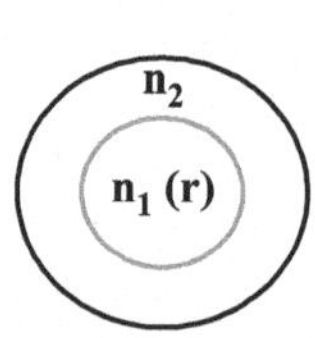

Fig. 5.6 (*b*) *Cross-sectional view of GRIN fibre*

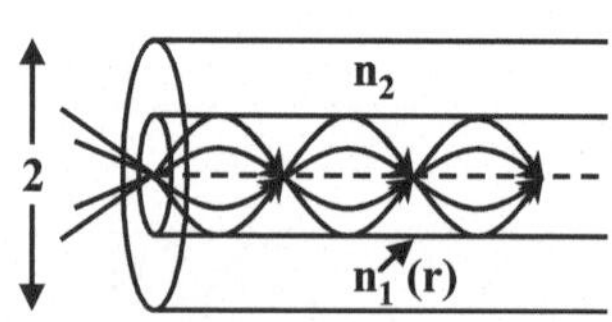

Fig. 5.6 (*c*) *Side view of GRIN fibre*

The refractive index varies with distance from the fibre axis. It has high value at the center and falls off rapidly along outer surface. Here entering rays are continuously bent and travel periodic paths along axis. The rays entering at different angles follow different paths but with the same period both in space and time.

For, GRIN fibre maximum number of modes, N_m, supported is

$$N_m \cong \frac{1}{4}.V^2.$$

These fibres exhibit less intermodal dispersion than multimode step index fibre due to their refractive index profile. Multimode graded index with parabolic or near parabolic index profile of core, have transmission bandwidth which may be order of magnitude greater than multimode step index fibre bandwidth.

Conventional notation for optical fibre: X/Y

Here, *X* gives the core diameter in μm and *Y* gives the cladding diameter in μm.

Normally, for *SI* fibre: 50/125, 100/40, 200/230, etc.

For GRIN fibre: 50/125, 62.5/125, 85/125, etc.

3. APPLICATIONS

I. Medical

In medical field the application of fibre optics can be broadly classified into following categories.

(*a*) **Clinical applications:** It has find applications in almost all clinical equipments. Optical fibres lend flexibility to the systems besides making it convenient to separate the source of light from the subject to be illuminated. This makes easier to illuminate the areas in the human body which are otherwise not accessible.

(*b*) **Laser medicine:** Since laser invention, it has found niches in both diagnostic and therapeutic medical applications. Optical fibres with their properties, allow a high degree of selectivity ensuring that the target is preferentially treated with minimal impact on surrounding tissues. Optical fibre has become the preferred medium for many applications because of its flexibility and its ability to be inserted interstitially. Laser medicine is also used in removing wrinkles and unwanted hair.

(*c*) **Endoscopy:** It is an optical instrument used for viewing internal organs through natural opening (ear, throat, rectum etc) or through a small incision in the skin. A flexible endoscopy is referred as a Fibreoscope. Optical fibres are being used in microscopes to illuminate subjects sensitive to light/heat. An approach known as IR spectroscopy takes advantage of the fact that the brain tissue is transparent to light in the 700nm to 1000nm wavelength range.

II. Military

The military uses the aspect of security to benefit from the fibre optic technology. Fibre optics has been used in submarines, ships, satellite, earth stations and transmission lines for tactical command post communications. Ships and submarines use fibre optics to reduce shocks, fire and spark. Ground stations that link satellite use fibre optic cable for secure lines. Fibre guided missiles are used in wars. Use of fibres in place of copper wires reduces weight and maintains true communication silence to the enemy.

III. Entertainment

Fibre optic link carries voice, video and data which links computers, *TV*, cable television etc. A coherent optical fibre bundle is used to enlarge the image display on *TV* screen in place of conventional system.

IV. Sensors

An optical sensor can be defined as a device in which an external measurement *e.g.*, temperature, pressure, or strain, modulates an optical signal which when demodulated yields precise quantitative information about measurement. The inherent property of variation of refractive index of fibre under the influence of external forces leads to use it as transducers.

V. Communication

The fibre optic communication system block diagram is shown in Fig. 5.7. The electrical circuit converts non-electrical signal into the electrical signal (measurable signal) first. This signal is then converted into the light signal by the light emitting diode (*LED*). In a communication system light waves are the carrier waves. Now this light signal is launched into optical fibre cable. The output of fibre cable is converted into electrical signal by a device known as photodetector. The optical signal traveling through the fibre will progressively attenuate and distorted because of various losses and dispersion occurring in the fibre. Therefore, repeaters or amplifiers are used in the transmission line to reshape the signal. The signal is then fed into a suitable transducer to convert into an audio or video form.

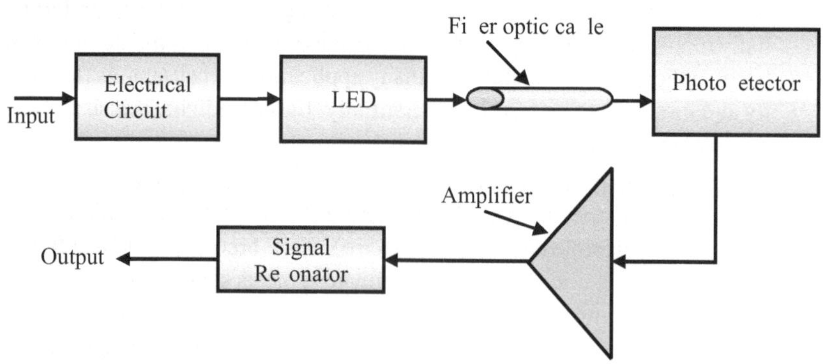

Fig. 5.7 *Block diagram of a typical fibre optic communication system*

- An optical fibre data bus offers a great reduction in cost and enormously increases information handling capacity compared to a parallel multiwire data bus.
- Wired city concept was experimented in Japan using an interactive optical fibre system.
- Telephone cables connecting various countries use mainly optical fibre network system.

4. ADVANTAGES

(*i*) Cheaper: made from Silica.

(*ii*) Small in size, light in weight, flexible but mechanically strong.

(*iii*) Not hazardous.

(*iv*) Immune to electro magnetic interference and radio frequency interference.

(*v*) Reduce cross talk possibility.

(*vi*) Wider bandwidth.

(*vii*) Have low loss per unit length.

5. DISADVANTAGES

(*i*) Some types of cables are harder to join. Splicing (the technology used to joint two fibre cables) is time consuming and thus increasing the cost of installation of cables.

(*ii*) Fibre optic cable is nonconductive. Sometime separately conducting medium is needed to the cable connections.

(*iii*) Lack of standards and knowledge are problem. This makes implementation of network and installation of technique difficult.

6

Simple Harmonic Motion

The motion of any body generally consists of three types, *e.g.*,

(*a*) translational, where position of the body varies linearly with time,

(*b*) rotational, where the body rotates about an axis, and,

(*c*) vibrational, where the body moves back & forth and repeats itself after a fixed time interval.

Normally both the rotational and vibrational motions are periodic motion as they both repeats after a regular time interval.

Periodic Motion

A motion in which the body describes the same path in same way continuously after equal interval of time is known as the periodic motion. Simple Harmonic Motion (SHM) is a special part of periodic motion with the following characteristics.

- This is a type of periodic motion.
- The acceleration of the body is directly proportional to its displacement from a fixed point and always directed towards the fixed point.
- The restoring force developed due to the friction between the vibrating body and the medium tries to bring the particle in its equilibrium position.

Characteristics of SHM

- It is a periodic motion.
- When displaced from mean position, a restoring force (which tends to bring it there and directed towards it) must occur
- The restoring force is proportional to the displacement from its mean position.

EXAMPLES

I. Simple Pendulum

It consists of an ideally massless inextensible string hanging from a rigid support *A* with a point mass (*i.e.*, the bob of the pendulum) connected to its other end *B*.

This is shown in Fig. 6.1. When the bob is displaced to a new position *P* and then is released, oscillations start from the equilibrium position *B* and it is in simple harmonic motion. If θ is the angle made by the string at *P* with the equilibrium condition of the string *AB*, *m* and *g* are the mass of the bob and the acceleration due to gravity, then, the radial component of *mg* (*i.e.*, *mg* cos θ) balances the tension *T* across the string. The tangential component '*mg* sin θ' tries to bring the bob at *B*. These way oscillations continue till it stops due to air friction.

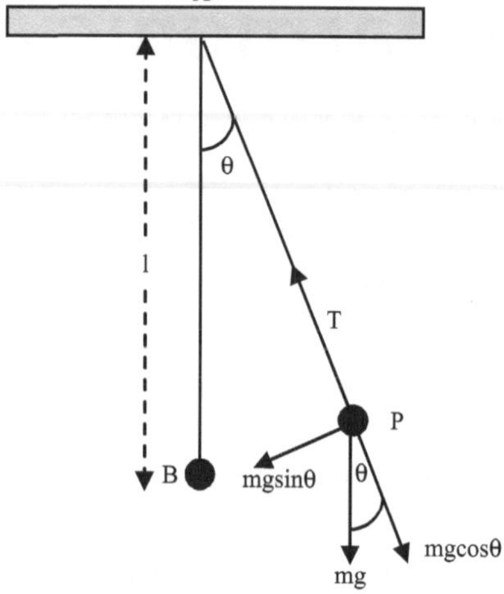

Fig. 6.1 *Simple pendulum*

II. Spring - Mass system

Here a mass block is connected to a spring either vertically or horizontally from rigid supports as shown in Fig. 6.2 (*a*) and (*b*). On displacing the mass from its equilibrium position and then releasing it, again simple harmonic oscillations set in.

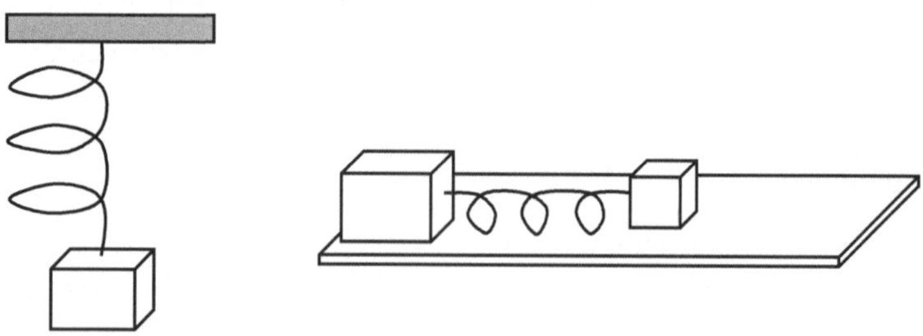

Fig. 6.2 (*a*) *Vertically connected* **Fig. 6.2** (*b*) *Horizontally connected*

Fig. 6.2 *SHM by mass block connected with spring*

1. SOME PARAMETERS OF SHM

In order to describe SHM of a particle we are considering a reference circle centered at *O* (as shown in Fig. 6.3) with radius *OP*. Here *X′OX* & *Y′OY* are the axes.

P is the present position of the particle. So *OP* is making any angle θ with *OX*. We draw perpendicular from *P* on *Y*-axis at *N*.

I. Displacement

This is the distance of the particle executing SHM along the path of motion from its mean position at a given instant.

From the Fig. 6.3 the displacement of the particle at *P* is given by

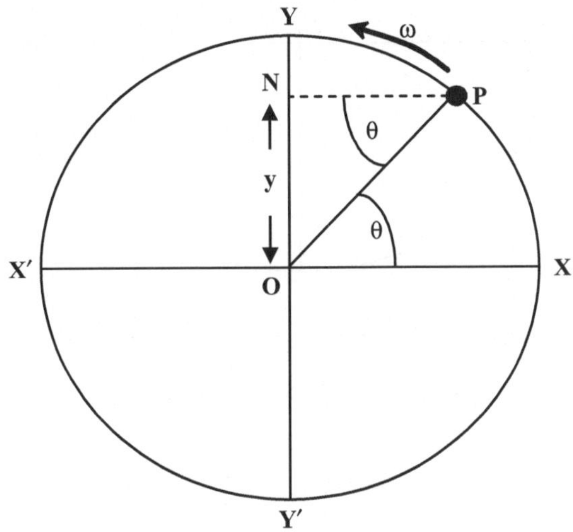

Fig. 6.3 *SHM in circular motion*

$$ON = y = OP \sin\theta \quad \Rightarrow \quad y = r \sin\theta$$

As the position of *P* varies, the position of *N* changes but it will be limited along *Y′OY*. *N* first moves to *Y*, then comes back & crosses *O* to reach *Y′* and then it turns back.When *P* completes one complete revolution, the point *N* again takes the same path. So, we observe that the point *N* suffers SHM when *P* moves in circular path. If 'ω' be the angular velocity of the particle, then the displacement at time *t* is

$$y = r \sin \omega t$$

Here, $\theta = \omega t.$

II. Amplitude

This is the maximum distance covered by the particle on either side of the mean position. In Fig. 6.4 for the particle executing SHM the amplitude is *OY*.

III. Velocity

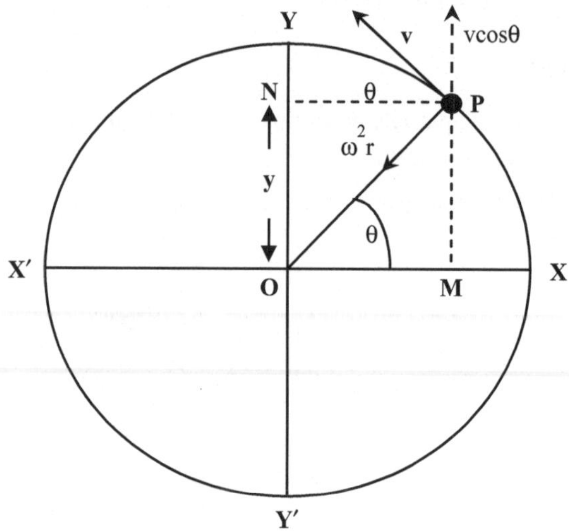

Fig. 6.4 *Velocity component in circular motion*

This is the velocity of particle along the path of motion at a given instant.

As shown in Fig. 6.4 the velocity of the particle at point P is equal to the component of velocity along diameter YOY'. The component of velocity v along YOY' is $v \cos \theta$.

At a time t, the velocity of the particle at P, then $V = v \cos \theta = \omega r \cos \omega t$

$$= \omega r \sqrt{1 - \sin^2 \omega t}$$

$\Rightarrow$ $$V = \omega r \sqrt{1 - \left(\frac{PM}{OP}\right)^2} \qquad \text{[from } \Delta MOP]$$

$\Rightarrow$ $$V = \omega r \sqrt{1 - \left(\frac{ON}{OP}\right)^2} \qquad \text{[as } PM = ON]$$

$\Rightarrow$ $$V = \omega r \sqrt{1 - \left(\frac{y}{r}\right)^2} = \omega \sqrt{r^2 - y^2}$$

So, at $y = 0$, $V = V_{max} = \omega r$ and

at $y = r$ (at the end of vibrations), $V = V_{min} = 0$.

IV. Acceleration

Acceleration of the particle executing SHM at any instant is equal to the component of acceleration of the reference particle along YOY'.

We now consider a particle at P moving with velocity v in a circle of radius r.

So, acceleration is given by $a = \dfrac{v^2}{r} = \omega^2 r$

This acceleration is directed towards center of the circle (as shown in Fig. 6.4).

So, acceleration of the particle at P along YOY' is given by

$$a = -\omega^2 r \sin\theta = -\omega^2 y.$$

The reason of negative sign is that its direction is towards center of the circle, *i.e.*, against the direction of increasing y.

V. Time Period

The time required to complete one vibration by a particle executing SHM is known as Time Period.

From the definition of ω and time period T, we have

$$\omega = \frac{2\pi}{T}$$

$$\Rightarrow \quad T = \frac{2\pi}{\omega} = \frac{2\pi}{\sqrt{a/r}} \qquad [\text{as } a = \omega^2 r]$$

$$\Rightarrow \quad T = \frac{2\pi}{\omega} = \frac{2\pi}{\sqrt{\text{Acceleration/Displacement}}}$$

$$= 2\pi\sqrt{\frac{\text{Displacement}}{\text{Acceleration}}}$$

VI. Frequency

This is the number of vibrations made in one second by a particle executing SHM. The frequency n is given by

$$n = \frac{1}{T}$$

VII. Phase and Phase Difference

Phase is defined as the state or conditions of the particle as regard to its position and direction of motion with respect to the mean position.

Phase difference between two SHMs indicates how much the two motions are out of step, or, by how much time one is ahead of the other.

As per Fig. 6.5 for a particle at position P_2, its displacement with respect to axis OX is given by

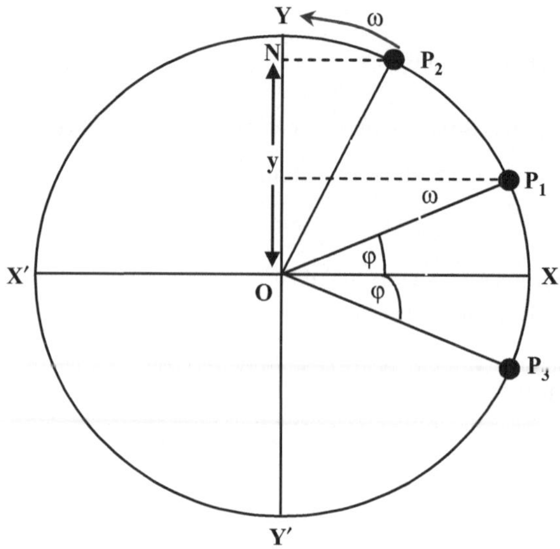

Fig. 6.5 *Measurement of phase difference in circular motion*

$$y = r \sin \omega t$$

Now, using OP_1 (which is an angle φ above X-axis) as the starting position, we have this displacement as

$$y = r \sin (\omega t - \varphi)$$

Here ϕ is the angle made by OP_1 with OX. Similarly, using OP_3 (which is an angle φ below X-axis) as the starting position, we have this displacement as

$$y = r \sin (\omega t + \varphi)$$

So, the general displacement of a particle is given by

$$y = r \sin (\omega t \pm \varphi)$$

Here, φ is known as the initial phase of the particle.

2. FORCED VIBRATION

For any oscillating body the oscillations continue in absence of any restoring force, *i.e.*, if the oscillations continues in free space. This is free oscillations. However, under an external periodic force with frequency not being equal to the natural frequency of the vibrating body, the oscillating body slowly changes its vibrational frequency from its own natural one to that of the external periodic force. Such oscillations under external periodic force are known as forced oscillation.

This is the vibrations in which the body vibrates with a frequency other than its natural frequency under the action of an external periodic force.

Theory

For a body or a particle under forced vibrations the forces acts upon it are:

(*a*) **a restoring force (F_r).** This force is proportional to the displacement, but oppositely directed,

i.e., $F_r = -\mu y$, where, μ is Force constant,

(*b*) **a frictional force (F_f).** This force is proportional to the velocity, but oppositely directed,

i.e., $F_r = -\beta \dfrac{dy}{dt}$, where, β is frictional force per unit velocity, and,

(*c*) **an external periodic force.** This force is represented by $F \sin pt$, where, F is the maximum force and $p/2\pi$ is its frequency.

So the net force acting on the particle is

$$-\mu y - \beta \frac{dy}{dt} + F \sin pt$$

Hence, for the particle mass '*m*' and instantaneous acceleration '$\dfrac{d^2y}{dt^2}$', we have,

$$m\frac{d^2y}{dt^2} = -\mu y - \beta \frac{dy}{dt} + F \sin pt$$

$$\Rightarrow \quad \frac{d^2y}{dt^2} + \frac{\beta}{m}\frac{dy}{dt} + \frac{\mu}{m} y = \frac{F}{m} \sin pt$$

$$\Rightarrow \quad \frac{d^2y}{dt^2} + 2b\frac{dy}{dt} + \omega^2 y = f \sin pt \qquad \ldots(6.1)$$

Here, $$2b = \frac{\beta}{m},\ \omega^2 = \frac{\mu}{m},\ \text{and},\ f = \frac{F}{m}$$

Now at steady state, the particle will vibrate with the frequency of applied force, so the solution of differential equation (6.1) must be of type as,

$$y = A \sin (pt - \theta) \qquad \ldots(6.2)$$

Here, A is steady state amplitude and θ is the angle by which y lags behind the applied force $F \sin pt$. So, using equation (6.2)

$$\frac{dy}{dt} = Ap \cos (pt - \theta) \text{ and}$$

$$\frac{d^2y}{dt^2} = -Ap^2 \sin (pt - \theta).$$

Using these into equation (6.1), we have

$$-Ap^2 \sin (pt - \theta) + 2\,bAp \cos (pt - \theta) + \omega^2 A \sin (pt - \theta) = f \sin pt$$

$$\Rightarrow -Ap^2 \sin(pt-\theta) + 2bAp\cos(pt-\theta) + \omega^2 A\sin(pt-\theta) = f\sin\left(pt-\theta+\theta\right)$$

$$\Rightarrow \quad A\left[\omega^2 - p^2\right]\sin(pt-\theta) + 2bAp\cos(pt-\theta) = f\left[\sin\left(pt-\theta\right)\cos\theta + \cos\left(pt-\theta\right)\sin\theta\right]$$

The above relation holds well for all values of t, so the coefficient of sin $(pt-\theta)$ and cos $(pt-\theta)$ terms of both sides on above equation must be equal.

So, comparing the coefficient of sin $(pt-\theta)$, we have

$$A\left[\omega^2 - p^2\right] = f\cos\theta \qquad \text{...(6.3)}$$

Also, comparing the coefficient of cos $(pt-\theta)$, we have

$$2Abp = f\sin\theta \qquad \text{...(6.4)}$$

Now, from eqns. (6.3) and (6.4), we have

$$A^2\left[\omega^2 - p^2\right]^2 + [2Abp]^2 = [f\cos\theta]^2 + [f\sin\theta]^2$$

$$\Rightarrow \quad A^2\left[\left[\omega^2 - p^2\right]^2 + [2bp]^2\right] = f^2$$

$$\Rightarrow \qquad A = \frac{f}{\sqrt{\left[\omega^2 - p^2\right]^2 + [2bp]^2}} \qquad \text{...(6.5)}$$

Also, using eqns. (6.3) and (6.4), we have

$$\theta = \tan^{-1}\left[\frac{2bp}{\omega^2 - p^2}\right] \qquad \text{...(6.6)}$$

Equations (6.5) and (6.6) give the amplitude A and phase θ of a particle vibrating under forced vibration.

3. RESONANCE

The phenomenon of making a body vibrate with its natural frequency under the influence of another vibrating body with the same frequency is called Resonance.

Condition of Amplitude Resonance

From equation (6.5), we observe the variation of A with p. Now for a particular value of p, A becomes maximum. This phenomenon is known as Amplitude Resonance.

Using equation (6.5), A is maximum only when,

$$\sqrt{\left[\omega^2 - p^2\right]^2 + \left[2bp\right]^2} \text{ is minimum.}$$

So, then from the condition of minimum,

$$\frac{d}{dp}\left[\left[\omega^2 - p^2\right]^2 + 4b^2p^2\right] = 0$$

$$\Rightarrow 2\left[\omega^2 - p^2\right](-2p) + 4b^2.2p = 0$$

$$\Rightarrow \qquad 2pb^2 = p\left[\omega^2 - p^2\right] \text{ [as } p \text{ ¹ } 0]$$

$$\Rightarrow \qquad 2b = \omega^2 - p^2$$

$$\Rightarrow \qquad p^2 = \omega^2 - 2b^2 \qquad \ldots(6.7)$$

So at resonance condition, we should have

$p = p_r$ and

$$p_r = \sqrt{\omega^2 - 2b^2}\,.$$

Hence, the resonant frequency is $\frac{p_r}{2\pi}$ and is given by,

$$\frac{1}{2\pi}\sqrt{\omega^2 - 2b^2}\,.$$

For no or negligible damping, $b \approx 0, p_r = \omega$.

So the maximum amplitude is given by the value of A at $p = p_r$ and using eqn. (6.5) with the value of p_r, we have

$$A_{max} = \frac{f}{\sqrt{\left[\omega^2 - p_r^2\right]^2 + 4b^2p_r^2}}$$

$$\Rightarrow \qquad A_{max} = \frac{f}{\sqrt{\left[\omega^2 - \left[\omega^2 - 2b^2\right]\right]^2 + 4b^2\left[\omega^2 - 2b^2\right]}}$$

$$= \frac{f}{\sqrt{4b^4 + 4b^2\omega^2 - 8b^4}} = \frac{f}{2b\sqrt{\omega^2 - b^2}}$$

Also,

$$A_{max} = \frac{f}{2b\sqrt{\left[p_r^2 + 2b^2\right] - b^2}}$$

$$= \frac{f}{2b\sqrt{p_r^2 + b^2}} \qquad \text{[as } \omega^2 = p^2 + 2b^2 \text{ from eqn.(6.7)]}$$

For negligible damping, $b \approx 0$ and

$$A_{max} \approx \frac{f}{2bp_r}.$$

As $b \to 0$, Amax $\to \propto$.

4. SHARPNESS OF RESONANCE

The rate of change (fall) of amplitude with the change of forcing frequency on each side of resonant frequency is known as sharpness of resonance.

Figure 6.6 shows the variation of A with forcing frequency at different conditions of damping.

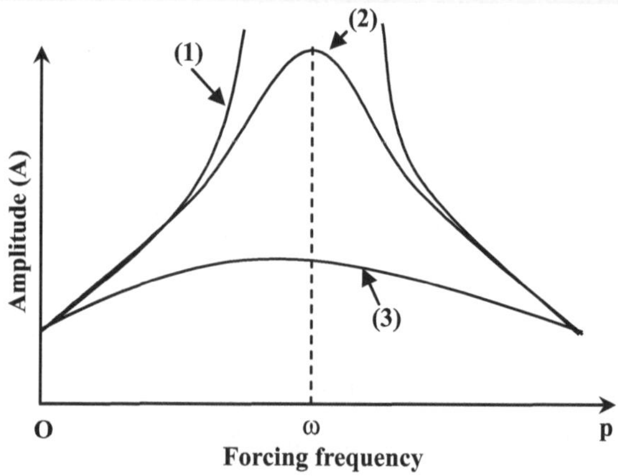

Fig. 6.6 *Sharpness of resonance curve*

(*i*) **Ideal case :** $b = 0$

So, $A \to \infty$ at $p = \omega$

This is shown in curve (1) in the Fig. 6.6

(*ii*) **Small *b***

Sharp resonance. This is shown in curve (2) in the Fig. 6.6

(*iii*) **Large *b***

Flat or no resonance . This is shown in curve (3) in the Fig. 6.6

5. EXAMPLES

I. Pendulums Hanging from a Common String

In the adjoining Fig. 6.7, there are five pendulums having same bob mass are hanging from different positions of a string between fixed supports. The pendulums have different length except two (the 2nd one and the pendulum P). When we set the pendulum P in SHM, slowly all the pendulums start oscillating because of their common coupling with the same frequency. However, the amplitude of pendulum 2 is maximum as it is in resonance with the forcing pendulum P.

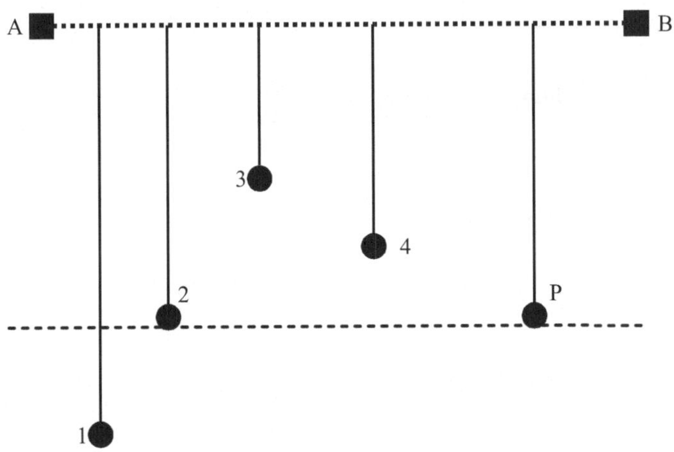

Fig. 6.7 *Experiment of forced vibration*

II. Radio Receivers

Each radio stations have their own broadcasting frequencies. When we tune our radio set, the moment the radio tuner attains exactly the frequency of any broadcasting station, we start hearing the loud and clear sound of that particular radio station due to the resonance.

III. Others

There are other examples also. The loud speaker diaphragm vibrates according to the amplifier circuit. Air column in resonance tube vibrates as per the vibrations of the tuning fork.

6. ENERGY OF THE FORCED OSCILLATOR

The total energy of the oscillating system at any particular instant is given by

E = Kinetic Energy (KE) + Potential Energy (PE)

Now, the instantaneous KE in the steady state here is given by

$$\text{KE} = \frac{1}{2} m \left(\frac{dy}{dt}\right)^2$$

From eqn.(6.2),

$$y = A \sin(pt - \theta)$$

So,

$$KE = \frac{1}{2} m \left(\frac{dy}{dt}\right)^2 = \frac{1}{2} m \left(pA \cos(pt - \theta)\right)^2$$

$$= \frac{1}{2} m \frac{f^2 p^2}{\left[\omega^2 - p^2\right]^2 + [2bp]^2} \cos^2 (pt - \theta) \qquad \ldots(6.8)$$

Also here the PE is given by

$$\text{PE}=\frac{1}{2}\mu y^2=\frac{1}{2}\mu\frac{f^2}{\left[\omega^2-p^2\right]^2+[2bp]^2}\sin^2(pt-\theta) \quad \text{... (6.9)}$$

Hence, the total energy at any instant is

$$\text{E} = \text{KE} + \text{PE}$$

$$=\frac{1}{2}m\frac{f^2p^2}{\left[\omega^2-p^2\right]^2+[2bp]^2}\cos^2(pt-\theta)$$

$$+\frac{1}{2}\mu\frac{f^2}{\left[\omega^2-p^2\right]^2+[2bp]^2}\sin^2(pt-\theta) \quad \text{...(6.10)}$$

Here, energy is time dependent, so we can take the time average value of it over a complete period. The time average of any cosine function and sine function are same and are given by

$$\left\langle\sin^2(\beta t+\phi)\right\rangle=\frac{1}{T}\int_0^T\sin^2(\beta t+\phi)\,dt=\frac{1}{2}$$

$$\left\langle\cos^2(\beta t+\phi)\right\rangle=\frac{1}{2}$$

So using eqn.(6.10), the time average energy of the forced oscillator is given by

$$\langle E\rangle=\langle KE\rangle+\langle PE\rangle$$

$$=\frac{1}{2}m\frac{f^2p^2}{\left[\omega^2-p^2\right]^2+[2bp]^2}\left\langle\cos^2(pt-\theta)\right\rangle$$

$$+\frac{1}{2}\mu\frac{f^2}{\left[\omega^2-p^2\right]^2+[2bp]^2}\left\langle\sin^2(pt-\theta)\right\rangle$$

$$=\frac{1}{4}m\frac{f^2p^2}{\left[\omega^2-p^2\right]^2+[2bp]^2}+\frac{1}{4}\mu\frac{f^2}{\left[\omega^2-p^2\right]^2+[2bp]^2}$$

$$\Rightarrow \qquad \langle E\rangle=\frac{1}{4}\left(mp^2+\mu\right)\frac{f^2}{\left[\omega^2-p^2\right]^2+[2bp]^2}$$

7. DAMPED VIBRATION

An oscillating body when put into a medium, every time during its oscillation through the medium, the body experiences a frictional resistive force. The magnitude of the resistance depends on the nature of the medium. This resistance forces the oscillation to be stopped and to regain its equilibrium position. The vibration thus continues with reducing amplitude. The mechanical energy of the system decreases with time and eventually comes to rest. This type of vibration is known as damped vibration.

There is no external force acting on the vibrating body in this case in order to compensate the energy lost in overcoming the friction in the medium.

Using equation (6.1) and by making the forced part zero there, we get

$$\frac{d^2y}{dt^2} + 2b\frac{dy}{dt} + \omega^2 y = 0 \qquad \text{...(6.11)}$$

Now we consider the solution of above second order differential equation as

$$y = A.e^{\alpha t} \qquad \text{...(6.12)}$$

Here, A and α are arbitrary constants. Using equation (6.9), we have

$$\frac{dy}{dt} = A\alpha e^{\alpha t} \text{ and}$$

$$\frac{d^2y}{dt^2} = A\alpha^2 e^{\alpha t}$$

Now using these into equation (6.11), we get

$$Ae^{\alpha t}\left[\alpha^2 + 2b\alpha + \omega^2\right] = 0$$

As $Ae^{\alpha t} \neq 0$, we have

$$\left[\alpha^2 + 2b\alpha + \omega^2\right] = 0$$

$$\Rightarrow \qquad \alpha = -b \pm \sqrt{b^2 - \omega^2}$$

So the general solution of equation (6.11) is

$$y = A_1 e^{\left[-b-\sqrt{b^2-\omega^2}\right]t} + A_2 e^{\left[-b+\sqrt{b^2-\omega^2}\right]t} \qquad \text{...(6.13)}$$

Here, A_1 and A_2 are arbitrary constants.

Depending on the values of the practical values of b and ω, we have three cases:

I. $b^2 > \omega^2$

So, $\sqrt{b^2 - \omega^2}$ is real and $\sqrt{b^2 - \omega^2} < b$.

Thus, $\left[-b+\sqrt{b^2-\omega^2}\right]$ and $\left[-b-\sqrt{b^2-\omega^2}\right]$ are both negative. So, the displacement, y, will be dying off exponentially to zero without any oscillations as shown in Fig. 6.8. This oscillation is known as Over Damped or Dead Beat Oscillations.

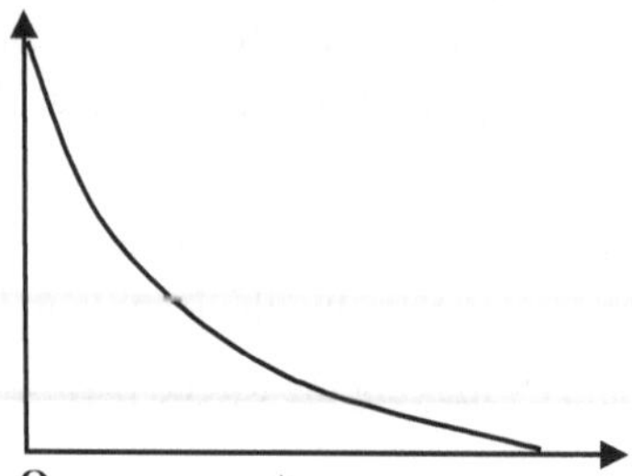

Fig. 6.8 *Amplitude decay under dead beat oscillations*

II. b^2 is slightly greater than ω^2

Using this condition in equation (6.13), this solution does not satisfy equation (6.11).

Here we assume that,

$$\sqrt{b^2-\omega^2}=h, \quad \text{where } h\to 0$$

So from equation (6.13), we have

$$y= A_1e^{[-b-h]t}+A_2e^{[-b+h]t}=e^{-bt}\left[A_1e^{-ht}+A_2e^{ht}\right]$$

$$y= e^{-bt}\left[A_1\left(1-ht+\frac{h^2t^2}{2!}-\text{K}\right)+A_2\left(1+ht+\frac{h^2t^2}{2!}+\text{K}\right)\right]$$

$$= e^{-bt}\left[\left[A_1+A_2\right]+\left[A_2-A_1\right]ht+\left[A_1+A_2\right]h^2t^2+\text{K}\right]$$

$$\approx e^{-bt}\left[\left[A_1+A_2\right]+\left[A_2-A_1\right]ht\right] \qquad (\text{as } h\to 0, \text{ then, } h^2, h^3, \ldots \to 0)$$

$$= e^{-bt}\left[P+Qt\right] \qquad \ldots(6.14)$$

Here,

$$P=[A_1+A_2], \text{ and, } Q=h[A_2-A_1]$$

Equation (6.14) gives the possible solution of equation (6.11) in this case. From equation (6.14), we see that as t increases, $(P+Qt)$ also increases. With higher values of t, the e^{-bt} term increases more than the $(P+Qt)$ term. So displacement y slowly approaches zero at a certain value of t. Then again it decreases more rapidly.

This damping condition is known as Critically damped oscillations. Here the particle acquire its position of equilibrium more rapidly than in earlier case.

Example: Pointer movement in Ammeter and Voltmeter.

III. $b^2 < \omega^2$

So, $\sqrt{b^2 - \omega^2}$ is imaginary and let $\sqrt{b^2 - \omega^2} = i\beta$,

so, from equation (10), we have,

$$y = A_1 e^{[-b - i\beta]t} + A_2 e^{[-b + i\beta]t}$$

$$= e^{-bt}\left[A_1\left[\cos\beta t - i\sin\beta t\right] + A_2\left[\cos\beta t + i\sin\beta t\right]\right]$$

$\Rightarrow$
$$y = e^{-bt}\left[\left[A_1 + A_2\right]\cos\beta t + i\left[A_2 - A_1\right]\sin\beta t\right]$$

$$= e^{-bt}\left[a\sin\phi\cos\beta t + a\cos\phi\sin\beta t\right]$$

[where, $a\sin\phi = A_1 + A_2$, and, $a\cos\phi = i\left[A_2 - A_1\right]$]

$$= ae^{-bt}.\sin\left[\beta t + \phi\right]$$

$$y = ae^{-bt}.\sin\left[t\sqrt{\omega^2 - b^2} + \phi\right] \qquad \ldots(6.15)$$

Equation (6.15) represents SHM with variable amplitude ae^{-bt} due to damping and thus y varies from + ae^{-bt} to – ae^{-bt}. This is shown in Fig. 6.9. The decay of amplitude depends on b. This damping vibration is known as under damped oscillations.

It has the time period, T, given by,

$$T = \frac{2\pi}{\beta} = \frac{2\pi}{\sqrt{\omega^2 - b^2}}$$

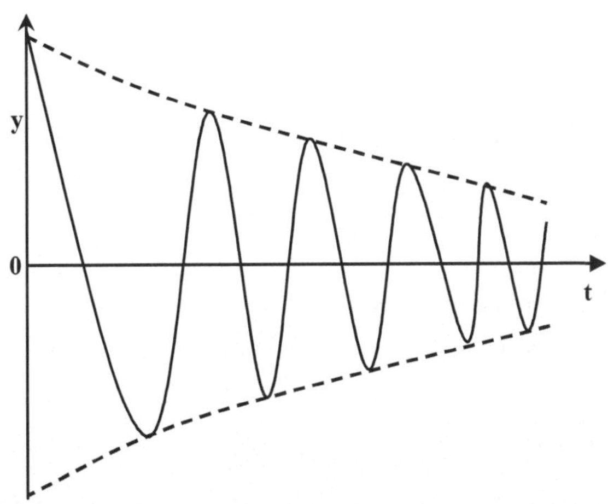

Fig. 6.9 *Decay of amplitude in under damped oscillations*

8. ENERGY OF DAMPED HARMONIC OSCILLATOR

From eqn.(6.15) the displacement is given by

$$y = ae^{-bt} \cdot \sin\left[t\sqrt{\omega^2 - b^2} + \phi\right]$$

Now, the instantaneous KE in the steady state here is given by

$$KE = \frac{1}{2} m \left(\frac{dy}{dt}\right)^2$$

So, $$KE = \frac{1}{2} m \left(\frac{dy}{dt}\right)^2$$

$$= \frac{1}{2} m \left[-abe^{-bt} \sin(\beta t + \phi) + a\beta e^{-bt} \cos(\beta t + \phi)\right]^2 \quad ...(6.16)$$

The PE at any instant is given by

$$PE = \frac{1}{2} \mu (y)^2$$

$$= \frac{1}{2} m\omega^2 a^2 e^{-2bt} \cdot \sin^2[\beta t + \phi] \quad ...(6.17)$$

Hence, the total instantaneous energy is given by

$$E = KE + PE$$

$$= \frac{1}{2} m\left[a^2b^2e^{-2bt}\sin^2(\beta t + \phi) + a^2\beta^2e^{-2bt}\cos^2(\beta t + \phi) - 2a^2b\beta e^{-2bt}\right.$$

$$\left.\sin(\beta t + \phi)\cos(\beta t + \phi)\right] + \frac{1}{2} m\omega^2 a^2 e^{-2bt} \cdot \sin^2[\beta t + \phi]$$

$$= \frac{1}{2} ma^2e^{-2bt}\left[b^2\sin^2(\beta t + \phi) + \beta^2\cos^2(\beta t + \phi) - 2b\beta\sin(\beta t + \phi)\right.$$

$$\left.\cos(\beta t + \phi) + \omega^2\sin^2(\beta t + \phi)\right]$$

$$\Rightarrow E = \frac{1}{2} ma^2e^{-2bt}\left[b^2\sin^2(\beta t + \phi) + \beta^2\cos^2(\beta t + \phi) - b\beta\sin 2(\beta t + \phi)\right.$$

$$\left. + \omega^2\sin^2(\beta t + \phi)\right] \quad ...(6.18)$$

Here damping is small, so we can have, $b << \omega$, so e^{-2bt} term does not change much over one time period of the oscillation.

Hence, the net time average of the energy of the oscillator over one period is given by

$$\langle E\rangle = \frac{1}{2} ma^2e^{-2bt}\left[b^2\langle\sin^2(\beta t + \phi)\rangle + \beta^2\langle\cos^2(\beta t + \phi)\rangle - b\beta\langle\sin 2(\beta t + \phi)\rangle\right.$$

$$\left. + \omega^2\langle\sin^2(\beta t + \phi)\rangle\right] \quad ...(6.19)$$

However as earlier,

$$\left\langle \sin^2 (\beta t + \phi) \right\rangle = \frac{1}{2}, \text{ and,}$$

$$\left\langle \cos^2 (\beta t + \phi) \right\rangle = \frac{1}{2}$$

Also, $\left\langle \sin 2 (\beta t + \phi) \right\rangle = \frac{1}{T} \int_0^T \sin 2 (\beta t + \phi)\, dt = 0$

Using these into eqn. (6.19), we have,

$$\langle E \rangle = \frac{1}{2} m a^2 e^{-2bt} \left[b^2 \left(\frac{1}{2}\right) + \beta^2 \left(\frac{1}{2}\right) - b\beta (0) + \omega^2 \left(\frac{1}{2}\right) \right]$$

$$= \frac{1}{2} m a^2 e^{-2bt} \left[b^2 \left(\frac{1}{2}\right) + \left(\omega^2 - b^2\right) \frac{1}{2} + \omega^2 \left(\frac{1}{2}\right) \right]$$

$$= \frac{1}{2} m a^2 e^{-2bt} \left[2\omega^2 \left(\frac{1}{2}\right) \right]$$

So, $$\langle E \rangle = \frac{1}{2} m \omega^2 a^2 e^{-2bt} = E_0 e^{-2bt} \quad \ldots(6.20)$$

Here, $$E_0 = \frac{1}{2} m \omega^2 a^2$$

This is the total energy of the damped harmonic oscillator.

The decay of the energy of the system is shown in the Figure over a complete cycle.

9. LOGARITHMIC DECREMENT (λ)

This is a important parameter to give how the amplitude of vibration is dying away.

This is the natural logarithm of the ratio between two successive maximum amplitudes of oscillations of a body undergoing damped oscillations which are separated by one period.

We have the amplitude of damped harmonic oscillator is ae^{-bt}.

So at $t = 0$, $a_0 = a$. Let $a_1, a_2, a_3, \ldots$ are the amplitudes at $t = T, 2T, 3T, \ldots$

Hence, at $t = T$, $a_1 = ae^{-bT}$
at $t = 2T$, $a_2 = ae^{-2bT}$,
at $t = 3T$, $a_3 = ae^{-3bT}$, and so on.

Hence we have,

$$\frac{a}{a_1} = \frac{a_1}{a_2} = \frac{a_2}{a_3} = \text{KK} = e^{bT} = e^{\lambda}.$$

Here, $\lambda = bT$ and is known as the logarithmic decrement of the oscillation and is given by

$$\lambda = \log_e\left(\frac{a_0}{a_1}\right) = \log_e\left(\frac{a_1}{a_2}\right) = \log_e\left(\frac{a_2}{a_3}\right) = \ldots$$

10. RELAXATION TIME (τ)

Relaxation time, τ, is the time taken for the total mechanical energy to decay to (1/e)th of its original value.

Mechanical energy of damped harmonic oscillator is

$$E = E_0 e^{-2bt}.$$

After the relaxation time, τ, *i.e.*, at $t = \tau$, $E = E_0 e^{-2b\tau}$.

But at $t = \tau$, $E = E_0/e$.

Hence from above,

$$\frac{E_0}{e} = E_0 e^{-2b\tau}.$$

$$\Rightarrow \quad e^{-1} = e^{-2b\tau}$$

$$\Rightarrow \quad 2b\tau = 1 \quad \Rightarrow \quad \tau = \frac{1}{2b}$$

So, $$\text{E} = E_0 e^{-\frac{t}{\tau}}.$$

11. QUALITY FACTOR (Q)

It is defined as 2π times the ratio of the energy stored in the system to the energy lost per period.

Hence, $$Q = 2\pi \times \frac{\text{Average energy stored in system}}{\text{Average energy lost per period}} = 2\pi \cdot \frac{E}{PT}.$$

Here, P is the average power dissipated and is given by

$$P = \frac{E}{\tau}.$$

So, $$Q = 2\pi \cdot \frac{E}{\left(\frac{E}{\tau}\right)T} = \frac{2\pi}{T} \cdot \tau = \omega\tau$$

This clearly indicates that for higher Q, we require high value of τ and low value of damping. It is a measure of decay rate of energy of the under damped system.

7

Dielectrics

The experiment of Faraday showed that the capacitance of a capacitor is increased when an insulator is put in between the capacitor plates. This increase is the maximum if the insulator fills the total space in between the plates and is depending on the nature of the insulating material.

The insulator materials which when placed in electric field produce induced charges are called Dielectrics. Dielectrics do not change electric charges, only allow the inductive effect to pass through them, when placed in an electric field. It does not contain any free electron or their number is negligibly small to constitute an electric moment. The electrons here are tightly bound to the nucleus of the atom.

The property of the dielectric is represented by a constant K known as the 'dielectric constant'. Materials such as glass, ceramics, polymers, papers, etc., are the common examples of dielectrics. Due to their insulating properties they conduct negligible current and have resistivity $\approx 10^{10} - 10^{20}\ \Omega$-cm.

There are generally two kinds of Dielectrics.

1.(A) POLAR DIELECTRIC

Although the molecule is electrically neutral their spatial arrangement of charges varies. We can consider a single positive charge of the molecule if this charge is placed at the centre of gravity of all the positive charges. Similarly, we can do this with negative charges also. The point of action of these positive and negative charges may or may not coincide. When they do not coincide molecule will possess a permanent intrinsic dipole moment. They are called as 'Polar dielectrics'. Here, in these molecules the centre of gravity of positive charges does not coincide with the centre of gravity of negative charges. Examples are water, HCl, NH_3, etc.

In absence of electric field molecules are randomly oriented as shown in Fig. 7.1 (a). When an electric field, $\vec{E}$, is applied, individual permanent dipoles experience a torque. It turn the dipole smoothly towards the direction of $\vec{E}$. However, initially thermal motion does not allow complete alignment (Fig. 7.1 (*b*)). So with higher value of $\vec{E}$ at constant temperature dipoles overcome thermal resistance and align themselves along $\vec{E}$ (Fig. 7.1 (*c*)). Applied electric field separates positive charge from the negative charge. The dielectric in this state is said to be polarized and this enforcement to all dipoles to align them throughout the volume of the dielectric is known as Polarization.

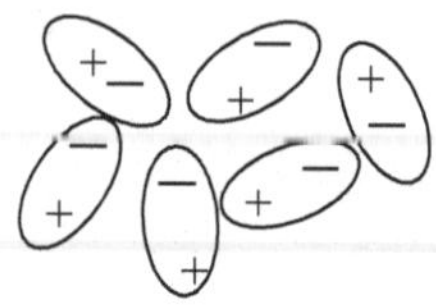

Fig. 7.1 (*a*) *Polar dielectric without applied electric field*

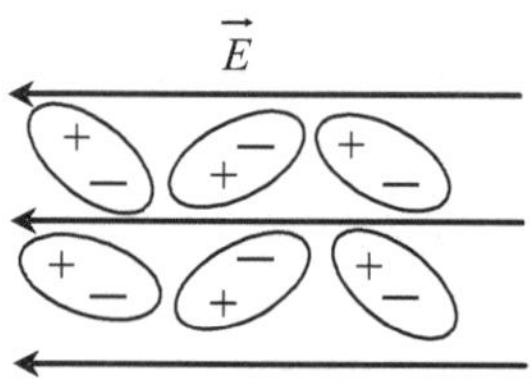

Fig. 7.1 (*b*) *Polar dielectric with weak applied electric field*

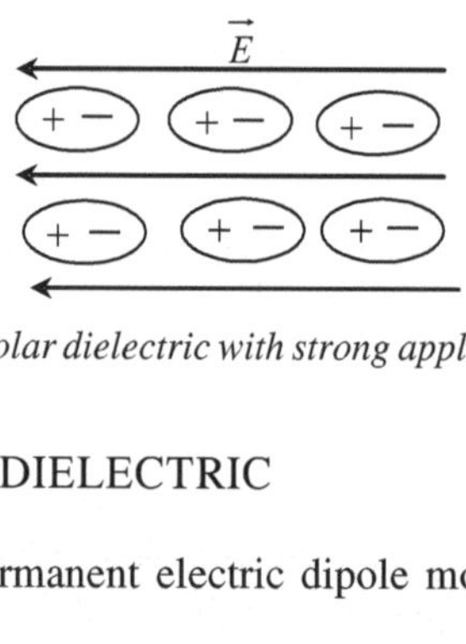

Fig. 7.1 (*c*) *Polar dielectric with strong applied electric field*

1. (B) NON-POLAR DIELECTRIC

They do not have any permanent electric dipole moment for their symmetric structures.

This is shown in Fig. 7.2 (*a*). When electric field is applied the molecules are deformed as in Fig. 7.2 (*b*). The positive charges are then displaced a little in the direction of applied field and negative charges in the other way. Thus, a dipole moment is acquired by the molecules of the dielectric by induction. Such molecule is called Induced Dipole and its moment is called Induced Dipole Moment.

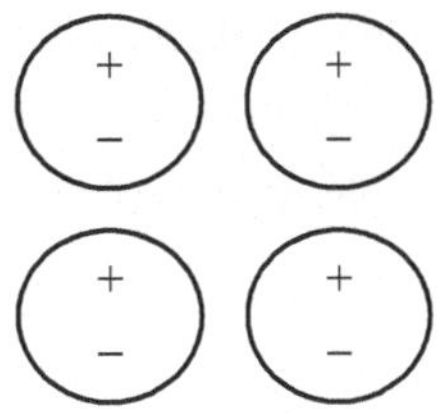

Fig. 7.2 *(a) Non-polar dielectric without applied electric field*

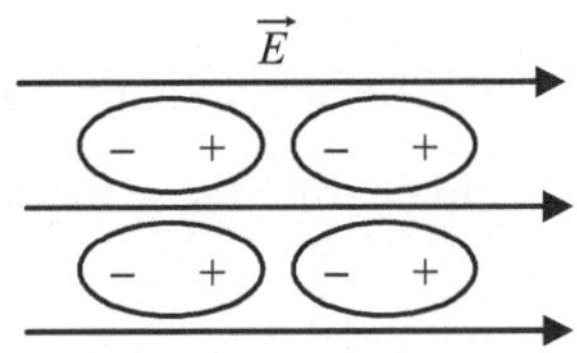

Fig. 7.2 *(b) Non-polar dielectric with strong applied electric field*

On withdrawal of electric field, initial situation again comes up. Examples are H_2, N_2, O_2, CCl_4, etc.

2. POLARIZATION

We place a dielectric bar in an external electric field (between plates of charged capacitor here) as shown in Fig. 7.3. So, non-polar atoms are converted into the polar atoms. If polar atoms exist in addition, they will also align. The result will be a total alignment of dipoles along electric field $\vec{E}_0$, resulting in a net dipole moment of the dielectric bar. On the surfaces of the bar at right angle to the external field $\vec{E}_0$, there will be an appearance of a net negative and positive charge.

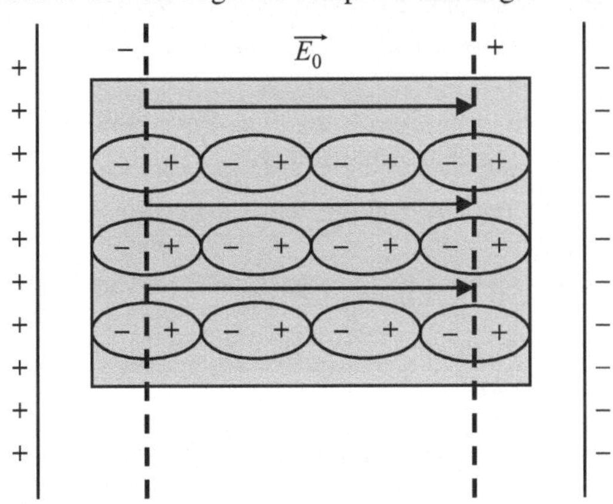

Fig. 7.3 *Dielectric bar placed under electric*

This phenomenon of appearance of charges on the surfaces of the dielectric bar, induced by an external electric field, is called Polarization.

The Dipole Moment per Unit Volume of the Dielectric is termed as Polarization.

The Polarization is represented by P and its unit is Coulomb per square meter.

3. DIFFERENT RELATIONS BETWEEN ELECTRICAL PARAMETERS

I. Between Polarization and Surface density of charges (σ_p)

Let the dipole moment of each dipole is p. If n be the number of dipoles per unit volume, then, $P = np$.

We now consider a dielectric bar (Fig. 7.4) of cross-section A and length L. So its volume is AL. When placed in $\overrightarrow{E_0}$, let the surface density of charges becomes σ_p.

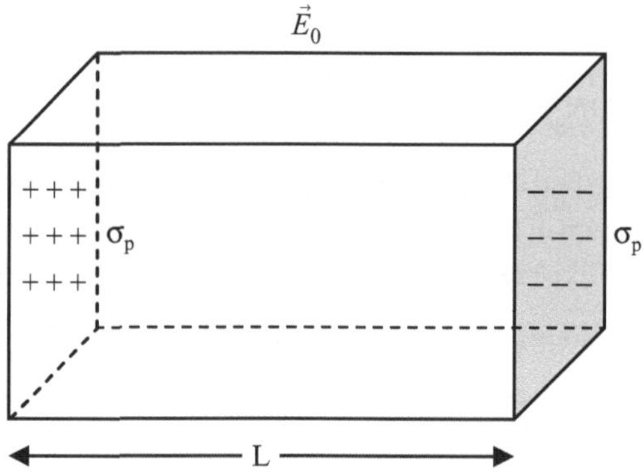

Fig. 7.4 *A polarized dielectric bar under electric field*

So, the total charge on the surface is given by $\sigma_p A$, and, net dipole moment of the bar is given by $\sigma_p AL$ (= charge × distance)

If P is dipole moment per unit volume, then the net dipole moment of the bar is $P.AL$.

Hence, $P.A.L = \sigma_p.A.L \Rightarrow P = \sigma_p$

Hence, dielectric polarization is numerically equal to the surface density of charges appearing at the ends.

II. Dielectric Constant (K)

On introduction of dielectric slab, the potential difference V between the plates of a capacitor also falls. If E_0 and E are the magnitudes of electric field in the absence and presence of dielectric respectively, then, the dielectric constant, K, is given by,

$$K = \frac{E_0}{E}$$

Also, $K = \frac{V_0}{V}$ for same capacitor.

We also know, $V = Ed$

So, as $E_0 > E$ we have $K > 1$

III. Displacement vector $(\vec{D})$

Let $\vec{E_0}$ be the electric field between plates of charged capacitor having charges $+q$ and $-q$, length between them L, and A be the area of each plate.

So, $$E_0 = \frac{q}{A\varepsilon_0} \qquad \text{...(7.1)}$$

where, ε_0 is the permittivity of free space.

Now, as dielectric is introduced between plates of the capacitor and $+q'$ and $-q'$ are the charges induced on the faces of dielectric. These charges change original electric field $\vec{E_0}$ to new electric field $\vec{E}$ inside the dielectric and so,

$$E = \frac{q - q'}{A\varepsilon_0} = \frac{q}{A\varepsilon_0} - \frac{q'}{A\varepsilon_0}$$

$$\Rightarrow \qquad \varepsilon_0 E = \frac{q}{A} - \frac{q'}{A} \Rightarrow \frac{q}{A} = \varepsilon_0 E + \frac{q'}{A}$$

As, $P = \text{Total dipole moment / Volume} = \frac{q'L}{AL} = \frac{q'}{A}$,

we have, $$\frac{q}{A} = \varepsilon_0 E + P \qquad \text{...(7.2)}$$

The right hand side of equation (7.2) is given the name as Electric Displacement (D).

So, $$\frac{q}{A} = \varepsilon_0 E + P = D \Rightarrow D = \varepsilon_0 E + P \qquad \text{...(7.3)}$$

Vectorically, $\vec{D} = \varepsilon_0 \vec{E} + \vec{P}$ (in SI system)

$\vec{D} = \vec{E} + 4\pi \vec{P}$ (in CGS system)

IV. Relation between ε and K

Using eqn.(7.1), $$K = \frac{E_0}{E} = \frac{q/A\varepsilon_0}{q/A\varepsilon} = \frac{\varepsilon}{\varepsilon_0}$$

where, ε is the permittivity of the medium.

V. Relation between $\vec{D}$ and $\vec{E}$

We know,

$$K = \frac{E_0}{E} \Rightarrow E = \frac{E_0}{K} = \frac{1}{K} \times \frac{q}{A\varepsilon_0} \qquad \text{[using eqn. (7.1)]}$$

So using eqn. (7.3), $D = \frac{q}{A} = K\varepsilon_0 \left[\frac{q}{K\varepsilon_0 A}\right] = K\varepsilon_0 E$ ...(7.4)

As, $K = \frac{E_0}{E} = \frac{\varepsilon}{\varepsilon_0}$,

$D = \varepsilon E$, vectorically, $\vec{D} = \varepsilon\vec{E}$.

VI. Relation between $\vec{P}$ and $\vec{E}$

We know from eqn.(7.3), $\vec{D} = \varepsilon_0\vec{E} + \vec{P}$,

and, from eqn.(7.4), $\vec{D} = K\varepsilon_0\vec{E}$.

So, using them, $\vec{D} = \varepsilon_0\vec{E} + \vec{P} = K\varepsilon_0\vec{E}$

$\Rightarrow$ $\vec{P} = (K-1)\,\varepsilon_0\vec{E}$ ...(7.5)

For free space, no dielectric medium, so. $K = 1$ and $\vec{P} = 0$

4. ELECTRIC SUSCEPTIBILITY (X)

For homogeneous and isotropic medium, polarization is linearly dependent on applied electric field. Hence,

$$\vec{P} \propto \vec{E} \Rightarrow \vec{P} = \varepsilon_0 X\vec{E}.$$

For magnitude only,

$$P = \varepsilon_0 XE \Rightarrow X = \frac{P}{\varepsilon_0 E} \qquad \text{...(7.6)}$$

Here, X represents the Electric Susceptibility of the dielectric medium. ε_0 is included only for simplification has no unit and so no dimension.

I. Relation between X and K

We now consider a capacitor carrying charges $+q$ and $-q$, and, A be the area of each plate. So, the electric field in absence of dielectric (as in Fig. 7.5(a)), E_0, is given by

$$E_0 = \frac{q}{\varepsilon_0 A} \qquad \text{[using eqn.(7.1)]}$$

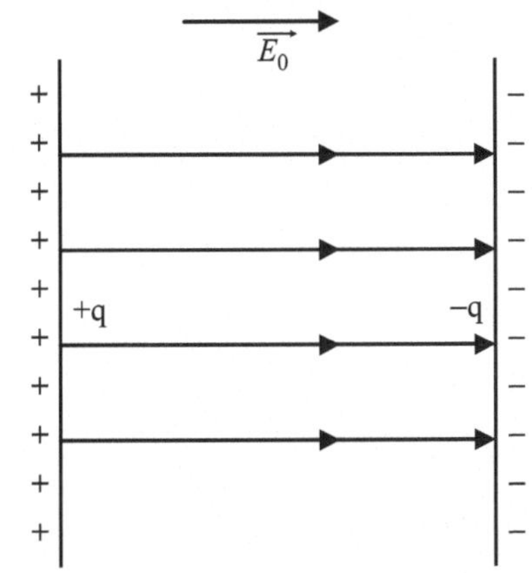

Fig. 5. *(a) Field inside a charged capacitor*

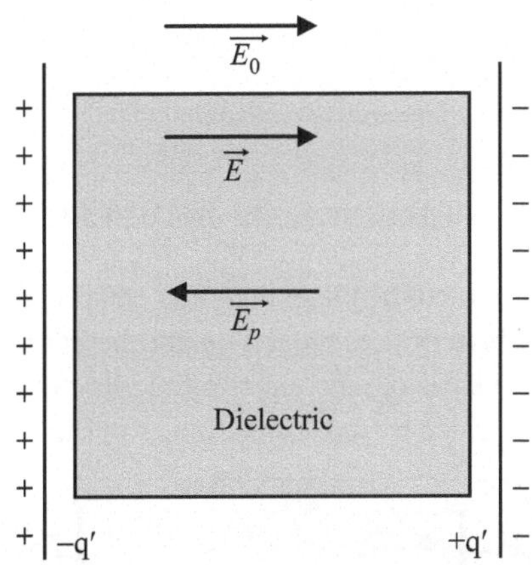

Fig. 7.5 *(b) Field inside dielectric placed inside charged capacitor*

We now place a bar of dielectric of dielectric constant K (Fig. 7.5(b)), between plates of capacitor. $\vec{P}$ is the polarization vector and $+q'$ and $-q'$ are the induced charges on the surface of dielectric. So, E_p is the induced electric field due to polarization of dielectric and is given by,

$$E_p = \frac{q'}{\varepsilon_0 A} = \frac{P}{\varepsilon_0}$$

and the resultant electric field is

$$E = E_0 - E_p = E_0 - \frac{\sigma_p}{\varepsilon_0} = E_0 - \frac{P}{\varepsilon_0}$$

So, $$E_0 = E + \frac{P}{\varepsilon_0}$$

$$\Rightarrow \qquad \frac{E_0}{E} = 1 + \frac{P}{\varepsilon_0 E} = 1 + X \qquad \text{[using eqn.(7.6)]}$$

But,
$$\frac{E_0}{E} = \frac{\left[\dfrac{q}{\varepsilon_0 A}\right]}{\left[\dfrac{q}{\varepsilon A}\right]} = \frac{\varepsilon}{\varepsilon_0} = K \qquad \text{...(7.7)}$$

Hence, from above, $K = 1 + X$ ...(7.8)

II. Relation between ε and X

Using eqns.(7.7) and (7.8), we have

$$K = \frac{\varepsilon}{\varepsilon_0} = 1 + X$$

$$\Rightarrow \qquad X = \frac{\varepsilon}{\varepsilon_0} - 1 \qquad \text{...(7.9)}$$

5. GAUSS'S LAW IN PRESENCE OF A DIELECTRIC

We consider here a pair of metallic plates and their ends connected to positive and negative terminals of a battery. Let $+q$ and $-q$ are the charges on the plates and these are free charges. We denote the electric field due to above charges as E_0. We draw the Gaussian surface by the dotted line as in Fig. 7.6 (*a*).

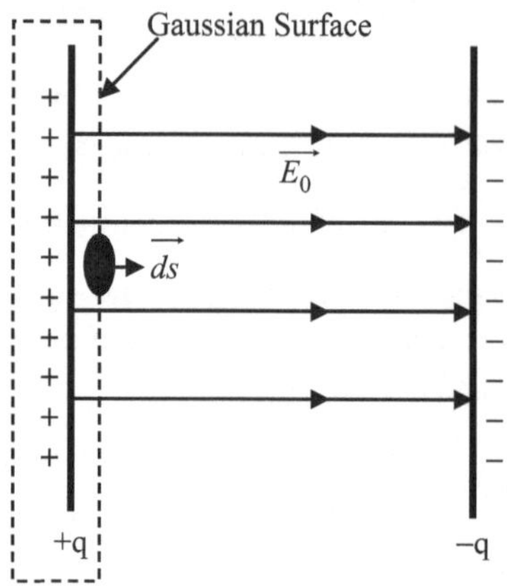

Fig. 7.6 (*a*) *Electric field inside a charged capacitor with Gaussian surface*

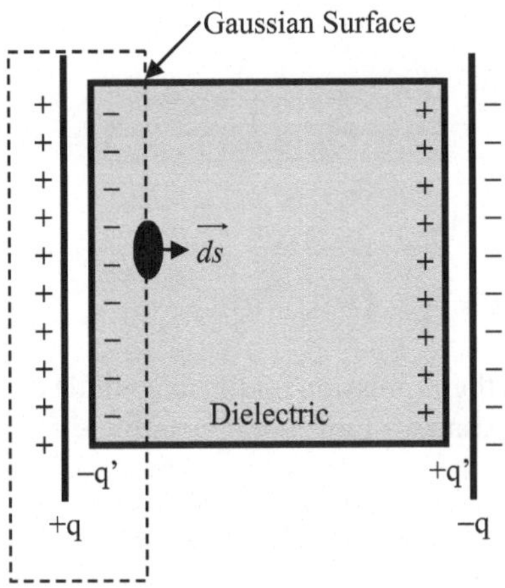

Fig. 7.6 *(b) Dielectric placed inside a charged capacitor with Gaussian surface*

From Gauss's law, electric flux, ϕ, through any closed surface is $\left(1/\varepsilon_0 \right)$ times the net charge, q, enclosed by the surface, *i.e.*,

$$\phi = \oint_S \vec{E_0} \cdot \vec{dS} = \frac{q}{\varepsilon_0}$$, where $\vec{dS}$ is the surface area vector.

So, from above

$$\oint_S E_0 dS \cos 0° = \frac{q}{\varepsilon_0} \quad \Rightarrow \quad E_0 \oint_S dS = \frac{q}{\varepsilon_0}$$

$$\Rightarrow \qquad E_0 A = \frac{q}{\varepsilon_0} \quad \Rightarrow \quad E_0 = \frac{q}{\varepsilon_0 A} \qquad \ldots(7.10)$$

Equation (7.10) gives the magnitude of electric field in free space in terms of charges on capacitor. Now, we introduce a dielectric bar with dielectric constant, K, between plates of the capacitor as in Fig. 7.6 (*b*). We also assume that the induced charges $+q'$ and $-q'$ appear on ends of dielectric as in Fig. 7.6 (*b*). These charges produce own electric field which opposes $\vec{E_0}$.

So total charge on the Gaussian surface = $q - q'$.

If $\vec{E}$ is net electric field inside dielectric, then applying Gauss's law to the same Gaussian surface, we have

$$\oint_S \vec{E} \cdot \vec{dS} = \frac{q - q'}{\varepsilon_0}$$

$$\Rightarrow \qquad \oint_S E dS \cos 0° = \frac{q - q'}{\varepsilon_0} \quad \Rightarrow \quad E = \frac{q - q'}{\varepsilon_0}$$

However, $$K = \frac{E_0}{E} = \frac{q/\varepsilon_0 A}{(q-q')/\varepsilon_0 A} = \frac{q}{q-q'}$$

$$\Rightarrow \quad Kq - Kq' = q$$

$$\Rightarrow \quad q' = \frac{q}{K}(K-1) = q\left(1 - \frac{1}{K}\right) \qquad \ldots(7.11)$$

So, Eqn.(7.11) gives the expression for the magnitude of induced charges q' in terms of q and K. Thus, Gausses Law modifies into

$$\oint_S \vec{E}\,.\,\vec{dS} = \frac{q-q'}{\varepsilon_0} = \frac{q/K}{\varepsilon_0} \quad \left[\because \text{Eqn.(7.11)} \Rightarrow q - q' = q - q\left(1-\frac{1}{K}\right) = \frac{q}{K}\right]$$

$$\Rightarrow K\oint_S \vec{E}\,.\,d\vec{S} = \frac{q}{\varepsilon_0}$$

This is the Gauss's Law in a Dielectric with constant K. This is also called the Integral form of Gauss's Law for Dielectric.

6. ENERGY STORED IN DIELECTRIC

We now consider a capacitor whose one plate is given a charge $+q$ and the other is earthed. Let, V be the potential difference between capacitor plates and C be the capacitance of it. We are now supplying an additional charge dq to it. The charge present in capacitor, q, will repel this dq and work is said to be done to charge the capacitor further. So, the work done is, $dW = \frac{q}{C}\,dq$.

Hence, total work done is given by

$$W = \int dW = \int_0^q \frac{q}{C}\,dq = \frac{1}{2}\frac{q^2}{C}.$$

The energy stored by the charged capacitor is then

$$U = W = \frac{1}{2}\frac{q^2}{C} = \frac{1}{2}CV^2 \qquad [\because \; q = CV]$$

Now, if A and d are the area of each plate of capacitor and the distance between plates respectively, then,

$$C = \frac{\varepsilon_0 A}{d}$$

Hence, $$U = \frac{1}{2}CV^2 = \frac{1}{2} \times V^2 \times \frac{\varepsilon_0 A}{d}$$

$$= \frac{1}{2} \times \frac{\varepsilon_0 A}{d} \times (Ed)^2 \qquad [\because \ V = Ed]$$

$$\Rightarrow \qquad U = \frac{1}{2} \varepsilon_0 \ AdE^2$$

So, energy stored per unit volume is

$$\frac{U}{Ad} = \frac{\frac{1}{2} \varepsilon_0 AdE^2}{Ad} = \frac{1}{2} \varepsilon_0 E^2 \propto E^2.$$

After an introduction of dielectric with constant '*K*', we have the, the energy stored per unit volume is

$$\frac{1}{2} \varepsilon_0 KE^2 \qquad \left[\because \ C = \frac{\varepsilon A}{d} = \frac{(k\varepsilon_0) A}{d}\right].$$

So an extra energy (potential) per unit volume arises from the polarization of molecules of dielectric by an amount of

$$\frac{1}{2} \varepsilon_0 KE^2 - \frac{1}{2} \varepsilon_0 E^2 = \frac{1}{2} \varepsilon_0 (K-1) E^2.$$

This is the amount of energy stored inside a dielectric by its introduction.

7. TYPES OF POLARIZATION

There are basically three types of polarization as the followings.

I. Electronic Polarization (P_e)

For an atom, molecule or ion electron cloud surrounds positively charged heavy nucleus. On application of electric field the electron cloud shifts to a distance less than the atomic dimension. This is shown in Fig. 7.7 (*a*) and 7.7 (*b*). This polarization by the distortion of electron cloud from spherical symmetry is known as the electronic polarization. It sets over a time of $10^{-14} - 10^{-15}$ sec and is independent of temperature.

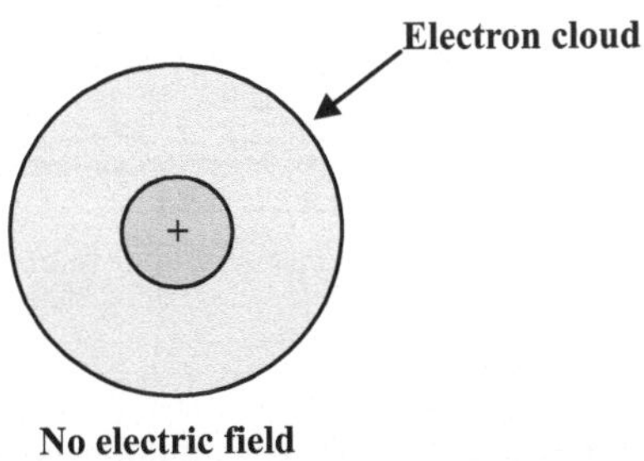

Fig. 7.7 (*a*) *Non-polar dielectric without externally applied electric field*

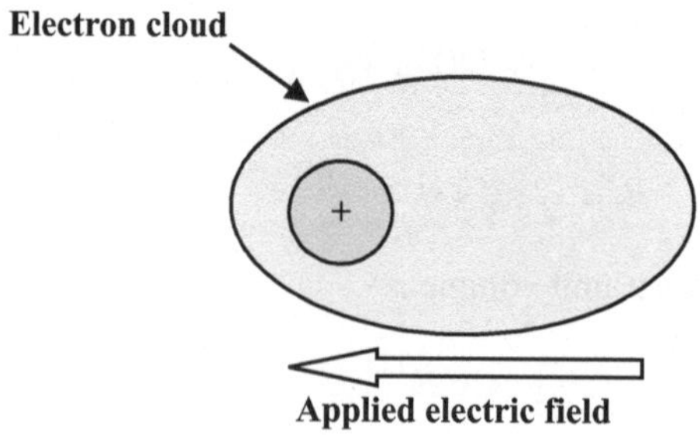

Fig. 7.7 *(b) Shifting of electron cloud with externally applied electric field*

II. Ionic Polarization (P_i)

It appears only in ionic crystals. On application of an electric field to the ionic molecules both the cations and anions are displaced in opposite directions till the ionic bonding forces them to stop. So there is a increase in dipole moment of the molecules. This is shown in Fig. 7.8 (*a*) and 7.8 (*b*) Here restoring force comes from the electrostatic forces between the two individual dipoles.

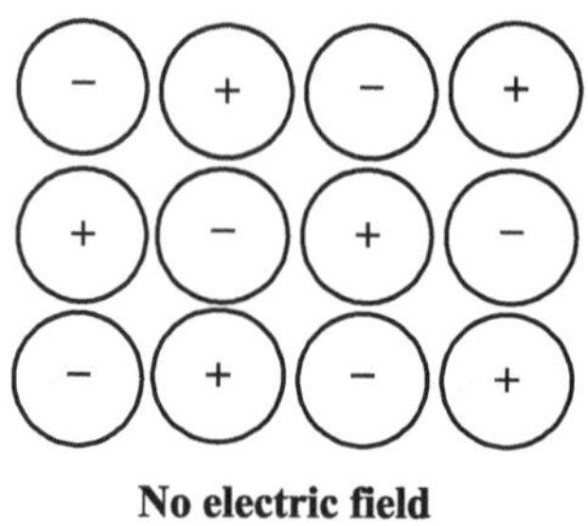

Fig. 7.8 *(a) Dielectric of ionic crystals without applied electric field*

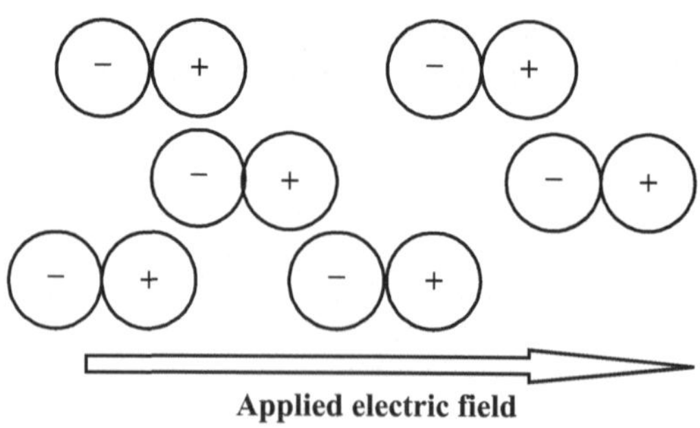

Fig. 7.8 *(b) Generation of ionic polarization with applied electric field*

III. Orientational Polarization (P_o)

This is a characteristic of polar dielectrics. The randomness of individual dipoles in absence of any external electric field yields zero net dipole moment. Under electric field they try to align as per electric field orientation which results a nonzero net dipole moment. This is clear from Fig. 7.9. Then the molecular dipoles rotate about their axis to align themselves along applied electric field. However, here no restoring force exists. Only problem comes from thermal agitation, due to which their alignments are disturbed even under external electric field.

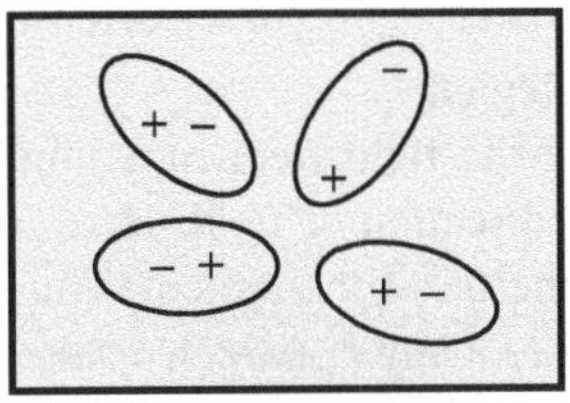

No Electric Field

Fig. 7.9 *(a) Polar dielectric without applied electric field*

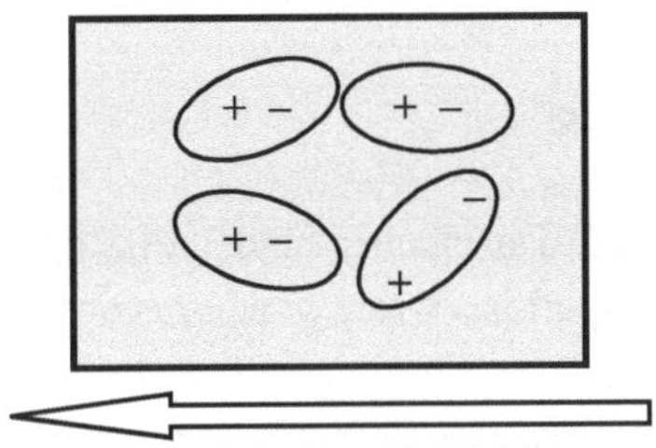

Applied Electric Field

Fig. 7.9 *(b) Generation of Orientational polarization with applied electric field*

So, net polarization in a dielectric, P, is given by,

$$P = P_e + P_i + P_o.$$

8. FREQUENCY DEPENDENCE OF DIELECTRIC (Behaviour of Dielectric in a.c. Field)

An a.c. field changes its direction with time. With each direction reversal, the polarization components are required to follow the field reversal. So the total polarization depends on the ability of dipoles to orient themselves in the direction of the field during each alternation of the field. The relative permittivity which is a measure of polarization shows marked differences in behavior at different frequencies.

(*i*) **At Low Frequency Region**

Here the dipoles will get sufficient time to orient themselves completely along the instantaneous direction of the field. The average time taken by the dipoles to reorient in the field direction is known as Relaxation Time, τ, and its reverse, the Relaxation frequency, ν. Thus, the polarization remains the maximum. If the frequency of the applied field is greater than ν, then the dipoles can not reverse fast enough.

(*ii*) **Radio Frequency or Microwave Region**

In this region permanent dipoles fail to follow the field reversals due to very high field frequency as heavy molecule does not follow the instantaneous orientation of the electric field. Thus P_o becomes zero in this frequency range. Only P_i and P_e exist there. So K decreases considerably.

(*iii*) **Infrared Frequency Region**

On further increasing the field frequency ions of large masses, fails to follow instantaneous direction of field. Here, P_i fails to follow the field reversals due to the inertia of the system. Thus the contribution of ionic polarizibility also ceases. Only P_e (which results only from displacement of electron cloud) then contributes.

(*iv*) **Ultra-violet Frequency Region**

Here with a electric field fast changing orientation, even electron cloud fail to follow the field alternations and thus the total polarization becomes zero.

9. DIELECTRIC LOSSES

When a material is subjected to an electric field, it dissipates a part of its electrical energy to heat energy. This part is a waste or loss, as no useful work is done by it. 'Power Loss' denotes the average electrical power dissipated in a material during a certain interval of time. [Generally, Power Loss $\propto$ (applied voltage $)^2$.]

I. In d.c. Field

A conductor of resistance R connected to a d.c.voltage V has a power loss as

$$P_{dc} = \frac{V^2}{R} = I^2 R .$$

Similarly, when it is connected to an voltage of same effective magnitude as d.c. voltage V, the power loss is given by

$$P_{ac} = \frac{V^2}{R} = I^2 R .$$

The power loss also called I^2R loss or 'Jouliean Heat Loss'.

So in conductor, $P_{ac} = P_{dc}$.

In dielectric, I^2R loss in d.c. will be very small due to high R and low current. However, power loss in a.c. field will be quite large.

II. In a.c. Field

The absorption of electric energy by dielectric subjected to a.c. field is known as 'Dielectric Loss'.

The dielectric loss caused by an a.c.field also results in dissipation of electrical energy as heat in the materials.

When a dielectric is placed in an a.c., there is an energy loss in dielectric due to the variation of polarization, $\vec{P}$ and displacement, $\vec{D}$ with time. Generally $\vec{P}$ & $\vec{D}$ may lag behind $\vec{E}$ in phase. This phenomenon of lagging of $\vec{D}$ behind $\vec{E}$ is known as 'hystersis', same as magnetic hystersis where flux $\vec{B}$ lag behind $\vec{H}$. Thus, the curve between $\vec{E}$ & $\vec{D}$ is called hystersis curve and area under the curve measures the loss of energy called the 'Dielectric Loss'.

Theory

Let an a.c. electric field applied to a dielectric is given by

$$E = E_0 \cos \omega t$$

If $\vec{D}$ is lagging behind $\vec{P}$, *i.e.*, $\vec{E}$ by an amount of phase δ.

Hence,

$$D = D_0 \cos(\omega t - \delta) = D_1 \cos \omega t + D_2 \sin \omega t \qquad \ldots(7.12)$$

Here,

$$D_1 = D_0 \cos \delta \;\&\; D_2 = D_0 \sin \delta, \; \delta \text{ is the phase lag.}$$

For most dielectrics $D_0 \propto E_0$, but the ratio (D_0/E_0) is dependent on frequency.

We now introduce two frequency dependent dielectric constants as

$$k' = \frac{D_1}{E_0} = \frac{D_0 \cos \delta}{E_0}$$

And,

$$k'' = \frac{D_2}{E_0} = \frac{D_0 \sin \delta}{E_0}.$$

These two constants can be combined in single complex dielectric constants as

$$k^* = k' + ik''$$

and,

$$D = k^* E_0 e^{i\omega t} = k^* E_0 (\cos \omega t + i \sin \omega t)$$

Also,

$$\tan \delta = k''/k'$$

Hence, δ is also frequency dependent.

The current density, I, in the capacitor is given by,

$$I = \frac{dD}{dt} = \omega\left[-D_1 \sin \omega t + D_2 \cos \omega t\right] \text{[using eqn.(7.12)]}$$

Hence, the energy density dissipated during one time period in the dielectric is given by

$$W = \frac{1}{T}\int_0^T IE\,dt$$

$$= \frac{\omega}{2\pi}\left[-\omega\int_0^{2\pi/\omega} E_0 D_1 \sin\omega t\cos\omega t dt + \omega\int_0^{2\pi/\omega} E_0 D_2 \cos^2\omega t dt\right]$$

$$= 0 + \frac{\omega}{2\pi}\times\omega\, E_0 D_2 \times \frac{(2\pi/\omega)}{2}$$

$$= \frac{\omega}{2} E_0 D_2$$

So,

$$\mathrm{W} = \frac{\omega}{2} E_0 D_0 \sin\delta = \frac{1}{2}\omega k'' E_0^2$$

Thus, energy losses are proportional to k'', or, $\sin\delta$. $\sin\delta$ is known as the Loss Factor. For small δ, $\tan\delta \approx \sin\delta \approx \delta$.

(*i*) **For Low frequency:** Contribution of W is mainly from d.c. resistivity.

(*ii*) **For High frequency:** Contribution of W is mainly due to dipole rotation.

Application of Dielectric Loss: Dielectric heating is a popular example of it. Here, insulators can be heated up by subjecting them to a high voltage of suitable frequency, *i.e.*, at which frequency the dielectric loss is a maximum. For example, in microwave oven microwave produced by a source are distributed uniformly inside the chamber by reflection from metal walls. They pass through glass or ceramic cooking utensils and are absorbed by water molecules. During absorption of microwaves by water molecules, heat is evolved and this heat helps in cooking food.

8

Electromagnetic Theory

Basic Ideas

The varying electrical and magnetic fields propagating in space in form of waves are called Electromagnetic waves (E.M.waves).

The direction of motion of e.m.waves is perpendicular to its both electric and magnetic field components. Examples are light waves, X-ray, etc. Thus, the propagation of E.M.waves implies variation in the electrical and magnetic fields.

As electrical and magnetic fields can exist in vacuum, so E.M.waves can propagate through it.

A comparison between the mechanical wave and e.m.waves

Sr.No.	*Mechanical waves*	*E.M. waves*
1	Generated through mechanical oscillations.	Generated through electrical oscillations.
2	Required material medium for propagation.	Do not require any medium for propagation.
3	Velocity of propagation is $v = \frac{I}{\sqrt{T/\rho}}$, where, I, T and ρ are inertia, tension in the medium and density respectively.	Velocity of propagation is $v = \frac{1}{\sqrt{\mu\varepsilon}}$, where, μ and ε are permeability and permittivity of the medium respectively.
4	Potential Energy (*P.E.*) is stored through T and Kinetic Energy (*K.E.*) is stored through ρ. Here ρ gives the frictional force comes due to kinetic energy.	Potential Energy (*P.E.*), *i.e.*, electrical energy is stored through ε and Kinetic Energy (*K.E.*), *i.e.*, magnetic energy is stored through μ.
5	Particles of the medium execute oscillatory	The electric and magnetic fields
6.	Motion and the variable is the displacement of the particles.	Execute oscillations and the variables are $\vec{E}$, $\vec{H}$ or $\vec{B}$.

Some important parameters with symbols and SI units

Sr. No.	Physical quantity	Symbol	Unit
1	Electric field strength	$\vec{E}$	Volt / meter
2	Magnetic field strength	$\vec{H}$	Ampere / meter
3	Electric displacement vector	$\vec{D}$	Coulomb / meter^2
4	Magnetic induction	$\vec{B}$	Tesla or Weber / meter^2
5	Permittivity	ε	Faraday / meter
6	Permittivity of free space	ε_0	Faraday / meter
7	Relative permittivity	ε_r, K	No dimension
8	Permeability	Ì	Henry / meter
9	Permeability of free space	μ_0	Henry / meter
10	Relative permeability	μ_r	No dimension
11	Current density	$\vec{J}$	Ampere / meter^2
12	Charge density	P	Coulomb / meter^3
13	Electrical conductivity	Σ	Siemens / meter
14	Displacement current	$\frac{d\vec{D}}{dt}$	Ampere / meter^2

Some useful relations

1. $\varepsilon_r = \frac{\varepsilon}{\varepsilon_0}$
2. $\mu_r = \frac{\mu}{\mu_0}$
3. $\vec{D} = \varepsilon\vec{E}$
4. $\vec{B} = \mu\vec{H}$
5. $\vec{J} = \sigma\vec{E}$.

Some sources

1. ρ is the source of $\vec{E}$.
2. $\vec{J}$ and $\frac{\partial\vec{D}}{\partial t}$ are the sources of $\vec{H}$
3. $\frac{\partial\vec{E}}{\partial t}$ is the sources of $\vec{B}$.
4. $\frac{\partial\vec{B}}{\partial t}$ is the sources of $\vec{E}$.

Values

$\varepsilon_0 = 8.8542 \times 10^{-12}$ Faraday / meter and $\mu_0 = 4\pi \times 10^{-7}$ Henry / meter

1. BASIC CONCEPTS

I. Electric Field Strength ($\vec{E}$)

When an electrical charge q is placed, then about the charge there is said to be an electric field of strength $\vec{E}$. Its magnitude at any point is force per unit charge at that point. The direction of $\vec{E}$ is the direction of the force on a positive probe charge and is along the radial from the positive charge q.

II. Electric Displacement 'Ψ' and Displacement Density 'D'

We see that $\vec{E}$ depends on magnitude and position of charge q and also depends on ε of the medium in which the field is measured. Hence, another electrical quantity (excluding q) is also added with q which is independent of the medium and is known as electric displacement or electric flux Ψ.

EXAMPLE

A sphere with charge Q is placed within a larger hollow sphere (without touching). The outer sphere was 'earthed' momentarily and then the inner sphere is removed. The charge on the outer sphere is now measured to be equal (and opposite) to the charge on the inner sphere. This is true for all inner spheres of different sizes and different types of dielectric medium. Hence, as an explanation we consider that there was an electric displacement through the dielectric media to the outer sphere from the charges on inner to outer sphere. This displacement is only depending on the magnitude of Q.

Hence, in *SI* $\quad \Psi = Q$ Coul.

The electric displacement per unit area *i.e.*, electric displacement density, $\vec{D}$, at any point on a spherical surface of radius r centered at the isolated charge q is

$$D = \frac{\Psi}{4\pi r^2} = \frac{q}{4\pi r^2} \text{ Coul./mt}^2.$$

D at any point depends upon the direction of the area vector, *i.e.*,

$$\vec{D} = \frac{q}{4\pi r^2}\hat{r}$$

III. Lines of Force and Lines of Flux

A line of electric force is a curve drawn in such a way that at every point on it shows the direction of electric field. The total numbers of lines per unit area is proportional to $|\vec{E}|$.

A line of electric flux is a curve where every point on it has the direction of the electric flux density or displacement density. The total numbers of flux lines per unit area is equal to $\left|\vec{D}\right|$. In homogeneous isotropic media both the lines of force and lines of flux always have the same direction.

The scalar product of a vector field $\vec{B}$ at any point and the elementary area $\vec{dS}$ drawn around the point is called Flux of $\vec{B}$ through the area at that point.

So, $$\text{flux} = d\phi = \vec{B}\,.\,\vec{dS} \;\Rightarrow\; \text{A scalar quantity.}$$

Physically, for the flow of liquid, if $\vec{v}$ be the velocity of flow of liquid at any point, then, $\left(\vec{v}\,.\,\vec{dS}\right)$ gives the volume of liquid flowing through the elementary area in one second.

IV. Gradient of scalar field

The gradient of a scalar field (*e.g.*, grad ϕ, say) at a point is defined as the maximum value of the spatial rate of change of the scalar field at that point. Mathematically,

$$\text{grad}\,\phi = \vec{\nabla}\phi = \left(\frac{d\phi}{dr}\right)_{\max} . \hat{r} \;\Rightarrow\; \text{A vector quantity.}$$

Direction of grad is normal to the level surface.

V. Divergence of a vector field

The total normal flux per unit volume through a closed surface drawn around a point, when the volume enclosed by the surface tends to zero, is called the divergence of the field at that point, *i.e.*, divergence of $\vec{E}$, or,

$$\text{div}\,\vec{E} = \vec{\nabla}\,.\,\vec{E} = \underset{V\to 0}{Lt}\left[\frac{1}{V}\oint_S \vec{E}\,.\,\vec{dS}\right] \;\Rightarrow\; \text{A scalar quantity}.$$

Here, V is the volume enclosed by the surface.

VI. Curl of a vector field

The maximum value of circulation per unit area drawn around a point, when the area enclosed by the closed curve tends to zero, gives the magnitude of the curl of the vector field at that point, *i.e.*,

$$\left[curl\,\vec{B}\right]_{magnitude} = \left|\left[\vec{\nabla}\times\vec{B}\right]\right| = \underset{S\to 0}{Lt}\left[\frac{1}{S}\oint_S \vec{B}\,.\,\vec{dl}\right].$$

Curl of a vector is also a vector. The direction of curl $\vec{B}$ is same as direction of circulation, *i.e.*, same as direction of area vector S. This is also called as Rotation.

VII. Gauss's Law

The total displacement or electric flux through any closed surface surrounding charges is equal to the amount of charge enclosed, *i.e.*,

$$\oint \vec{D}\,.\,\vec{dS} = \int_V \rho dV\,,$$

where, ρ is the charge density, *i.e.*, charge per unit volume.

Proof

We consider a point charge (q) located in a homogeneous isotropic medium with dielectric constant ε.

Here, $$\vec{E} = \frac{q}{4\pi\varepsilon r^2}\,\hat{r} \qquad \text{...(8.1)}$$

and, $$\vec{D} = \varepsilon\vec{E} = \frac{q}{4\pi r^2}\,\hat{r} \qquad \text{...(8.2)}$$

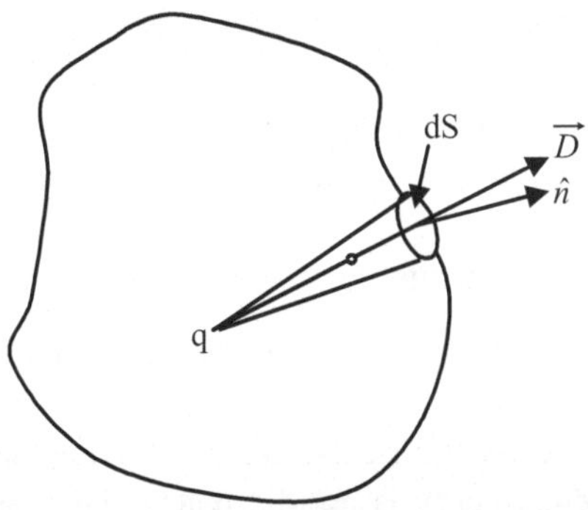

Fig. 8.1 *An arbitrary volume containing charge*

We now consider the displacement through some surface enclosing the charge (q) as in Fig. 8.1. The displacement or electric flux through the element of surface $\vec{dS}$ is

$$d\Psi = D\,.\,dS\cos\theta \qquad \text{...(8.3)}$$

where, θ is the angle between $\vec{D}$ and $\hat{n}$ (direction of $\vec{dS}$) [as dY = $\vec{D}\,.\,\vec{dS}$ = $D\,.\,dS\cos\theta$].

From the definition of solid angle $d\Omega$ subtended at q by dS is

$$D\,.\,dS\cos\theta = r^2\,d\Omega \qquad \text{...(8.4)}$$

Hence, $$\text{Y} = \oint d\Psi = \oint D\,.\,dS\,.\cos\theta \qquad \text{(using eqn.(8.3))}$$

$$= \oint Dr^2\,d\Omega \qquad \text{(using eqn.(8.4))}$$

$$= \frac{q}{4\pi} \oint d\Omega \qquad \text{...(8.5) (using eqn.(8.2))}$$

But, $$W = \oint d\Omega = 4\pi \text{ solid angle} \qquad \text{...(8.6)}$$

From eqns.(8.5) and (8.6), $\Psi = \frac{q}{4\pi} \times 4\pi = q$ ...(8.7)

For a homogeneous medium with charge density 'ρ', we have,

$$\Psi = q = \int_V \rho dV \qquad \text{...(8.8)}$$

Hence, using eqn.(8.3),

$$Y = \oint d\Psi = \oint D \,.\, dS \,.\, \cos\theta = \oint \vec{D} \,.\, \overrightarrow{dS} \qquad \text{...(8.9)}$$

So, from eqns.(8.8) and (8.9),

$$\oint \vec{D} \,.\, \overrightarrow{dS} = \int_V \rho dV$$

This is Gauss's law. It indicates that the net outward displacement through a closed surface is equal to the charge contained in the volume by the surface.

VIII. The Potential Function

An electric field is a field of force. When a body is acted upon by a force and is moved from one position to another, work will be done on or by the body. A charge moving in a static electric field or a mass moving under gravity without friction in both of them, energy is not dissipated. They are the examples of conservative fields. If some point is taken as a reference (normally taken as a point at infinity) or zero point, the field of force can be described by the work that must be done in moving the body from the reference point up to any point in the field.

We consider a small body of charge q and a second body of charge dq is moved from infinity along a radius line to a point P at a distance R from q, then, the work done on the system in moving the test charge against the force F will be

$$W = -\int_{\infty}^{R} (dF)\, dr \qquad \text{...(8.10)}$$

Now, $$dF = \frac{qdq}{4\pi\varepsilon r^2} \qquad \text{...(8.11)}$$

So from eqns.(8.10) and (8.11),

$$W = -\int_{\infty}^{R} \left(\frac{qdq}{4\pi\varepsilon r^2} \right) dr = \frac{qdq}{4\pi\varepsilon R}$$

So the work done on the test charge dq per unit charge is

$$V = \frac{W}{dq} = \frac{q}{4\pi\varepsilon R} \qquad ...(8.12)$$

This V is known as the 'Potential' at the point P due to the charge q. This is scalar and thus it is known as 'Scalar potential'.

For two points separated by an infinitesimal distance dl the work done by the external force in moving a unit positive charge from one point to the other is

$$dW = dV = -\vec{E}\,.\,\vec{dl} \qquad ...(8.13)$$

But, V is a function of (x,y,z), so we can write the eqn.(8.13) as

$$\frac{\partial V}{\partial x}\,.\,dx + \frac{\partial V}{\partial y}\,.\,dy + \frac{\partial V}{\partial z}\,.\,dz = -\vec{E}\,.\,\vec{dl}$$

$$\Rightarrow \left(\hat{i}\frac{\partial V}{\partial x} + \hat{j}\frac{\partial V}{\partial y} + \hat{k}\frac{\partial V}{\partial z}\right)\left(\hat{i}dx + \hat{j}dy + \hat{k}dz\right) = -\vec{E}\,.\,\vec{dl}$$

$$\Rightarrow \qquad \vec{\nabla}V\,.\,\vec{dl} = -\vec{E}\,.\,\vec{dl}$$

Hence, $$\vec{E} = -\vec{\nabla}V$$

Thus, the electric field at any point is just the negative gradient of potential. The direction of the electric field is in the direction, in which the gradient is greatest, *i.e.*, in which direction the potential changes most rapidly.

IX. Gauss's Divergence Theorem

It states that the surface integral of a vector field $\vec{E}$ over a closed surface S is equal to the volume integral of the divergence of the vector over the volume enclosed by the Gaussian surface, *i.e.*,

$$\oint_S \vec{E}\,.\,\vec{dS} = \int_V \left(\vec{\nabla}\,.\,\vec{E}\right) dV$$

X. Stoke's theorem

It states that the line integral of a vector field over a curved path C is equal to the surface integral of curl of the vector over the open surface bounded by the curve C, *i.e.*,

$$\oint_C \vec{E}\,.\,\vec{dl} = \oint_S \left(\vec{\nabla}\times\vec{E}\right).\,\vec{ds}$$

XI. Equation of Continuity

(Conservation of number of charges)

In absence of steady state, the change of increase of charge equals to the net flow of charge into the volume. Let us consider here a tube in which charges q are flowing as shown in Fig. 8.2. As rate of flow of charge defines the current, therefore, current I flowing from the volume can be written as

$$I = -\frac{dq}{dt} \quad \ldots(a)$$

Negative sign is taken as there is decrease of charges within the volume as the flow is outwards. Similarly current flow in Fig. 8.2 can be expressed in terms of current density, J, as

$$I = \oint_S \vec{J} \cdot d\vec{S} \quad \ldots(b)$$

This current is flowing through the cross sectional area of the tube. By divergence theorem, converting surface integral into volume integral, *i.e.*,

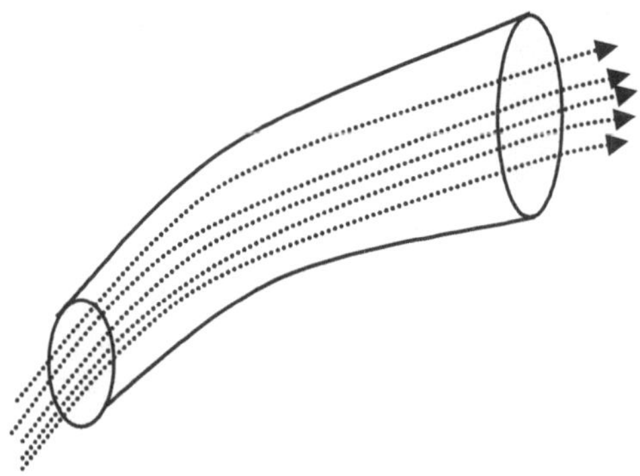

Fig. 8.2 *Flow of charge through a tube of arbitrary volume*

$$\oint_S \vec{J} \cdot d\vec{S} = \int_V \left(\vec{\nabla} \cdot \vec{J}\right) dV \quad \ldots(c)$$

So, eqn.(b) can be written as

$$I = \int_V \left(\vec{\nabla} \cdot \vec{J}\right) dV$$

Comparing eqns.(a) and (c), we get

$$\int_V \left(\vec{\nabla} \cdot \vec{J}\right) dV = -\frac{dq}{dt} \quad \ldots(d)$$

If ρ is the charge density of charge flowing inside the tube, then

$$q = \int \rho dV$$

Thus, eqn.(d) can be written as

$$\int \left(\vec{\nabla} \cdot \vec{J}\right) dV = -\frac{d}{dt}\left(\int \rho dV\right) = -\int \frac{\partial \rho}{\partial t} \cdot dV$$

$$\Rightarrow \quad \int\left(\left(\vec{\nabla}\,.\,\vec{J}\right)+\frac{\partial\rho}{\partial t}\right) = 0$$

Since the equation is true for all values, so integrand must vanishes, *i.e.*,

$$\vec{\nabla}\,.\,\vec{J}+\frac{\partial\rho}{\partial t} = 0 \qquad \text{...(8.14)}$$

Equation (8.14) is called as equation of continuity.

XII. Some vector algebric relations

(*a*) div (curl $\vec{A}$) = 0 $\Rightarrow$ $\vec{\nabla}\,.\left(\vec{\nabla}\times\vec{A}\right)=0$

(*b*) curl (grad ϕ) = 0 $\Rightarrow$ $\vec{\nabla}\times\left(\vec{\nabla}\,.\,\phi\right)=0$ (ϕ is scalar).

(*c*) $\vec{\nabla}\times\left(\vec{\nabla}\times\vec{A}\right)=\vec{\nabla}\left(\vec{\nabla}\,.\,\vec{A}\right)-\nabla^2\vec{A}$

(*d*) For irrotational vector $\vec{A}$, $\vec{\nabla}\times\vec{A}=0$

(*e*) For solenoidal vector $\vec{A}$, $\vec{\nabla}\,.\,\vec{A}=0$

2. MAXWELL'S EQUATIONS

There are four fundamental equations by Maxwell on which the subject of electromagnetism developed.

I. First Equation

$$div\,\vec{D} = \vec{\nabla}\,.\,\vec{D}=\rho$$

This is the Differential form of Gauss's Law in Electrostatics.

II. Second Equation

$$div\,\vec{B} = \vec{\nabla}\,.\,\vec{B}=0$$

This is the Differential form of Gauss's Law in Magnetostatics.

III. Third Equation

$$curl\,\vec{E} = \vec{\nabla}\times\vec{E}=-\frac{\partial\vec{B}}{\partial t}$$

This is the Differential form of Faraday's Law of Electromagnetic Induction.

IV. Fourth Equation

$$curl\,\vec{H} = \vec{\nabla}\times\vec{H}=\vec{J}+\frac{\partial\vec{D}}{\partial t}$$

This is the Maxwell's Modification of Ampere's Law.

All these Maxwell's Equations are the mathematical expressions of a certain experimental results.

3. DERIVATION

(Maxwell's Equations in Differential Form)

I. First Equation: $\vec{\nabla} \cdot \vec{D} = \rho$

We consider a surface S bounding a volume V in a dielectric medium. In the dielectric total charge consists of free charges plus polarization charges. Let ρ and ρ_p are the free charge density and polarization charge density at a point in a small volume element dV, then, from Gauss's law of electrostatics,

$$\oint_S \vec{E} \cdot \vec{dS} = \frac{1}{\varepsilon} \int_V (\rho + \rho_p)\, dV \qquad \text{...(8.15)}$$

But, $$\rho_p = -\vec{\nabla} \cdot \vec{P}.$$

Hence, using (8.15), $$\varepsilon \oint_S \vec{E} \cdot \vec{dS} = \int_V \left(\rho - \vec{\nabla} \cdot \vec{P}\right) dV \qquad \text{...(8.16)}$$

Using Gauss's divergence theorem (to change surface integral to a volume integral), we have,

$$\varepsilon \oint_S \vec{E} \cdot \vec{dS} = \int_V \left(\rho - \vec{\nabla} \cdot \vec{P}\right) dV$$

$$\Rightarrow \quad \varepsilon \int_V \left(\vec{\nabla} \cdot \vec{E}\right) dV = \int_V \rho dV - \int_V \left(\vec{\nabla} \cdot \vec{P}\right) dV$$

$$\Rightarrow \quad \int_V \vec{\nabla} \cdot \left(\varepsilon \vec{E} + \vec{P}\right) dV = \int_V \rho dV$$

$$\Rightarrow \quad \int_V \left[\vec{\nabla} \cdot \vec{D} - \rho\right] dV = 0 \qquad \text{...(8.17) [as we know, } \vec{D} = \vec{P} + \varepsilon \vec{E}\text{]}$$

Since this equation (8.17) is true for all volumes, then, the integrand must vanish, *i.e.*,

$$\vec{\nabla} \cdot \vec{D} - \rho = 0 \;\Rightarrow\; \vec{\nabla} \cdot \vec{D} = \rho$$

This is Maxwell's first equation.

II. Second Equation: $div \vec{B} = \vec{\nabla} \cdot \vec{B} = 0$

The isolated magnetic poles and magnetic currents due to them has no physical significance. So, magnetic lines of force are either curved or go to infinity.

Consequently, the number of magnetic lines of force entering any closed surface is same as leaving it. It means that the flux of magnetic induction $\vec{B}$ across any closed surface is always zero, *i.e.*,

$$\oint_S \vec{B} \cdot \overrightarrow{dS} = 0.$$

On applying Gauss's divergence theorem here we get,

$$\int_V \vec{\nabla} \cdot \vec{B} dV = 0.$$

As the surface bounding the volume is arbitrary, this holds true for all volume and integrand must vanish.

Hence, $\quad div\, \vec{B} = \vec{\nabla} \cdot \vec{B} = 0.$

This is Maxwell's second equation.

III. Third Equation: $curl\, \vec{E} = \vec{\nabla} \times \vec{E} = -\dfrac{\partial \vec{B}}{\partial t}$

From Faraday's law of electromagnetic induction, we know that electromotive force (emf) induced in a closed loop is equal to the rate of change of magnetic flux, *i.e.*,

$$e = -\frac{d\phi}{dt} \qquad \text{...(8.18)}$$

However, the magnetic flux, ϕ, from its definition is given by

$$\phi = \oint_S \vec{B} \cdot \overrightarrow{dS}$$

Here, S is any surface having loop as boundary.

So, using eqn.(8.18),

$$e = -\frac{d\phi}{dt}$$

$$= -\oint_S \frac{\partial}{\partial t}\left(\vec{B} \cdot \overrightarrow{dS}\right) = -\oint_S \frac{\partial \vec{B}}{\partial t} \cdot \overrightarrow{dS} \qquad \text{...(8.19)}$$

Now, the emf e can be computed by calculating the work done in carrying a unit charge round a closed loop C. Thus, if electric field intensity, E, at a small element dl of loop is present, then, we have

$$e = \int_C \vec{E} \cdot \overrightarrow{dl}$$

$$\Rightarrow \qquad -\oint_S \frac{\partial \vec{B}}{\partial t} \cdot \overrightarrow{dS} = \int_C \vec{E} \cdot \overrightarrow{dl} \qquad \text{...(8.20) [using eqn.(8.19)]}$$

Using Stokes's theorem for the conversion of line integral to surface integral, we have from eqn.(8.20),

$$\oint_S (\vec{\nabla} \times \vec{E}) . \vec{dS} = -\oint_S \frac{\partial \vec{B}}{\partial t} . \vec{dS}$$

$$\Rightarrow \quad \oint_S \left[(\vec{\nabla} \times \vec{E}) + \frac{\partial \vec{B}}{\partial t} \right] . \vec{dS} = 0 \qquad \text{...(8.21)}$$

As surface is arbitrary, equation (8.21) holds only when integrand vanishes, so,

$$(\vec{\nabla} \times \vec{E}) + \frac{\partial \vec{B}}{\partial t} = 0$$

$$\Rightarrow \quad (\vec{\nabla} \times \vec{E}) = -\frac{\partial \vec{B}}{\partial t}$$

This is Maxwell's third equation.

IV. Fourth Equation: $curl\, \vec{H} = \vec{\nabla} \times \vec{H} = \vec{J} + \dfrac{\partial \vec{D}}{\partial t}$

Ampere's circuital law states that the line integral of the magnetic induction around any closed path is numerically equal to permeability multiplied by the current enclosed in the path. So, we have,

$$\oint_C \vec{B} . \vec{dl} = \mu I$$

$$\Rightarrow \quad \oint_C \vec{H} . \vec{dl} = I \qquad \text{...(8.22)}$$

However, the current is $\quad I = \oint_S \vec{J} . \vec{dS}$

So, using equation (8.22),

$$\oint_C \vec{H} . \vec{dl} = I = \oint_S \vec{J} . \vec{dS}$$

Using Stokes's theorem,

$$\oint_S (\vec{\nabla} \times \vec{H}) . \vec{dS} = \oint_S \vec{J} . \vec{dS} \Rightarrow \oint_S \left[(\vec{\nabla} \times \vec{H}) - \vec{J} \right] . \vec{dS} = 0$$

As the surface is arbitrary, integrand must vanish, so,

$$\left[(\vec{\nabla} \times \vec{H}) - \vec{J} \right] = 0 \Rightarrow \vec{\nabla} \times \vec{H} = \vec{J} \qquad \text{...(8.23)}$$

Now, we know that for any vector $\vec{A}$,

$$\text{div (curl } \vec{A}\text{)} = 0 \;\Rightarrow\; \vec{\nabla}.\left(\vec{\nabla}\times\vec{A}\right) = 0$$

So here from eqn.(8.23),

$$\text{div (curl } \vec{H}\text{)} = 0 = \text{div (} \vec{J}\text{)} \qquad ..(8.24)$$

However, from continuity equation for the conservation of charge from eqn.(8.14), we have,

$$\vec{\nabla}.\vec{J} + \frac{\partial \rho}{\partial t} = 0 \;\Rightarrow\; \vec{\nabla}.\vec{J} = -\frac{\partial \rho}{\partial t} \qquad ..(8.25)$$

So from equations (8.24) and (8.25), we see that div ($\vec{J}$) = 0 only when ñ is constant with time. Thus we found that Ampere's law is valid only for steady state condition. For time varying field this equation must be modified.

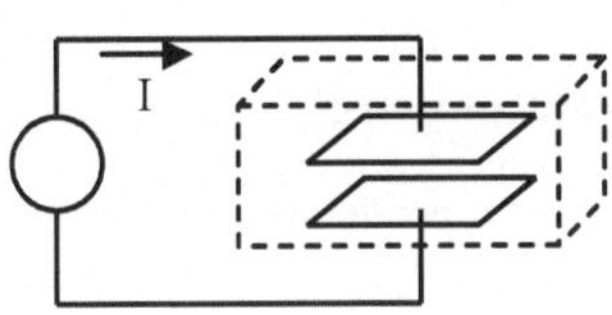

Fig. 8.3 (*a*) *Current flowing through the capacitor*

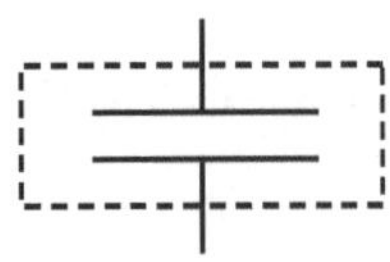

Fig. 8.3 (*b*) *A capacitor*

Fig. 8.3 above shows a surface enclosing one plate of a two plate capacitor. On applying eqn.(8.22) we found that during charge or discharge, the conduction current in the wire attached to the plate is equal to the displacement current passing between the plates.

Maxwell investigated this to make it consistent with the equation of continuity for all practical time varying field. He assumed that the definition of current density is insufficient and a part $\overrightarrow{J_d}$ must be added to it. So the total current density becomes $\left[\vec{J} + \overrightarrow{J_d}\right]$.

As from equation (8.23), $\vec{\nabla}\times\vec{H} = \vec{J}$, we have

$$\vec{\nabla}\times\vec{H} = \vec{J} + \overrightarrow{J_d}$$

$$\Rightarrow \quad div\left(\vec{\nabla}\times\vec{H}\right) = div\left(\vec{J} + \overrightarrow{J_d}\right) \;\Rightarrow\; 0 = div\left(\vec{J} + \overrightarrow{J_d}\right)$$

$$\Rightarrow \quad div\left(\overrightarrow{J_d}\right) = -\,div\,\vec{J} = \frac{\partial \rho}{\partial t} \qquad \text{(using equation of continuity)}$$

$$\Rightarrow \quad div\left(\overrightarrow{J_d}\right) = \frac{\partial}{\partial t}\left[\vec{\nabla}.\vec{D}\right] \qquad \text{(using Maxwell's first equation)}$$

$$\Rightarrow \quad div\left(\overrightarrow{J_d}\right) = \vec{\nabla}.\left[\frac{\partial}{\partial t}\vec{D}\right]$$

$$\Rightarrow \qquad \overrightarrow{J_d} = \left[\frac{\partial}{\partial t}\vec{D}\right] \qquad \text{...(8.26)}$$

Hence, total current density becomes

$$\vec{J} + \frac{\partial \vec{D}}{\partial t}.$$

So, we have,

$$\vec{\nabla} \times \vec{H} = \vec{J} + \overrightarrow{J_d} = \vec{J} + \frac{\partial \vec{D}}{\partial t}$$

$$\Rightarrow \qquad \vec{\nabla} \times \vec{H} = \vec{J} + \frac{\partial \vec{D}}{\partial t}$$

This is the Maxwell's fourth equation, as well as, the modified form of Ampere's law.

(*a*) **Characteristics of Displacement Current Density** $\left(\overrightarrow{J_d}\right)$

(*i*) This is a current only in sense that it produces a magnetic field. It is not linked with motion of charges. It is having a finite value even in a perfect vacuum.

(*ii*) $\overrightarrow{J_d} = \left[\frac{\partial}{\partial t}\vec{D}\right]$

(*iii*) $\overrightarrow{J_d}$ makes the total current continuous across the discontinuity in a conduction current. Example: A battery charging a capacitor produces a closed current loop with $\overrightarrow{J_{total}} = \vec{J} + \overrightarrow{J_d}$.

(*iv*) For a good conductor less than optical frequency (10^{15} hz), $\vec{J} >> \overrightarrow{J_d}$.

4. PHYSICAL SIGNIFICANCE OF MAXWELL'S EQUATIONS

(Maxwell's Equations in Integral Form)

I. First Equation

Here, $\qquad \vec{\nabla} . \vec{D} = \rho$

$$\Rightarrow \qquad \int_V \vec{\nabla} . \vec{D} . dV = \int_V \rho dV$$

$$\Rightarrow \qquad \oint_S \vec{D} . \overrightarrow{dS} = \int_V \rho dV = q \qquad \text{(using Gauss's divergence theorem)}$$

Net outward flux of displacement vector through the surface enclosing a volume, is equal to the net charge contained within the volume.

II. Second Equation

Here, $$\vec{\nabla}.\vec{B} = 0$$

$$\Rightarrow \oint_S \vec{B}.\vec{dS} = 0 \qquad \text{(using Gauss's divergence theorem)}$$

Net outward flux of magnetic induction vector through the surface enclosing a volume, is zero.

III. Third Equation

Here, $$\vec{\nabla}\times\vec{E} = -\frac{\partial\vec{B}}{\partial t}$$

$$\Rightarrow \oint_S \left(\vec{\nabla}\times\vec{E}\right).\vec{dS} = \oint_S -\frac{\partial\vec{B}}{\partial t}.\vec{dS} = -\frac{\partial}{\partial t}\oint_S \vec{B}.\vec{dS}$$

Using Stoke's theorem,

$$\int_C \vec{E}.\vec{dl} = -\frac{\partial}{\partial t}\oint_S \vec{B}.\vec{dS}$$

$$\Rightarrow \qquad e = -\frac{\partial}{\partial t}\phi$$

The emf (e) around a closed path is equal to negative rate of change of magnetic flux linked with the path.

IV. Fourth Equation

Here, $$\vec{\nabla}\times\vec{H} = \vec{J}+\frac{\partial\vec{D}}{\partial t}$$

$$\Rightarrow \oint_S \left(\vec{\nabla}\times\vec{H}\right).\vec{dS} = \oint_S \left[\vec{J}+\frac{\partial\vec{D}}{\partial t}\right].\vec{dS}$$

Using Stoke's theorem,

$$\int_C \vec{H}.\vec{dl} = \oint_S \left[\vec{J}+\frac{\partial\vec{D}}{\partial t}\right].\vec{dS}.$$

The magneto motive force (mmf), $\left[mmf = \int_C \vec{H}.\vec{dl}\right]$, around a closed path is equal to the conduction current density plus displacement current density through any surface bounded by the path.

5. PLANE WAVE EQUATION IN A HOMOGENEOUS MEDIUM

A homogeneous medium is one for which the quantities ε, μ, σ are constant throughout the medium. The medium is isotropic, if ε is a scalar constant so that '$\vec{D}, \vec{E}$' have same values everywhere in all direction. A source free region is that where there are impressed voltages or currents (no generators).

We now consider a uniform linear medium having permittivity ε, conductivity σ, but does not have any free charge or any current other than that determined by Ohm's law.

So, in this medium, we have

$$\vec{D} = \varepsilon\vec{E}, \vec{B} = \mu\vec{H}, \vec{J} = \sigma\vec{E}, \text{and } \rho = 0 \qquad \ldots(8.27)$$

Thus, using eqn.(8.27) all the Maxwell's equations now become

$$\vec{\nabla}.\vec{D} = \rho \;\Rightarrow\; \varepsilon\vec{\nabla}.\vec{E} = 0$$

$$\Rightarrow \quad \vec{\nabla}.\vec{E} = 0 \qquad \ldots(8.28)$$

$$\vec{\nabla}.\vec{B} = 0 \;\Rightarrow\; \mu\vec{\nabla}.\vec{H} = 0$$

$$\Rightarrow \quad \vec{\nabla}.\vec{H} = 0 \qquad \ldots(8.29)$$

$$\vec{\nabla}\times\vec{E} = -\frac{\partial\vec{B}}{\partial t}$$

$$\Rightarrow \quad \vec{\nabla}\times\vec{E} = -\mu\frac{\partial\vec{H}}{\partial t} \qquad \ldots(8.30)$$

$$\vec{\nabla}\times\vec{H} = \vec{J} + \frac{\partial\vec{D}}{\partial t}$$

$$\Rightarrow \quad \vec{\nabla}\times\vec{H} = \sigma\vec{E} + \varepsilon\frac{\partial\vec{E}}{\partial t} \qquad \ldots(8.31)$$

Taking the curl of equation (8.30), we have

$$\vec{\nabla}\times\vec{E} = -\frac{\partial\vec{B}}{\partial t} \;\Rightarrow\; \vec{\nabla}\times\left(\vec{\nabla}\times\vec{E}\right) = \vec{\nabla}\times\left[-\mu\frac{\partial\vec{H}}{\partial t}\right]$$

$$\Rightarrow \quad \vec{\nabla}.\left(\vec{\nabla}.\vec{E}\right) - \nabla^2\vec{E} = -\mu\frac{\partial}{\partial t}\left[\vec{\nabla}\times\vec{H}\right]$$

$$\Rightarrow \quad \vec{\nabla}.(0) - \nabla^2\vec{E} = -\mu\frac{\partial}{\partial t}\left[\sigma\vec{E} + \varepsilon\frac{\partial\vec{E}}{\partial t}\right]$$

$$\Rightarrow \quad -\nabla^2\vec{E} = -\mu\sigma\frac{\partial\vec{E}}{\partial t} - \varepsilon\mu\frac{\partial^2\vec{E}}{\partial t^2}$$

$$\Rightarrow \quad \nabla^2 \vec{E} - \mu\sigma \frac{\partial \vec{E}}{\partial t} - \varepsilon\mu \frac{\partial^2 \vec{E}}{\partial t^2} = 0 \qquad \ldots(8.32)$$

Similarly, taking curl of eqn. (8.31), we have

$$\vec{\nabla} \times \left(\vec{\nabla} \times \vec{H}\right) = \vec{\nabla} \times \left[\sigma \vec{E} + \varepsilon \frac{\partial \vec{E}}{\partial t}\right]$$

$$\Rightarrow \quad \vec{\nabla} . \left(\vec{\nabla} . \vec{H}\right) - \nabla^2 \vec{H} = \left[\sigma\left(\vec{\nabla} \times \vec{E}\right) + \varepsilon \frac{\partial\left(\vec{\nabla} \times \vec{E}\right)}{\partial t}\right]$$

$$\Rightarrow \quad \vec{\nabla} . (0) - \nabla^2 \vec{H} = \left[\sigma\left(-\mu \frac{\partial \vec{H}}{\partial t}\right) + \varepsilon \frac{\partial}{\partial t}\left(-\mu \frac{\partial \vec{H}}{\partial t}\right)\right]$$

$$\Rightarrow \quad -\nabla^2 \vec{H} = -\mu\sigma \frac{\partial \vec{H}}{\partial t} - \mu\varepsilon \frac{\partial^2 \vec{H}}{\partial t^2}$$

$$\Rightarrow \quad \nabla^2 \vec{H} - \mu\sigma \frac{\partial \vec{H}}{\partial t} - \mu\varepsilon \frac{\partial^2 \vec{H}}{\partial t^2} = 0 \qquad \ldots(8.33)$$

Equations (8.32) and (8.33) represent wave equations which govern the electromagnetic field in a homogeneous, linear medium in which the charge density is zero, whether the medium is conducting or non-conducting.

6. PLANE WAVE EQUATION IN FREE SPACE

For free space, we have

$$\varepsilon = \varepsilon_0, \mu = \mu_0, \sigma = 0, \text{ and, } \rho = 0 \qquad \ldots(8.34)$$

Hence, the wave eqn. (8.32) and (8.33) become

$$\nabla^2 \vec{E} - \varepsilon_0\mu_0 \frac{\partial^2 \vec{E}}{\partial t^2} = 0 \text{ and}$$

$$\nabla^2 \vec{H} - \varepsilon_0\mu_0 \frac{\partial^2 \vec{H}}{\partial t^2} = 0 \qquad \ldots(8.35)$$

The above equations are similar to the wave equation,

$$\nabla^2 u - \frac{1}{v^2} \frac{\partial^2 u}{\partial t^2} = 0 \qquad \ldots(8.36)$$

Here, u be any scalar and can stand for any component of $\vec{E}$ and $\vec{H}$, and v be velocity of the wave. Comparing eqns.(8.35) and (8.36), we have the velocity of the E.M. wave as,

$$v = \frac{1}{\sqrt{\mu_0 \varepsilon_0}} = \sqrt{\frac{4\pi}{\mu_0 \, . \, 4\pi\varepsilon_0}} = \sqrt{\frac{4\pi}{4\pi \times 10^{-7}} \times 9 \times 10^9}$$

$$= 3 \times 10^8 \text{ meter/sec}$$

$$\left[\text{Using the values of } \mu_0 = 4\pi \times 10^{-7} \, H/m, \, \frac{1}{4\pi\varepsilon_0} = 9 \times 10^9 \, m/F \right]$$

Hence, $\dfrac{1}{\sqrt{\mu_0 \varepsilon_0}} = c$, the velocity of light.

So, using equation (8.35), we have

$$\nabla^2 \vec{E} \left(\text{or, } \vec{H} \right) = \frac{1}{c^2} \frac{\partial^2 \vec{E} \left(\text{or, } \vec{H} \right)}{\partial t^2} = 0 .$$

A plane wave is defined as a wave whose amplitude is same at any point in a plane perpendicular to a specified direction.

7. THE WAVE EQUATIONS FOR A CONDUCTING MEDIUM

In free space we have solved Maxwell's equations. However, for a perfect dielectric like free space, there were neither charges nor conduction currents. We need to get the solution now for regions in which the conductivity is not zero and conduction currents may exist. This is the general case.

From Maxwell's fourth equation,

$$\vec{\nabla} \times \vec{H} = \vec{J} + \frac{\partial \vec{D}}{\partial t}$$

And, $\vec{J} = \sigma \vec{E}$, we get,

$$\vec{\nabla} \times \vec{H} = \sigma \vec{E} + \varepsilon \frac{\partial \vec{E}}{\partial t} \quad \ldots(8.37)$$

Taking curl of both sides of Maxwell's third equation

$$\vec{\nabla} \times \vec{E} = -\mu \frac{\partial \vec{H}}{\partial t}, \text{ we have,}$$

$$\vec{\nabla} \times \vec{\nabla} \times \vec{E} = -\mu \frac{\partial}{\partial t} \left(\vec{\nabla} \times \vec{H} \right)$$

$$\Rightarrow \quad \vec{\nabla} \, . \, \vec{\nabla} \, . \, \vec{E} - \nabla^2 \vec{E} = -\mu \frac{\partial}{\partial t} \left[\sigma \vec{E} + \varepsilon \frac{\partial}{\partial t} \vec{E} \right] \quad \text{(from eqn.(8.37))}$$

$$\Rightarrow \nabla^2 \vec{E} - \mu\varepsilon \frac{\partial^2}{\partial t^2} \vec{E} - \sigma\mu \frac{\partial}{\partial t} \vec{E} = \vec{\nabla} \, . \, \vec{\nabla} \, . \, \vec{E} \quad \ldots(8.38)$$

From Maxwell's first equation,

$$\vec{\nabla}.\vec{D} = \rho \quad \Rightarrow \quad \vec{\nabla}.\vec{E} = \rho/\varepsilon \qquad \text{...(8.39)}$$

Since there are no net charge within a conductor (although there may be a charge on the surface), $\rho = 0$. This is because within any macroscopic element of volume the positive and negative charges are equal in number (on the average).

Hence, from eqn.(8.39), $\vec{\nabla}.\vec{E} = 0 \Rightarrow \vec{\nabla}.\vec{D} = 0$.

Hence, eqn.(8.38) becomes

$$\nabla^2 \vec{E} - \mu\varepsilon \frac{\partial^2}{\partial t^2} \vec{E} - \sigma\mu \frac{\partial}{\partial t} \vec{E} = 0 \qquad \text{...(8.40)}$$

Similarly for $\vec{H}$, we have

$$\nabla^2 \vec{H} - \mu\varepsilon \frac{\partial^2}{\partial t^2} \vec{H} - \sigma\mu \frac{\partial}{\partial t} \vec{H} = 0 \qquad \text{...(8.41)}$$

These eqns.(8.40) and (8.41) are the wave equations for $\vec{E}$ & $\vec{H}$.

8. POYNTING'S THEOREM

While propagating an E.M.waves transfer energy from one point to the other. There exists relationship between the rate of energy transfer and the amplitudes of $\vec{E}, \vec{H}$ of the e.m.wave.

From Maxwell's fourth equation,

$$\vec{\nabla} \times \vec{H} = \vec{J} + \varepsilon \frac{\partial \vec{E}}{\partial t}$$

$$\Rightarrow \qquad \vec{J} = \vec{\nabla} \times \vec{H} - \varepsilon \frac{\partial \vec{E}}{\partial t} \qquad \text{...(8.42)}$$

So, $$\vec{E}.\vec{J} = \vec{E}.\left(\vec{\nabla} \times \vec{H}\right) - \varepsilon \vec{E}.\frac{\partial \vec{E}}{\partial t} \qquad \text{...(8.43)}$$

Hence, $$\vec{E}.\vec{J} = \left[\vec{H}.\left(\vec{\nabla} \times \vec{E}\right) - \vec{\nabla}.\left(\vec{E} \times \vec{H}\right)\right] - \varepsilon \vec{E}.\frac{\partial \vec{E}}{\partial t} \qquad \text{...(8.44)}$$

$$\left(\text{using } \vec{\nabla}.\left(\vec{E} \times \vec{H}\right) = \vec{H}.\left(\vec{\nabla} \times \vec{E}\right) - \vec{E}.\left(\vec{\nabla} \times \vec{H}\right)\right)$$

From Maxwell's third equation, $\vec{\nabla} \times \vec{E} = -\mu \dfrac{\partial \vec{H}}{\partial t}$

Using eqn.(8.44) into eqn.(8.43), we get

$$\vec{E}.\vec{J} = \left[\vec{H}.\left(\vec{\nabla} \times \vec{E}\right) - \vec{\nabla}.\left(\vec{E} \times \vec{H}\right)\right] - \varepsilon \vec{E}.\frac{\partial \vec{E}}{\partial t}$$

$$= \left[\vec{H}\cdot\left(-\mu\frac{\partial\vec{H}}{\partial t}\right) - \vec{\nabla}\cdot\left(\vec{E}\times\vec{H}\right)\right] - \varepsilon\vec{E}\cdot\frac{\partial\vec{E}}{\partial t} \qquad \ldots(8.45)$$

Now, $\vec{E}\cdot\frac{\partial\vec{E}}{\partial t} = \frac{1}{2}\frac{\partial}{\partial t}(E)^2$ and

$$\vec{H}\cdot\frac{\partial\vec{H}}{\partial t} = \frac{1}{2}\frac{\partial}{\partial t}(H)^2 \qquad \ldots(8.46)$$

So from eqns.(8.45) and (8.46),

$$\vec{E}\cdot\vec{J} = \mu\cdot\frac{1}{2}\frac{\partial}{\partial t}(H)^2 - \vec{\nabla}\cdot\left(\vec{E}\times\vec{H}\right) - \varepsilon\cdot\frac{1}{2}\frac{\partial}{\partial t}(E)^2$$

Integrating over a volume V, we have

$$\int_V \vec{E}\cdot\vec{J}dV = -\frac{\partial}{\partial t}\int_V\left[\frac{\mu}{2}(H)^2 + \frac{\varepsilon}{2}(E)^2\right]dV - \int_V \vec{\nabla}\cdot\left(\vec{E}\times\vec{H}\right)dV$$

$$\Rightarrow \quad \int_V \vec{E}\cdot\vec{J}\,dV = -\frac{\partial}{\partial t}\int_V\left[\frac{\mu}{2}(H)^2 + \frac{\varepsilon}{2}(E)^2\right]dV - \oint_S\left(\vec{E}\times\vec{H}\right)dS \qquad \ldots(8.47)$$

I. Physical Interpretation of Poynting's Theorem

Eqn.(8.47) contains three terms. We can explain them in the following.

(*a*) $\int_V \vec{E}\cdot\vec{J}\,dV$

This is the instantaneous power dissipated in volume V. Hence, eqn.(8.47) is a generalization of Joule's law. We have consider a conductor of cross sectional area A carrying a current I and having a voltage drop E per unit length, will have a power loss of EI watts per unit length. So power dissipated per unit volume is

$$\frac{EI}{A} = EJ \text{ Watt/vol.}$$

Here $\vec{E}, \vec{J}$ are in the same direction. However, in other cases also $\vec{E}\cdot\vec{J}$ gives the power dissipated per unit volume and so for the whole volume the total power dissipated is given by

$$\int_V \vec{E}\cdot\vec{J}\,dV \qquad \ldots(8.48)$$

When $\vec{E}$ is required to produce $\vec{J}$ in the conducting medium, eqn.(8.47) represents ohmic loss or dissipation loss (I^2R). If $\vec{E}$ is due to a source of power, *e.g.*, battery, then eqn.(8.47) would be used up in driving the current against the battery voltage.

(*b*) $\frac{\partial}{\partial t}\int_V\left[\frac{\mu}{2}(H)^2+\frac{\varepsilon}{2}(E)^2\right]dV$

Here, $\frac{\varepsilon}{2}(E)^2$: This is the energy density or stored electric energy per unit volume of the electric field.

$\frac{\mu}{2}(H)^2$: This is the stored energy density of the magnetic field.

These quantities continue to represent stored energy densities when the field is changing with time, the integral represent the stored energy in the volume *V*. Hence, these terms with a negative sign represents the rate at which the stored energy in the volume is decreasing.

(*c*) $\oint_S(\vec{E}\times\vec{H})\,dS$

The interpretation of it follows from the law of conservation of energy. It indicates that the rate of energy dissipation in the volume *V* must equal the rate at which the stored energy in *V* is decreasing, plus, the rate at which energy is entering the volume *V* from outside, *i.e.*, 'Rate of energy dissipated equals the rate at which the stored energy is decreasing plus the rate at which energy is entering from outside '

Hence, '$-\oint_S(\vec{E}\times\vec{H})\,dS$' must give the rate of flow of energy outward through the surface enclosing the volume. The interpretation of eqn.(8.47) leads to the conclusion that '$(\vec{E}\times\vec{H})$' over any closed surface gives the rate of energy flow through that surface. This vector can be represented as

$$\vec{P}=(\vec{E}\times\vec{H}) \qquad \ldots(8.49)$$

This has the dimension of 'Watt/sq. mt'. This is Poynting's Theorem.

II. Poynting Vector

E.m.waves transport energy from one point to another by propagation of electric and magnetic field through space.

The rate of transmission of electromagnetic energy across a unit area perpendicular to the direction of propagation of energy is represented by Poynting Vector $(\vec{S})$ *i.e.*, $\vec{S}=\vec{E}\times\vec{H}$.

The cross product of electric and magnetic field $(\vec{E}\text{ and }\vec{H})$ is called as Poynting Vector.

Unit of Poynting Vector

In scalar, we have

$$S = EH = \frac{V}{mt} \times \frac{A}{mt} = \frac{VA}{mt^2} = \frac{W}{mt^2} = \frac{J/s}{mt^2} = Js^{-1}mt^{-2}$$

Now, $\vec{S} = \vec{E} \times \vec{H}$

Considering $\vec{E}$ along *X*-axis and $\vec{H}$ along *Y*-axis, we have then,

$$\vec{S} = \hat{i}E_x \times \hat{j}H_y = \hat{k}E_xH_y$$

But, E_xH_y is the energy per sec. per unit area. Hence, $\vec{S}$ represents the energy propagating along *Z*-axis.

Poynting Vector $(\vec{S})$ gives the time rate of flow of e.m.wave energy per unit area of the medium.

(*a*) **Illustration :** Fig. 8.4 shows a parallel plate capacitor in a circuit with battery with electric field $\vec{E}$ and a key, *K*. When K is *ON*, charging of capacitor begins. A conduction current, *I*, is set up in the circuit and a displacement current of density $\frac{\partial \vec{D}}{\partial t}$ is set up between the capacitor plates. The magnetic lines of force are circular and the electric field is directed downwards. So, the Poynting Vector $(\vec{E} \times \vec{H})$ is everywhere directed into the capacitor.

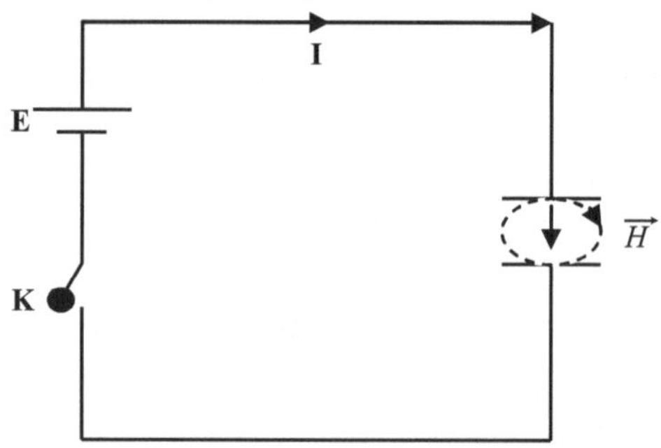

Fig. 8.4 *Example of a energy flow in a capacitor*

9. REPRESENTATION OF ELECTROMAGNETIC WAVE (E.M.WAVE)

E.M.wave propagates in a direction perpendicular to both $\vec{E}$ and $\vec{H}$. If $\vec{k}$ denotes the wave vector of the E.M.wave, then vectors $\vec{E}, \vec{H}$ and $\vec{k}$ form a set of orthogonal vectors which form a right handed co-ordinate system as shown in Fig. 8.5 aside.

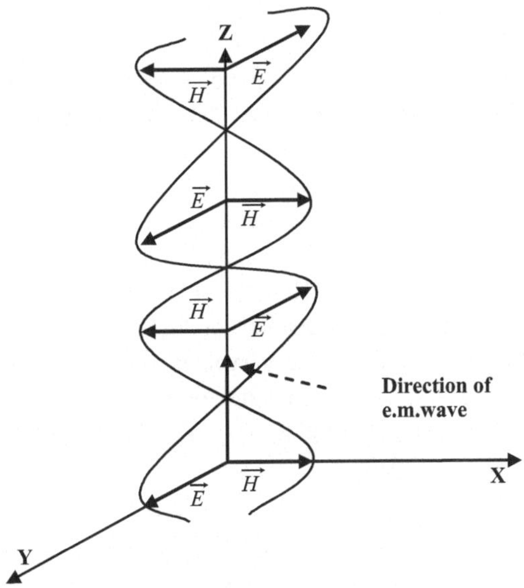

Fig. 8.5 *Propagation of e.m. wave alongwith its electric and magnetic vector components*

10. WAVEGUIDE

A waveguide is a hollow pipe of cylindrical, rectangular or spherical in shape. The microwave E.M. radiation is propagated along its cross-section uniformly along its length. The walls of the guide are perfectly conducting. A waveguide acts as a sort of high pass filter in the sense that only frequencies greater than cut off frequency can be propagated in the guide. Moreover, at any given frequency only a finite number of modes can propagate. So, it is often convenient to choose the dimensions of the guide so that at the operating frequency only the lowest mode can occur. Through the waveguide microwave e.m. radiations can be transmitted with relatively little attenuation. The dominating transmission modes in a waveguide are transverse electric (*TE*) and transverse magnetic (*TM*) modes. In *TE* mode, the electric vector has no component in a direction of propagation. Similarly, in *TM* mode the magnetic vector has no component in a direction of propagation. The waveguide is shown in Fig. 8.6. The *TEM* (Transverse Electromagnetic) mode cannot exist inside a single hollow conductor of large conductivity as the electric field vanishes on the surface as it is an equipotential one. So, two or more cylindrical surfaces are required to support the *TE*M mode in the waveguide.

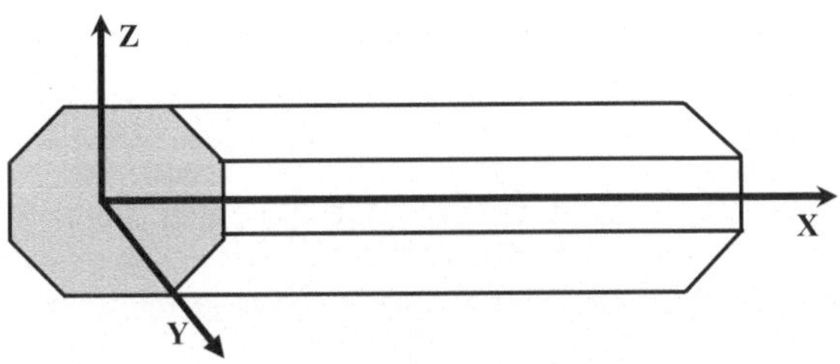

Fig. 8.6 *A typical waveguide*

11. CO-AXIAL CABLE

A co-axial cable is a conducting medium for transmission of radio frequency signals between the transmitter and receiver. The outer conductor is in the form of mesh and is further surrounded by an outer insulator as shown in Fig. 8.7. The inner and outer conductors are separated by low dielectric insulator, *i.e.*, polythene, teflon, etc. The central and outer conductors are coaxial, *i.e.*, has a common axis. Hence, the name is given as coaxial cable. They are used to transmit high frequency signal (1 – 20 GHz). Here the outer conductor forms a shield that confines the wave to the space between the conductors so that the radiation from the line is negligible. However, the line is basically unbalanced since the external capacitance is between the outer conductor only and the ground. Due to the complete insulation of the transmitted signals from the outside environment their transmission efficiency is excellent.

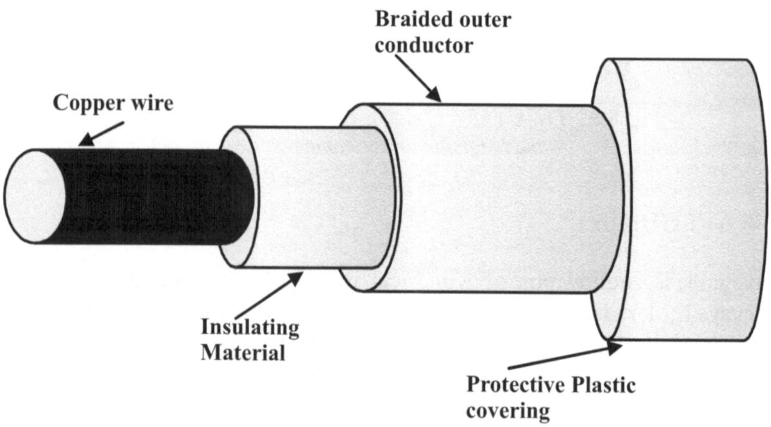

Fig. 8.7 *A typical co-axial waveguide*

For a co-axial cable, the value of inductance, L and capacitance, C, per meter length is given by,

$$L = \frac{\mu}{2\pi} \log_e \left[\frac{D}{d}\right] \text{ H/cm and}$$

$$C = \frac{2\pi\varepsilon}{\log_e \left[\frac{D}{d}\right]} \text{ H/cm.}$$

Here, D, d, μ and ε are diameter of outer conductor, diameter of inner conductor, permeability and permittivity of the medium respectively.

9

Special Theory of Relativity

With the acceptance of the electromagnetic theory of light, only two general problems remained. One of these was that of luminiferous ether, a hypothetical medium suggested as the carrier of light waves, just as air or water carries sound waves. The ether was assumed to have some very unusual properties, e.g., being massless but having high elasticity. A number of experiments performed to give evidence of the ether, most notably by *A. A.* Michelson in 1881 and by Michelson and *E. W.* Morley in 1887, failed to support the ether hypothesis. With the publication of the special theory of relativity in 1905 by Albert Einstein, the ether was shown to be unnecessary to the electromagnetic theory. Much before Einstein it was considered the velocity of light, like other material's particle velocity, should depend on the relative motion between the observer and the source of light. However, experiments failed to disapprove the constancy of velocity of light to generate a new branch of Physics. According to classical mechanics space and time are absolute and they are not interconnected. This absolute nature is broken by relativity which generally starts that both space and time are a function of observer's velocity as well as observable's velocity. At high velocities both space and time combined to form space-time continuum.

The theory which deals with the relativity of motion and rest is called the theory of relativity. Einstein was the pioneer in this field and did a lot which helps us to decode the mysteries of the universe.

It has two parts.

(*a*) **Special Theory :** It deals with objects and systems which are either moving at a constant speed with respect to one another or at rest.

(*b*) **General Theory :** It deals with objects and systems which are speeding up or slowing down with respect to one another.

Here, we are only dealing with Special Theory of Relativity only.

1. BASIC CONCEPTS

I. Frames of reference

We have seen that the motion of the body has no meaning unless it is described with respect to some well defined co-ordinate system. The co-ordinate system is known as the Frame of Reference. Thus a frame of reference is a coordinate system relative to which the position and motion of the body is described. The reference frame may be at rest or under motion. Its choice is totally at our own convenience. Suppose we want to study about anything from earth. Thus we are considering our reference frame as the earth. However, earth is not a stationary object. Similarly, a passenger sitting in a train observes outside trees moving backwards, which is not the fact. Actually in the reference system attached to the earth the train is moving forward.

There are two types of Frames of reference.

(a) **Inertial (unaccelarated) :** This obeys Newton's laws of inertia and other laws of Newtonian mechanics. In this frame, a body not acted upon by external force, is at rest, or moves with a constant velocity. Actually for different types of frames of references the laws of mechanics becomes different. We can approximate a reference frame which moves with constant velocity relative to the distant stars as a best approximation of inertial frame.

(b) **Non-inertial (accelerated) :** This does not obey Newton's laws of inertia and other laws of Newtonian mechanics. In this frame, a body not acted upon by external force, is accelerated. A coordinate system fixed in the Earth is a non-inertial frame if the object motion is observed from the outside of the earth as the earth itself rotates about its axis and also about the Sun.

II. Ether

Newton's concept of stationary frame of reference gave the idea of the 'ether' frame in 19th century. As we measure the speed of any moving object with respect to any stationary object, the velocity of light also requires a stationary frame for its velocity measurement. Hence, it seemed a fairly logical postulate to think for the ether medium having unusual properties like zero density and 100% transmittivity. These properties are essential for its undetectibility. It was assumed to fill all the space by ether and called as a universal frame of reference, relative to which any object motion can be defined.

2. GALILEAN TRANSFORMATION (OLD SYSTEM)

In this system the frames were taken as inertial ones, which obey Newton' laws. Also any reference frame moving with uniform velocity with respect to another inertial frame is also an inertial frame. Our experience shows that any experiment performed in a uniformly moving vehicle will be the same if we perform the same experiment in any stationary laboratory. It implies the universality of every inertial frame. Galileo's principle of relativity tells that laws of mechanics are the same in all inertial frames of reference. The transformation from one inertial frame of reference to another is called the Galilean Transformation.

We now consider here two frames of reference S and S' as in Fig. 9.1. The

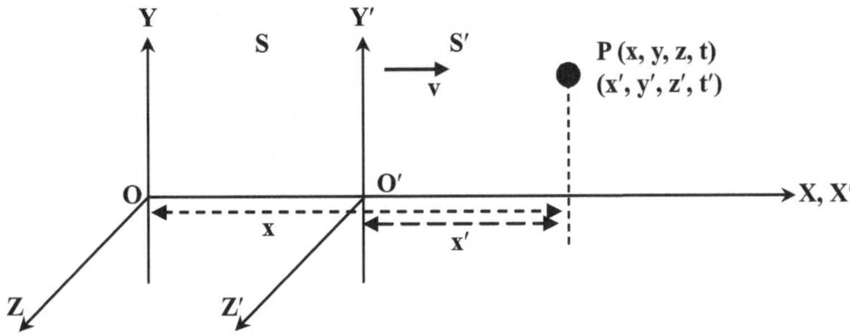

Fig. 9.1 *Stationary and moving frames of reference*

Galilean transformation establishes a relationship between the coordinates (x, y, z, t) of an event at S and the coordinates (x', y', z', t') of the same event at S'. We assume S' moves in such a way that x and x' are in one straight line and have same positive directions. For convenience we are restricting the motion of the moving frame in X-direction only. Also O and O' coincide at $t = 0$ and S' moves with a velocity v with respect to S.

I. Coordinate Transformation

From Fig. 9.1,

$$x = x' + vt \quad \Rightarrow \quad x' = x - vt$$

also, $$y' = y; z' = z; t' = t$$

So the Galilean Transformation for coordinate is

$$x' = x - vt$$

$$y' = y$$

$$z' = z$$

$$t' = t$$

II. Velocity Transformation

We now consider the movement of a moving body. Let the body be at the position P_1 at t_1 and P_2 at t_2. Now at P_1 the coordinate in system S is (x_1, y_1, z_1, t_1) and at P_2 the coordinate in system S is (x_2, y_2, z_2, t_2).

Therefore, for S' system, using coordinate transformation as above,

$$x_1' = x_1 - vt_1;\ y_1' = y_1;\ z_1' = z_1$$

and, $$x_2' = x_2 - vt_2;\ y_2' = y_2;\ z_2' = z_2$$

So, $$x_2' - x_1' = (x_2 - x_1) - v(t_2 - t_1)$$

$\Rightarrow$ $$\frac{x_2' - x_1'}{(t_2 - t_1)} = \frac{(x_2 - x_1)}{(t_2 - t_1)} - v$$

Let, $$u'_x = \frac{(x'_2 - x'_1)}{(t_2 - t_1)};\ u_x = \frac{(x_2 - x_1)}{(t_2 - t_1)}$$

So the Galilean Transformation for velocity is

$$u'_x = u_x - v$$

$$u'_y = u_y$$

$$u'_z = u_z$$

We have here simply the classical velocity addition theorem.

III. Acceleration Transformation

We now consider the body in S moves along X-axis with acceleration a and so acceleration a' with reference to S'. So at some moment of time,

$$u = u' + v \qquad \ldots(9.1)$$

After time Δt, the velocity in both systems increase and the new velocities are

$$(u + \Delta u) = (u' + \Delta u') + v$$

Using equation (9.1), we have

$$((u' + v) + \Delta u) = (u' + \Delta u') + v$$

$$\Rightarrow \qquad \Delta u = \Delta u' \quad \Rightarrow \quad \frac{\Delta u}{\Delta t} = \frac{\Delta u'}{\Delta t}$$

$$\Rightarrow \qquad a = a'$$

This is Galilean Transformation for acceleration.

If F and F' are the forces in S frame and S' frame respectively, then

$$F = ma = ma' = F'.$$

It shows that the acceleration and forces do not change from one inertial frame to another inertial frame. Hence, this is an Invariant.

3. MICHELSON - MORLEY EXPERIMENT

The requirement to measure the speed of light necessitates the need of reference frame or coordinate frame with respect to which we can measure the velocity. It was assured initially that ether is present in the space through which light travels. This ether is perfectly transparent to light and the material bodies may pass through it without any resistance. For earth its orbital motion with velocity 30 km/s is so with respect to ether. If an ether exists, the spinning and orbital motion of the earth should be trough it. An observer on the earth would sense an 'ether wind' whose velocity is v relative to the earth. Hence, as ether was assumed to be at rest, it was considered possible to measure absolute velocity of light moving through it. A number of scientists namely Fizeau, Michelson & Morley had tried it. *A*. Michelson, the inventor of the optical interferometer with excellent sensitivity, first performed the experiment in 1881 and then again in 1887 with *E*. Morley. They carried out one of the more precise and breakthrough experiment in the history of science and eventually it leads to the foundation of the theory of relativity. Eventually Michelson was the first American to be awarded the Nobel Prize in Physics in 1907. The experiment set up by Michelson and Morley is described below. The experimental arrangement is shown Fig. 9.2.

Michelson and Morley mounted the interferometer on a massive stone slab for stability and floated the apparatus in mercury so that we can rotate the apparatus smoothly about a central pin. The light path was increased by making it travel through mirrors on the slab to reflect the beams back and forth through eight round trips. Light from a monochromatic source *S* after being made parallel by a collimating lens L falls on semi-silvered glass plate *G* inclined at 45° to the beam. It is then divided into two parts. First part, ray 1 travels towards mirror M_1 and second part ray 2 travels towards mirror M_2. Then they reflected back along their original paths meet again at the semi-silvered face of *G* and enter a short focus

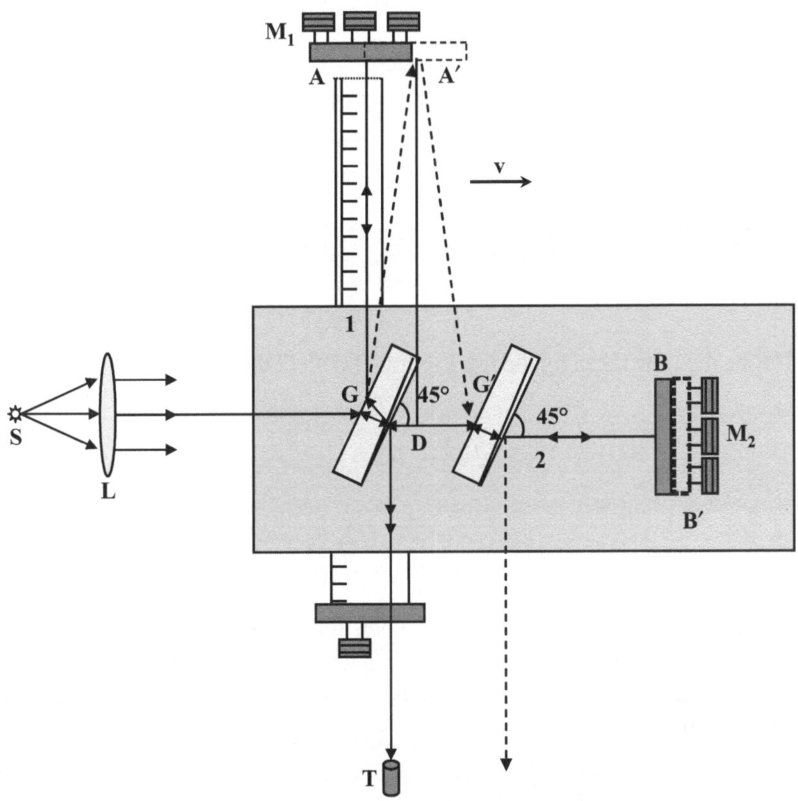

Fig. 9.2 *Schematic diagram of Michelson-Morley experiment*

telescope T. These two rays produce interference fringes in the field of view of T. A desired path difference can be introduced between the two reflected rays by moving M_1. The optical distances of M_1 and M_2 from G are made equal by using a compensating glass plate G'. If the apparatus is at rest, the two reflected rays would take equal time to return the glass plate G. But actually the whole apparatus is moving along with the earth. Let us suppose, the direction of motion of earth is in the direction of the initial direction of the initial beam. Due to the motion of the earth the optical paths traversed by both the rays are not the same. The reflections at M_1 and M_2 do not take place at A and B, but at A' and B' respectively as shown in Fig. 9.2. Thus, the time taken by the two rays to travel to the mirrors and back to G will be different in this case.

Theory

Let M_1 and M_2 be at distances l from G. Also let c and v are velocities of light and the apparatus or earth respectively. Now the reflected ray 1 from G strikes M_1 at A' and not at A due to earth's motion. The total path of the ray 1 from G to A' and back to G' will be GAG'.

By laws of reflection,

$$GA' = A'G'.$$

So, $$GA'G' = GA' + A'G' = 2GA'.$$

However, $$(GA')^2 = (GD)^2 + (A'D)^2 = (AA')^2 + (A'D)^2 \quad \ldots(9.2)$$

If t be the time taken by the ray to move from G to A', then,

$$GA' = ct;\ AA' = vt \text{ and } A'D = l$$

So from eqn.(9.2),

$$c^2t^2 = v^2t^2 + l^2 \quad \Rightarrow \quad t = \frac{l}{\sqrt{c^2 - v^2}}.$$

If t_1 be the time taken by the ray 1 to travel $GA'G'$, then,

$$t_1 = 2t = \frac{2l}{\sqrt{c^2 - v^2}}$$

$$= \frac{2l}{c}\left[1 - \frac{v^2}{c^2}\right]^{-\frac{1}{2}} \approx \frac{2l}{c}\left[1 + \frac{v^2}{2c^2}\right] \quad \ldots(9.3)$$

Now, we see the transmitted ray 2 moves longitudinally towards M_2. It has velocity $(c - v)$ relative to the apparatus when it is moving from G to B. During its return its velocity relative to the apparatus is $(c + v)$. If t_2 be the time taken by the ray 2 to travel to GM_2G, then

$$t_2 = \frac{l}{c - v} + \frac{l}{c + v}$$

$$= l\left[\frac{(c+v)+(c-v)}{(c+v)(c-v)}\right] = \frac{2lc}{c^2-v^2}$$

$$= \frac{2lc}{c^2}\left[1-\frac{v^2}{c^2}\right]^{-1} = \frac{2l}{c}\left[1-\frac{v^2}{c^2}\right]^{-1}$$

$$\approx \frac{2l}{c}\left[1+\frac{v^2}{c^2}\right] \qquad \ldots(9.4)$$

Thus, from eqns. (9.3) and (9.4), the difference in times of travel of longitudinal and transverse journey is

$$\Delta t = t_2 - t_1$$

$$= \frac{2l}{c}\left[1+\frac{v^2}{c^2}-1-\frac{v^2}{2c^2}\right] = \frac{lv^2}{c^3} \qquad \ldots(9.5)$$

The number of fringes shift between the two rays is given by

$$N = \text{Path Difference / Wavelength}$$

$$= \frac{c\Delta t}{\lambda} = \frac{lv^2}{\lambda c^2} \qquad \ldots(9.6)$$

This shift will be observed if earth is made at rest. However, this is not possible. Hence Michelson and Morley performed the experiment in two steps, **firstly**, by setting up the apparatus as shown in Fig. 9.2 and **secondly**, by turning the apparatus through 90°.

When the apparatus was turned through 90°, the positions of M_1 and M_2 are changed. Now the path difference is in opposite direction, *i.e.*, the path difference is $-\dfrac{lv^2}{\lambda c^2}$ fringes.

The resultant path difference now becomes $\dfrac{lv^2}{\lambda c^2} - \left(-\dfrac{lv^2}{\lambda c^2}\right) = \dfrac{2lv^2}{\lambda c^2}$ fringes.

So a fringe shift of $\dfrac{2lv^2}{\lambda c^2}$ is expected.

Using l = 10 meter, $\lambda = 5000$ Å, $v = 3 \times 10^6$ cm/sec and $c = 3 \times 10^{10}$ cm/sec, we get a fringe shift, n, as $n = \dfrac{2lv^2}{\lambda c^2} = 0.4$ fringe.

Thus a fringe shift of less than half of a fringe was expected.

However, Michelson and Morley could observe a shift of about 0.01 fringe only.

They repeated the experiment at different locations in Europe and USA in day and night and at different seasons, but could not detect any measurable shift. So, it was a complete NEGATIVE RESULT.

This result suggests it is impossible to measure the speed of earth relative to the fixed frame of reference, *i.e.*, ether, no existence of the ether was detected, also the speed of light in vacuum is the same in all frames of reference which are in uniform relative motion.

4. FAILURE OF GALILEAN TRANSFORMATION

We can also look into this example, we have two equal and like charges stationery in stationery frame of reference S. As per Coulomb's law they repel each other. Suppose an observer watches the charges from a moving frame of reference $S2$ moving with velocity u along a line normal to the line joining the charges. So this observer find the charges moving with a relative velocity $-u$. Hence these moving charges constitute parallel currents along the same direction and thus they attract each other. However, the observer in moving frame finds that these like charges exert both a mutual repulsive force and also an attractive force at the same time which is not possible. Here either Galilean Transformation is wrong or the Maxwell's equations are wrong. However Maxwell's equations are experimentally verified one and hence the Galilean Transformation fails.

5. POSTULATES OF SPECIAL THEORY OF RELATIVITY

The negative result of Michelson – Morley experiment gave rise to postulates of Special Theory of relativity by A. Einstein. In his paper "On the Electrodynamics of Moving Bodies" Einstein wrote "... no properties of observed facts correspond to a concept of absolute rest; ... for all coordinate systems for which the mechanical equations hold, the equivalent electrodynamical and optical equations hold also ... In the following we make these assumptions (which we shall subsequently call the Principle of Relativity) and introduce the further assumption – an assumption which is at the first sight irreconcilable with the former one – the light is propagated in vacant space, with a velocity c which is independent of the nature of motion of the emitting body. These two assumptions are quite sufficient to give us a simple and consistent theory of electrodynamics of moving bodies on the basis of the Maxwellian theory of bodies at rest."

First postulate

The fundamental laws of Physics have the same form for all inertial frames. No preferred inertial system exists. This is the principle of relativity.

Second postulate

The velocity of light in vacuum is independent of the relative motion of the source and the observer. This is the principle of constancy of the speed of light.

Above first postulate is direct outcome of Newtonian Mechanics. However, for second postulate some modifications were required other than the Galilean Transformation.

6. LORENTZ TRANSFORMATION (NEW SYSTEM)

Lorentz tried to replace Galilean transformation to give results in agreement with Michelson and Morley experiment and second postulate of the special theory of relativity and developed his transformation equations in 1890. The transformation equations apply to speed upto c.

I. Coordinate Transformation

Here, we consider two inertial systems S and S' as in Fig. 9.3. We assume S' moves in such a way that x and x' are in one straight line and have same positive directions. Also O and O' coincide at $t = 0$ and S' moves with a velocity v with respect to S. For our convenience, we take the axes of two coordinate systems coincide at $t = t' = 0$.

Let a pulse of light be generated at $t = 0$ at the origin which glows in the space.

Now we consider the situation when the pulse reaches at point P. Let (x, y, z, t) and (x', y', z', t') are the position and time coordinates of P measured by observers O and O' in frames S and S' respectively. When the pulse is observed from S, we have, the velocity of light as distance divided by the time taken; hence,

$$c = \frac{\sqrt{x^2 + y^2 + z^2}}{t}$$

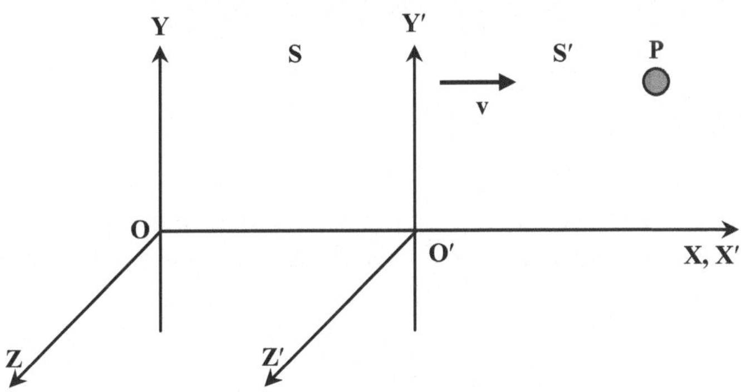

Fig. 9.3 *Stationary and moving frames of reference*

$$\Rightarrow \quad x^2 + y^2 + z^2 - c^2t^2 = 0 \qquad \ldots(9.7)$$

Similarly, when the pulse is observed from S′, we have

$$(x')^2 + (y')^2 + (z')^2 - c^2 (t')^2 = 0 \qquad \ldots(9.8)$$

However, $\quad y = y'; z = z' \qquad \ldots(9.9)$

Using eqn.(9.9) into eqns.(9.7) and (9.8), we have,

$$x^2+y^2+z^2-c^2t^2=(x')^2+(y')^2+(z')^2-c^2(t')^2$$

$$\Rightarrow \quad x^2+y^2+z^2-c^2t^2=(x')^2+(y)^2+(z)^2-c^2(t')^2$$

$$\Rightarrow \quad x^2-c^2t^2=(x')^2-c^2(t')^2 \qquad \ldots(9.10)$$

Let the transformation between x and x' can be represented by

$$x'=\lambda(x-vt) \qquad \ldots(9.11)$$

where, λ being independent of x and t. If we suppose that the system S is moving relative to S' with velocity $-v$ along X-direction (because the motion is relative), then,

$$x=\lambda'(x'+vt') \qquad \ldots(9.12)$$

where, λ' being independent of x and t.

From eqns.(9.11) and (9.12), we have

$$x=\lambda'(\lambda(x-vt)+vt')$$

$$\Rightarrow \quad vt'=\frac{x}{\lambda'}-\lambda(x-vt)$$

$$=\lambda\left[\frac{x}{\lambda\lambda'}-x+vt\right]=\lambda\left[vt-x\left(1-\frac{1}{\lambda\lambda'}\right)\right]$$

$$\Rightarrow \quad t'=\lambda\left[t-\frac{x}{v}\left(1-\frac{1}{\lambda\lambda'}\right)\right] \qquad \ldots(9.13)$$

Now, using eqns.(9.13) and (9.11) into eqn.(9.10), we have

$$\Rightarrow x^2-c^2t^2=\left(\lambda(x-vt)\right)^2-c^2\left[\lambda\left[t-\frac{x}{v}\left(1-\frac{1}{\lambda\lambda'}\right)\right]\right]^2$$

$$x^2-c^2t^2-\lambda^2\left[x^2-2xvt+v^2t^2\right]+c^2\lambda^2\left[t^2-\frac{2xt}{v}\left(1-\frac{1}{\lambda\lambda'}\right)+\left(\frac{x}{v}\right)^2\times\left(1-\frac{1}{\lambda\lambda'}\right)^2\right]$$

$$=0 \qquad \ldots(9.14)$$

This is an identity and hence coefficients of various powers of x and t must vanish separately. So, equating the coefficient of t^2 from both sides to zero, we have

$$-c^2-\lambda^2v^2+c^2\lambda^2=0$$

$$\Rightarrow \quad c^2=\lambda^2[c^2-v^2] \qquad \ldots(9.15)$$

$$\Rightarrow \quad \lambda=\frac{1}{\sqrt{1-\frac{v^2}{c^2}}} \qquad \ldots(9.16)$$

$$\Rightarrow \quad \lambda^2=\frac{c^2}{\left[c^2-v^2\right]} \qquad \ldots(9.17)$$

Now equating the coefficient of *xt* from both sides to zero, we have

$$+ 2v\lambda^2 + c^2\lambda^2 \left[-\frac{2}{v}\left(1 - \frac{1}{\lambda\lambda'}\right) \right] = 0$$

$$\Rightarrow \qquad 2v\lambda^2 = \frac{2c^2\lambda^2}{v}\left(1 - \frac{1}{\lambda\lambda'}\right)$$

$$\Rightarrow \qquad v^2\lambda\lambda' = c^2 [\lambda\lambda' - 1]$$

$$\Rightarrow \qquad \lambda\lambda' (c^2 - v^2) = c^2 \qquad \ldots(9.18)$$

Comparing equations (9.15) and (9.18), we have

$$\lambda\lambda' (c^2 - v^2) = c^2 = \lambda^2 [c^2 - v^2]$$

$$\Rightarrow \qquad \lambda\lambda' = \lambda^2 \;\Rightarrow\; \lambda = \lambda'$$

So, using eqn.(9.17), $\lambda = \lambda' = \dfrac{c}{\sqrt{c^2 - v^2}} = \dfrac{1}{\sqrt{1 - \dfrac{v^2}{c^2}}}$.

Substituting equation (9.16) into eqn. (9.13), we have

$$t' = \frac{1}{\sqrt{1 - \frac{v^2}{c^2}}}\left[t - \frac{x}{v}\left(1 - \frac{1}{\lambda^2}\right) \right]$$

$$= \frac{1}{\sqrt{1 - \frac{v^2}{c^2}}}\left[t - \frac{x}{v}\left[1 - \frac{c^2 - v^2}{c^2}\right] \right]$$

$$= \frac{1}{\sqrt{1 - \frac{v^2}{c^2}}}\left[t - \frac{x}{v} \times \frac{v^2}{c^2} \right]$$

$$\Rightarrow \qquad t' = \frac{t - \frac{vx}{c^2}}{\sqrt{1 - \frac{v^2}{c^2}}} \qquad \ldots(9.19)$$

Substituting equation (9.16) into eqn. (9.11), we have

$$\Rightarrow \qquad x' = \frac{x - vt}{\sqrt{1 - \frac{v^2}{c^2}}} \qquad \ldots(9.20)$$

Equations (9.19) and (9.20) are known as Lorentz Transformation Equations.

II. Velocity Transformation

Let the velocities of the particle in S & S' frames are $u\,(u_x, u_y, u_z)$ & $u'\,(u_x', u_y', u_z')$ respectively.

So, $u_x = dx/dt \;\&\; u_x' = dx'/dt$

However, from eqn. (9.20), $x = \lambda(x - vt)$,

where, $$\lambda = 1\Big/\sqrt{1 - \frac{v^2}{c^2}}$$

and, from eqn. (9.9), $y = y'; z = z'$

On differentiating, $dx' = \lambda(dx - vdt),\ dy' = dy,\ dz' = dz$

Also, from eqn. (9.19), $$t' = \lambda\left(t - \frac{xv}{c^2}\right)$$

On differentiating, $$dt' = \lambda\left(dt - \frac{v}{c^2}dx\right)$$

Hence, $$\frac{dx'}{dt'} = \frac{\lambda\,[dx - vdt]}{\lambda\left[dt - \frac{v}{c^2}dx\right]} = \frac{dx - vdt}{dt - \frac{v}{c^2}dx} \qquad \ldots(9.21)$$

$$\frac{dy'}{dt'} = \frac{dy}{\lambda\left[dt - \frac{v}{c^2}dx\right]} \qquad \ldots(9.22)$$

$$\frac{dz'}{dt'} = \frac{dz}{\lambda\left[dt - \frac{v}{c^2}dx\right]} \qquad \ldots(9.23)$$

Using eqn.(9.21), we have,

$$\frac{dx'}{dt'} = \frac{\frac{dx}{dt} - v}{1 - \frac{v}{c^2}\cdot\frac{dx}{dt}}$$ (by dividing numerator & denominator by dt)

$$\Rightarrow \qquad u_x' = \frac{u_x - v}{1 - \frac{v}{c^2}\cdot u_x} \qquad \ldots(9.24)$$

Similarly, $$u_y' = \frac{u_y}{\lambda\left[1 - \frac{v}{c^2}\cdot u_x\right]} \qquad \ldots(9.25)$$

and $$u_z' = \frac{u_z}{\lambda\left[1 - \frac{v}{c^2}\cdot u_x\right]} \qquad \ldots(9.26)$$

Also from eqn.(9.24),

$$u_x' - \frac{v}{c^2} u_x' u_x = u_x - v$$

$$\Rightarrow \qquad u_x = \frac{u_x' + v}{1 + \frac{v}{c^2} . u_x'} \qquad \text{...(9.27)}$$

From eqn.(9.25),

$$\lambda u_y' \left[1 - \frac{v}{c^2} u_x\right] = u_y$$

$$\Rightarrow \qquad \lambda u_y' \left[1 - \frac{v}{c^2}\left\{\frac{u_x' + v}{1 + \left(v/c^2\right) u_x'}\right\}\right] = \lambda u_y' \left[1 - \frac{\frac{v}{c^2} u_x' + \frac{v^2}{c^2}}{1 + \frac{v}{c^2} u_x}\right]$$

$$\Rightarrow \quad u_y = \lambda u_y' \left[\frac{1 + \frac{v}{c^2} u_x' - \frac{v}{c^2} u_x' - \frac{v^2}{c^2}}{1 + \frac{v}{c^2} u_x'}\right] = \lambda u_y' \left[\frac{1 - \frac{v^2}{c^2}}{1 + \frac{v}{c^2} u_x'}\right]$$

$$= \lambda u_y' \left[\frac{1/\lambda^2}{1 + \frac{v}{c^2} u_x'}\right] = u_y' \left[\frac{1/\lambda}{1 + \frac{v}{c^2} u_x'}\right]$$

$$\Rightarrow \qquad u_y = \frac{u_y' . \sqrt{1 - \left(v^2/c^2\right)}}{1 + \frac{v}{c^2} . u_x'} \qquad \text{...(9.28)}$$

Similarly,

$$u_z = \frac{u_z' . \sqrt{1 - \left(v^2/c^2\right)}}{1 + \frac{v}{c^2} . u_x'} \qquad \text{...(9.29)}$$

(*a*) **For non-relativistic case**

Here, $v \langle\langle c$ & $\lambda \approx 1$.

So, $\quad u_x = u_x' + v,\ u_{yx} = u_y',\ u_z = u_z'$

(*b*) **When the motion is along X-axis only**

Here, $\quad u_x = u,\ u_y = 0,\ u_z = 0$

$$\Rightarrow \qquad u_x' = u' = \frac{u - v}{1 - \frac{v}{c^2} u}$$

(*c*). **For relativistic case**

When, $\quad u = c$

then,

$$u' = \frac{c - v}{1 - \frac{v}{c^2} c} = \frac{c - v}{(c - v)/c} = c$$

So, $u = u' = c.$

It proves the constancy of the speed of light in any reference frame.

(*d*) **For $v > c$**

Here the left hand side becomes negative and it has no physical significance.

III. Consequences of Lorentz Transformation

(*a*) **Length Contraction**

We consider a rod parallel to *X*-axis in system *S* at rest.

The length of the rod in system S (stationary frame) is

$$L = x_2 - x_1$$

and in moving system *S′* it is measured as

$$L' = x_2' - x_1'$$

Using Lorentz transformation we have,

$$x = \frac{x' + vt'}{\sqrt{1 - \frac{v^2}{c^2}}}.$$

So, $$x_1 = \frac{x_1' + vt'}{\sqrt{1 - \frac{v^2}{c^2}}}$$ and

$$x_2 = \frac{x_2' + vt'}{\sqrt{1 - \frac{v^2}{c^2}}}$$

Hence, $$L = x_2 - x_1$$

$$= \frac{x_2' + vt'}{\sqrt{1 - \frac{v^2}{c^2}}} - \frac{x_1' + vt'}{\sqrt{1 - \frac{v^2}{c^2}}}$$

$$= \frac{x_2' - x_1'}{\sqrt{1 - \frac{v^2}{c^2}}} = \frac{L'}{\sqrt{1 - \frac{v^2}{c^2}}}$$

$$\Rightarrow \qquad L' = L\sqrt{1 - \frac{v^2}{c^2}}$$

This length is also known as Proper Length.

Here $\left[1 - \frac{v^2}{c^2}\right] < 1$, so, $L' \langle L$, hence the rod seems to be contracted from the measured value of length in moving frame when observed from a stationary frame. This is called as Length Contraction.

The body's length is thus measured to be the maximum when it is at rest relative to the observer. As soon as it starts moving with a velocity v relative to the observer, its measured length is contracted in the direction of its motion, whereas its dimensions perpendicular to the direction of motion remains unaffected. Fitzgerald and Lorentz proposed this idea of length contraction long before Einstein's Special Theory of relativity. So this is also known as Lorentz – Fitzgerald contraction. In present particle physics experiments in high energy accelerators the colliding particles (nuclei, protons, etc) have achieved a very high velocity approaching c, there the length contraction is regularly observed.

(*b*) **Time Dilation**

We consider here a rail car (*i.e.*, a moving frame) moving with a velocity v. A mirror is fixed to its roof inside the car and an observer is at rest inside the car. He starts a light pulse at an instant vertically towards the mirror and the light signal reflects immediately back to the observer with a velocity c.

This time interval is measured as

$$\Delta t' = \frac{2h}{c} \qquad \ldots(9.30)$$

Here, h is the vertical height of the mirror from the floor of the rail car as shown in Fig. 9.4. Also an observer on the ground outside the rail car (*i.e.*, a stationary frame) observes that both the mirror and the light source are moving with a velocity v which is now at the point C.

If Δt is the time taken for the light pulse to reach at C from A, then using geometry as per Fig.9.4, we have

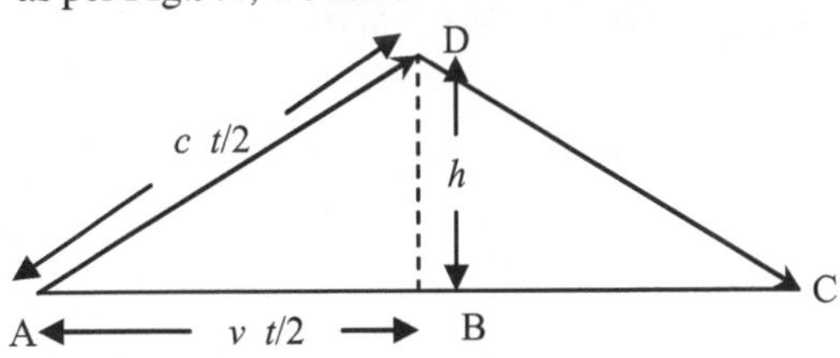

Fig. 9.4 *Representation of moving light pulse and mirror from outside the rail*

$$h^2 + \left(v\frac{\Delta t}{2}\right)^2 = \left(c\frac{\Delta t}{2}\right)^2$$

$$\Rightarrow \qquad \Delta t = \frac{2h/c}{\sqrt{1-\frac{v^2}{c^2}}} = \frac{\Delta t'}{\sqrt{1-\frac{v^2}{c^2}}} \qquad \text{(using eqn.(9.30))}$$

Hence,

$$\Delta t' = \Delta t\sqrt{1-\frac{v^2}{c^2}} \qquad \ldots(9.31)$$

So, time measured in the moving frame becomes lengthened or dilated. Hence from the point of view of the observer at the stationary frame the events will be found happening at a slower rate in the moving frame. This is known as Time Dilation effect. The effect has been experimentally verified by the decay of Muons in 1941 and also in 1971 in a jet plane travel.

(*c*) **Explanation of the Negative Result of Michelson-Morley Experiment**

The negative result of Michelson-Morley experiment was explained by considering following assumptions.

(*i*) **Constancy of speed of light:** If we consider, the speed of light as constant then there would not be path difference between reflected and transmitted ray and so no fringe shift will be observed *i.e.*, $\Delta N = 0$.

(*ii*) **Ether drag hypothesis:** If we consider, that material bodies are dragging with the ether medium then there won't be any relative motion between earth and ether. So, no path difference should be observed and there will be no fringe shift.

(*iii*) **Lorentz Fitzgerald contraction hypothesis:** In 1892, Lorentz Fitzgerald proposed a hypothesis that length of all material bodies moving with velocity v contract in the direction of motion by a factor of $\sqrt{\left(1-\frac{v^2}{c^2}\right)}$. This hypothesis retains the concept of ether frame.

If l_o is the length of the moving object at rest with respect to ether frame then length of body moving with velocity v can be written as

$$l = l_o \sqrt{\left(1-\frac{v^2}{c^2}\right)}$$

According to this hypothesis, t_2, from eqn.(9.4) can be modified as

$$t_2 = \frac{2l}{c}\left(1-\frac{v^2}{c^2}\right)^{1/2}\left(1+\frac{v^2}{c^2}\right)$$

$$= \frac{2l}{c}\left(1-\frac{v^2}{2c^2}\right)\left(1+\frac{v^2}{c^2}\right)$$

$$= \frac{2l}{c}\left(1+\frac{v^2}{c^2}-\frac{v^2}{2c^2}-\frac{v^4}{2c^4}\right)$$

$$= \frac{2l}{c}\left(1+\frac{v^2}{2c^2}\right) \quad \text{(neglecting high powers of } v/c\text{)}$$

This is equal to time t_1 as given by eqn. (9.3)

Thus, there is no path difference and hence, no fringe shift is observed in experiment.

7. VARIATION OF MASS WITH VELOCITY

According to Newtonian Mechanics the mass of a body does not change with velocity. However, conservation laws, especially here the law of conservation of momentum, hold for any inertial system. Hence in order to maintain the momentum conserved in any isolated system, mass of the body must be related to its velocity. So according to Einstein, the mass of the body in motion is different from the mass of the body at rest. We consider two systems of coordinates S and S' as in Fig. 9.5.

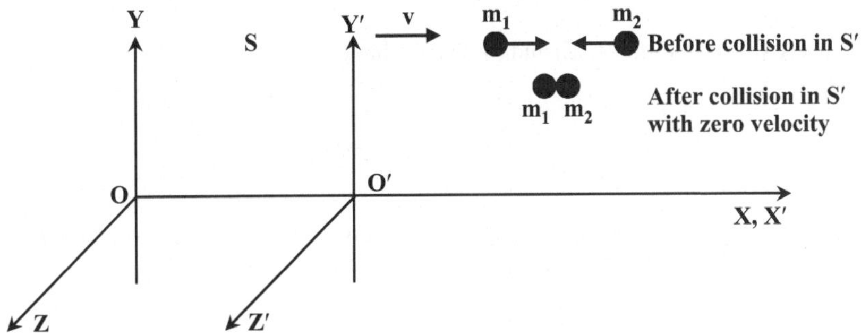

Fig. 9.5 *Collision between masses viewed from stationary and moving frames of reference*

We now consider the collision of two bodies in system S' and view it from the system S. Let the two bodies of masses m_1 and m_2 are traveling with velocity u' and $-u'$ parallel to X-axis in system S'. The two bodies collide and after collision they coalesce into one body.

In system S: Before Collision

Mass of bodies are m_1 and m_2. Let the velocities of them are u_1 and u_2 respectively.

In system S: After Collision

Mass of the coalesced body is $(m_1 + m_2)$ and the velocity is v.

Using Law of addition of velocities,

$$u_1 = \frac{u' + v}{1 + \frac{u'v}{c^2}}, \; u_2 = \frac{-u' + v}{1 - \frac{u'v}{c^2}} \qquad \ldots(9.32)$$

Applying the principle of conservation of momentum of the system before and after the collision, we have

$$m_1 u_1 + m_2 u_2 = (m_1 + m_2)v$$

$$\Rightarrow m_1 \left[\frac{u' + v}{1 + \frac{u'v}{c^2}} \right] + m_2 \left[\frac{-u' + v}{1 - \frac{u'v}{c^2}} \right] = (m_1 + m_2)v$$

$$\Rightarrow \qquad m_1\left[\frac{u'+v}{1+\frac{u'v}{c^2}}-v\right] = m_2\left[v-\frac{-u'+v}{1-\frac{u'v}{c^2}}\right]$$

$$\Rightarrow \qquad \frac{m_1}{m_2} = \left[\frac{1+\frac{u'v}{c^2}}{1-\frac{u'v}{c^2}}\right] \qquad \text{...(9.33)}$$

Now, using equation (9.32) and (9.33), we have

$$\Rightarrow \qquad \frac{m_1}{m_2} = \left[\frac{\sqrt{1-(u_2/c)^2}}{\sqrt{1-(u_1/c)^2}}\right] \qquad \text{...(9.34)}$$

Let the body of mass m_2 is moving with velocity zero in S before collision, *i.e.*, $u_2 = 0$, hence, using (9.34), we have

$$\frac{m_1}{m_2} = \left[\frac{1}{\sqrt{1-(u_1/c)^2}}\right] \qquad \text{...(9.35)}$$

Using common notation as $m_1 = m$, $m_2 = m_0$, $u_1 = v$, by using eqn.(9.35), we have

$$m = \frac{m_0}{\sqrt{1-\frac{v^2}{c^2}}} \qquad \text{...(9.36)}$$

This is the relativistic formula for variation of mass with velocity, where, m_0 is the rest mass and m is the effective mass of the body. There are a large numbers of experimental observations of this enhancement of mass of particles in high energy physics. Larger magnetic field is required to deflect a high velocity charged projectile as compared to the amount of magnetic field when the projectile is at stationary condition.

I. When $v >> c$

$$\therefore \quad v^2 << c^2, \frac{v^2}{c^2} \approx 0 \quad \Rightarrow \quad m = m_0$$

When velocity of the moving particle is much smaller as compared to velocity of light, effective mass equals the rest mass.

II. When $v \approx c$

$$\therefore \qquad v^2 \approx c^2, \frac{v^2}{c^2} \approx 1 \ \Rightarrow \ \left(1-\frac{v^2}{c^2}\right) < 1 \ \Rightarrow \ m > m_0$$

When velocity of the moving particle is comparable to velocity of light, effective mass of the body appears to be greater than the rest mass.

III. When $v = c$

$$\therefore \qquad v^2 = c^2, \frac{v^2}{c^2} = 1 \quad \Rightarrow \quad m = \infty$$

When velocity of the moving particle is exactly equal to velocity of light, effective mass of the body appears to be infinite and this is an impractical concept.

IV. When $v > c$

$$\therefore \quad v^2 >> c^2, \frac{v^2}{c^2} > 0 \quad \Rightarrow \quad m = \text{Imaginary}$$

When velocity of the moving particle is greater as compared to velocity of light, effective mass becomes imaginary and this is an impractical concept.

8. EQUIVALENCE OF MASS AND ENERGY

According to classical mechanics, force F is given by the time derivative of the product of mass (m) and velocity (v), *i,e.*, momentum of the body, and is given by

$$F = \frac{d}{dt}(mv).$$

However, according to the theory of relativity, both m and v are variable.

So, $$F = \frac{d}{dt}(mv) = m\frac{dv}{dt} + v\frac{dm}{dt} \qquad ..(9.37)$$

When a particle is displaced through a distance dx on application of a force F, then the increase in kinetic energy is given by,

$$dK = F \times dx$$

$$= \left[m\frac{dv}{dt} + v\frac{dm}{dt}\right] \times \frac{dx}{dt} \times dt$$

$$= \left[m\frac{dv}{dt} + v\frac{dm}{dt}\right] \times v \times dt$$

$$\Rightarrow \qquad dK = v^2\,dm + mvdv \qquad ...(9.38)$$

Now, we know

$$m = \frac{m_0}{\sqrt{1 - \left(\frac{v}{c}\right)^2}}$$

$$\Rightarrow \qquad m^2 = \frac{m_0^2}{\left(\frac{c^2 - v^2}{c^2}\right)} \quad \Rightarrow \quad m^2 = \frac{m_0^2 c^2}{c^2 - v^2}$$

$$\Rightarrow \qquad m^2c^2 - m^2v^2 = m_0^2c^2$$

On differentiation of both sides, we get

$$2m \times dm \times c^2 - [2m \times dm \times v^2 + 2v \times dv \times m^2] = 0$$

$$\Rightarrow \qquad 2m \times dm \times c^2 = 2m\,[dm \times v^2 + v \times m \times dv]$$

$$\Rightarrow \qquad c^2 dm = v^2 dm + vmdv \qquad \ldots(9.39)$$

By comparing equation (9.38) with (9.39), we have

$$dK = c^2 dm \qquad \ldots(9.40)$$

Now, we consider that the body initially at rest and on application of force it acquires a velocity v. Also the mass of the body increases from m_0 to m. Thereby, the total kinetic energy acquired by the body is given by,

$$\int dK = \int_{m_0}^{m} c^2 dm$$

$$\Rightarrow \qquad K = c^2 (m - m_0) \qquad \ldots(9.41)$$

This is the increase in kinetic energy due to the increase in mass.

Now the total energy of a moving body is the total of kinetic energy due to motion and the energy at rest.

So, $\qquad E = K + m_0c^2 = c^2 (m - m_0) + m_0c^2 = mc^2$

Hence, $\qquad E = mc^2$

This equation gives the universal equivalence between mass and energy. This is another major contribution of Einstein. It signifies the manifestation of energy into the matter and vive versa. Relativity shows that mass is not any independent quantity, however, it is a form of energy. Its experimental evidences are well known in nuclear fission and fusion reactions.

The conversion of mass and energy and the reverse are also illustrated in nature itself. When a particle collides with an anti-particle there is a mutual annihilation and the total mass is converted into radiant energy. Thus, the conservation of mass takes place into energy. Other way, when a radiant energy comes near a charged nucleus, particle and anti-particle are created. Thus, the energy is converted into mass.

10

Nuclear Physics

The electronic structure of the atom was understood before even the composition of its nucleus was known. The reason is that, forces that holds the nucleus together are vastly stronger than the electric forces that hold the electrons to the nucleus, and it is correspondingly harder to break apart a nucleus to find what is inside. Change in electron structure of an atom involves energies of only few electron volts. Change in nuclear structure on the other hand, involves energies in MeV range, a million times greater. Despite the strength of the forces that hold nucleons (particles residing inside the nucleus) together to form an atomic nucleus, many nucleus are unstable and spontaneously change into other nuclides by radioactive decay. All nuclei can be transformed by nuclear reaction with nucleons or other nuclei with them into different nuclei. The principle aspects of nuclear reactions, various nuclear detectors and their application are considered in this chapter.

1. NEUTRON CROSS SECTION

Cross section is a measure of a particular interation. So neutron crosssection deals with the interaction of neutron with nucleus or in other means it gives information about the probablity of scattereing and absorption of neutron when it approaches to nucleus. The number of scattered neutron is determined by scattering cross section of neutron while the number of absorbed neutron is determined by the absorption cross section.

If σ_{sc} is scattering cross section and σ_a is absorption cross section then the cross section σ_t of neutron is given by $\sigma_t = \sigma_{sc} + \sigma_a$

The neutron cross section for a particular role (scattering and absorption) depends on the energy of neutron, nature and size of target. If a collimated beam of monoenergetic neutrons is incident upon a slab of material of thickness dx; then its intensity, I, is reduced to dI. The decrease in intensity depends upon the incident intensity I and the number of nuclei present per unit area of the slab, ndx, where n is the number of nuclei per unit volume, so

$$dI \propto I(ndx) \quad \Rightarrow \quad dI = -\,\sigma_t\, Indx \qquad \ldots(10.1)$$

Here, negative sign is taken as intensity is decreasing with x, here σ_t is the total cross section for attenuation of beam.

The quantity, $\mu = n\sigma_t$, is called the attenuation coefficient.

If slab is of finite thickness x, I_o is intensity of incident beam and I is the intensity of transmitted beam, then from eqn.(10.1),

$$dI = -\,\mu I dx \quad \Rightarrow \quad \frac{dI}{I} = -\,\mu dx \qquad \ldots(10.2)$$

On integrating above equation we get,

$$\log \mathrm{I} = -\,\mu x + C \text{ at } x = 0, I = I_0 \quad \Rightarrow \quad C = \log I_0$$

Hence,
$$\log I = -\,\mu x + \log I_0 \quad \Rightarrow \quad I = I_0 e^{-\mu x} \qquad \ldots(10.3)$$

The equation (10.3) may be written as

$$I = I_0 e^{-(n\sigma_t)x}$$

$$\Rightarrow \qquad \sigma_t = \frac{1}{nx} \log_e \left(\frac{I_0}{I}\right) \qquad \ldots(10.4)$$

Hence, by measuring I, I_0, x and with known value of n, σ_t can be estimated. So neutron cross section may be defined as net cross sectional area of target nucleus atom offered for nuclear reaction to the incident neutron beam.

2. NUCLEAR FISSION

The phenomenon of splitting of a heavy nucleus into two or more lighter nuclei is known as nuclear fission. This phenomenon was discovered by Otto Hahn in 1938. In this phenomenon a nucleus of uranium isotope $_{92}U^{235}$ undergoes fission when it struck by a slow neutron. Such a nucleus is called a fissile or fissionable nucleus. This fissile splits into lighter nuclei $_{56}Ba^{141}$, $_{36}Kr^{92}$. The nuclear reaction involved in fission process are given as :

$$_{92}U^{235} + {_0}n^1 \longrightarrow [_{92}U^{236}] \longrightarrow {_{56}}a^{141} + {_{36}}r^{92} + 3\,{_0}n^1 + Q$$

o neutron   ompoun nuc eu un ta e

$$_{92}U^{235} + {_0}n^1 \longrightarrow [_{92}U^{236}] \longrightarrow {_{57}}a^{146} + {_{35}}r^{87} + 3\,{_0}n^1 + Q$$

Here $Q \approx 200$ MeV per fission, which is shared by the fission fragments, the neutrons and other particles emitted during the process. It may be pointed out that in each fission of uranium, the two fragments can be any stable isotope of middle weight atoms, with the emission of two or three neutrons per fission. Thus, on the average, each thermal neutron (slow neutron) fission of $_{92}U^{235}$ yields on an average of 2.5 neutrons. The distribution of mass number in the fragments from the fission of $_{92}U^{235}$ is shown in Fig. 10.1. The first satisfactory description of nuclear fission was given by Bohr and Wheeler in 1939, known as liquid drop model.

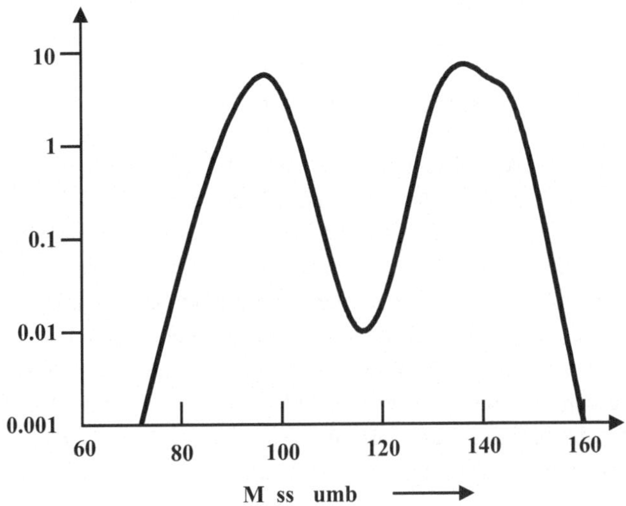

Fig. 10.1 *Variation of yield percentage with mass number*

According to this model the nucleus behaves like an uncompressible liquid drop whose shape is governed by the cohesive short range nuclear forces and the electrostatic repulsive forces between protons within the nucleus. The cohesive forces tend to make the nuclear drop into spherical and electrostatic forces tend to disrupt spherical shape. As nuclear forces are short range forces, considered much stronger than electrostatic repulsive forces, so the spherical nuclear shape is restored. If the energy is given to the nuclear drop by capture of slow neutrons, the nucleus is excited and oscillations are set up in in the nuclear drop distorting its spherical shape. The nuclear forces still tend to restore its spherical shape of nuclear drop, while excitation energy tends to distort its spherical shape. If the amplitude of oscillations becomes sufficiently large, the nuclear drop may attain the shape of dumbell and the electrostatic forces may then push two bell apart till it split into two fragments with the release of two or three neutrons. The sequence of step is shown in Fig. 10.2

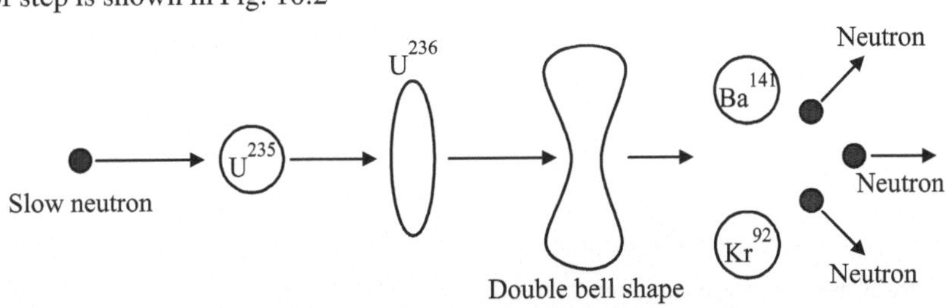

Fig. 10.2 *Liquid drop model of nuclear fission*

If the excitation energy is not sufficient to cause fission, the compound nucleus $_{92}U^{236}$ return to spherical shape by emitting γ -radiation. The minimum energy required for fission is known as activation energy. This activation energy has been calculated by Bohr and Wheeler theory which concludes that fission of $_{92}U^{235}$ is possible by slow neutrons while that of $_{92}U^{238}$ by fast neutron.

3. CHAIN REACTION AND REACTOR CRITICALITY

The chain reaction takes place only if the size of fissionable material is greater than a certain size, known as the critical size. If the size of fissionable material is less than critical size, then there will be less than one neutron per fission event and the chain reaction would stop. This criticality states whether the reaction is self sustaining or not. In a nuclear fission on an average 2.5 neutrons are emitted. All resulting secondary neutrons are never capable of producing fission, because some are absorbed by fission fragments, some by $_{92}U^{238}$ in fission of uranium and some may escape through the surface; but still on an average one neutron is produced in each fission capable of causing another fission provided the size of fissionable material is greator than critical size.

To have an idea of critical size, let the radius of fissionable material be *r*, *N* be the number of neutrons generated in fission, *A* be the number of neutron absorbed by fission fragments or other isotopes of fissionable material present and *L* be the number of neutrons escaping through the surface in given time interval. Then obviously *N* and *A* are proportional to the volume and *L* is proportional to the surface area of the fissionable material, *i.e.*,

$$N \propto r^3 \quad \Rightarrow \quad N = k_1 r^3$$
$$A \propto r^3 \quad \Rightarrow \quad A = k_2 r^3$$
$$L \propto r^3 \quad \Rightarrow \quad L = k_3 r^2$$

As chain reaction is possible only when rate of emission of neutron is greater than the rate of loss of neutrons, *i.e.*,

$$N > A + L \quad \Rightarrow \quad N - A > L$$

$$\Rightarrow \qquad \frac{N - A}{L} > 1$$

or

$$\frac{k_1 r^3 - k_2 r^3}{k_3 r^2} = \frac{(k_1 - k_2)\, r}{k_3} = kr > 1$$

where *k* is new constant and is given by

$$k = \frac{k_1 - k_2}{k_3}$$

and so, $$r > (1/k) \qquad \ldots(10.5)$$

Therefore, for the chain reaction to take place the size of fissionable material, *r*, must be greater than critical size, $1/k$. The condition of chain reaction in another form may be expressed as reproduction factor f, f is the rate of emission of neutrons per unit rate of loss of neutrons.

If $f < 1 \Rightarrow$ the state of reaction is sub critical and fission process is not self sustained, $f = 1 \Rightarrow$ the state of reaction is steady and the fission process is self sustained, and, if $f > 1 \Rightarrow$ the state of reaction is super critical and this leads to an explosion in due course of time.

There are two types of chain reaction, controlled and uncontrolled chain reaction.

In uncontrolled chain reaction the reproduction factor f is considerably greater than unity, so that the nuclear energy released in the fission process continuosly increases which causes system explosion. This type of reaction is the basic principle of atom bomb.

However, in controlled chain reaction factor f is unity or slightly greater than unity. In this reaction desired number of neutrons are absorbed in a controlled manner so that only one neutron per fission remains. Such a controlled chain reaction is the principle of nuclear reactor.

4. NUCLEAR FISSION REACTOR

Nuclear fission reactors are classified depending on the nature of fissionable materials, power level, number of neutrons. They are called thermal reactor, intermediate reactor or fast reactor,depending on whether fission is induced by slow neutrons, intermediate energy neutron or fast neutron respectively. Based on the utlity of the reactors, they are termed as research reactors, power reactors, production reactors and breeder reactors. The reactors which are used to supply neutrons for nuclear research and manufacture of radioactive isotopes, are called as research reactors. The reactors which are used to produce electricity from the heat energy generated in fission reactions, are called as power reactors. However production reactors are used to convert fertile materials ${}_{92}U^{238}$ into fissile materials and the breeder reactors produce more fissionable materials than it consumes and also provide useful power.

Principle

It is an assembly in which nuclear fission takes place in controlled manner such that huge amount of enery can be utilized properly. In this reactor, the nuclear reaction is self sustaining and does not become explosive. Here, chain reaction is controlled in such a way that only one neutron is made available for each fission in self-sustaining reaction. In general the fission reaction can be expressed as,

$${}_{92}U^{235} + {}_{0}n^{1} \rightarrow [{}_{92}U^{236}] \rightarrow X + Y + p\,{}_{0}n^{1} + Q$$

Here, X and Y are fission fragments, p is the number of fast neutrons produced. On an average in each fission of uranium 2.5 neutrons are produced and nearly 200 MeV of energy is released.

Construction

Nuclear reactors with its variuos componenets is shown schematically in the Fig. 10.3. These components are described below.

(*a*) **Nuclear fuel:** The fissionable material used in the reactor is known as nuclear fuel. Generally natural uranium is used as a fuel which consists of only 0. 72% of $_{92}U^{235}$ isotope and plutonium $_{94}Pu^{239}$. $_{92}U^{235}$ is useful isotope of uranium causing nuclear fisson reaction with slow neutron.

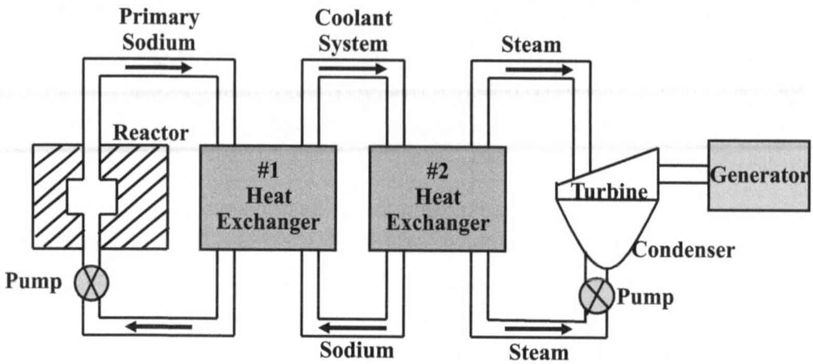

Fig. 10.3 *Schematic diagram of a Nuclear Fission reactor*

(*b*) **Modeartor:** In a nucleus fisssion nearly 2.5 neutrons are emitted, which are fast neutrons. However to self sustain the nuclear reaction slow neutrons are desirable. So to slow down these fast neutron, moderators are used. Moderator is material used to slow down the fast neutrons all along to maintain self sustain sequence of fission reactions. The moderator should have following properties:

(*i*) It should have low absorption cross section for neutron.

(*ii*) Moderator should have size equal to neutron since the collsion between neutron and moderator nuclei is elastic.

(*iii*) Materials should have low atomic weight.

(*iv*) It should reduce the kinetic energy of neutron.

The first six elements in order of increasing atomic weight are hydrogen, helium, lithium, berylium, boron and carbon. Of these there are no solid or liquid compound of helium and lithium. Hydrogen has a fairly large absorption cross section for slow neutrons. Boron has too large cross section for slow neutrons that it used as an absorber of neutron in a reactor, hence can not be used as moderator. That is why, heavy water (D_2O) and carbon (graphite) are the best choices as moderators.

(*c*) **Reactor coolants:** In the nuclear fission reaction a large amount of energy liberate in the reactor core. The materials used to remove this heat are called as coolant. These coolants may be liquid or gases. Liquid coolants are water, heavy water mercury while gas coolants are air, nitogen, CO_2. These coolants are circulated through the reactor core to absorb heat energy produced anf transfer it to the outside of core. These coolant give up its energy to heat exchanger to produce steam which is then then used to produce electric power. Normally water or heavy water are used as good coolant in a reactor operating at ordinary temperature.

(*d*) **Controller:** To control the rate of production of heat it is necessary to absorb neutrons in each fission so that neutrons are available for self sustaining controlled chain reaction. Controller should have fairly large absorbtion crosssection for neutrons as large amount of neutron may cause uncontrolled chain reaction which is not desirable in reactor. Cadmium and boron rods satisfy the criteria for controller in the reactor. However, mostly boron rods are used as controller. When the reactor reaches to required power level the control rods are inserted in the graphite between the uranium fuels in the reactor to absorb excess neutron. This maintains the effective reproduction factor as unity and the power level of the reactor becomes steady. Alternately control rods can be moved in and out according to the requirement of maintaining power level in the reactor.

(*e*) **Reactor shield:** The nuclear reactor are powerful source of neutrons and intense radioactive radiation which are very harmful to human working on these reactors. In order to protect persons working in these reactors, reactor core is surrounded by a radiation shield. The shield generally consists of concrete wall of about 6 to 8 feet thickness which is capable of absorbing neutron and ã-rays. In reators operating at high powers, the part of shield in immedaite contact gets heated considerably and special cooling facilities are needed to prevent it from heating. In such cases thermal shield is used to which is close to the core and consists of few inches of iron or steel.

Applications

Nuclear reactors are mainly used to generate electric power. This is used as an alternate to fossil fuel like oil, petrol, coal etc. However, breeder reactor produce more fossil fuel. The reactors may also be used to produce plutonium for enriching other fossil fuel. These reactor produce radioactive isotopes, which are helpful in nuclear research. Many artificial isotopes, thus, produced find applications in medicine, agriculture, science and industry as strong and stable source of γ-radiation. Radioactive tracers are very important tool in medical diagnosis and radiotherapy.

5. NUCLEAR FUSION

This process is an alternate to the nuclear fission as a source of energy. When two or more lighter nuclei combine together in such a way that they form a heavy nucleus with liberation of huge amount of energy. This phenomenon is known as Nuclear Fusion and the reaction is known as Nuclear Fusion Reaction. This fusion requires very high temperature which is not possible in laboratory. When two lighter nuclei come close together strong repulsive force start to act between them, due to positive charged nucleus. So, it's a major problem in this reaction to fuse two lighter nuclei. But if the nuclei are imparted a very high kinetic energy and then make them collide then the fusion reaction is possible on earth because in that situation it will overcome the coulomb repulsive force acting between them. In high temperature the product particle so formed in this reaction get sufficient kinetic energy so that the whole reaction becomes selfsustaining. Since this reaction occurs only at high temperature therefore it is called as thermonuclear reaction and the energy liberated is called as thermonuclear energy. It has been observed that necessary kinetic energy required for the fusion of two more lighter nuclei increases rapidly with atomic number. Therefore, three isotopes of hydrogen are preferred as reacting agent for this reaction. There are two ways of nuclear fusion reaction *i.e.*, proton –proton cycle and carbon-nitrogen cycle.

Proton-proton Cycle

In this reaction four protons react together to form helium nucleus and liberate 26.7 MeV energy. This cycle has following steps:

$$_1H^1 + {}_1H^1 \rightarrow {}_1H^2 + {}_1e^0 + \gamma$$

$$_1H^2 + {}_1H^1 \rightarrow {}_2He^3$$

$$_2He^3 + {}_1H^1 \rightarrow {}_2He^4 + {}_1e^0 + \gamma$$

the net result is $4\,{}_1H^1 \rightarrow {}_2He^4 + 2\,{}_1e^0 + 2\gamma$

Here ${}_1e^0$, í represent positron and neutrino repectively.

Carbon–nitrogen Cycle

Here also four protons react together to form helium nucleus with liberation of huge energy. But in these reaction carbon and nitrogen act as catalyst. This type of nuclear fusion reaction occur in following steps:

$$_6C^{12} + {}_1H^1 \rightarrow {}_7N^{13}$$

$$_7N^{13} \rightarrow {}_6C^{13} + {}_1e^0 + \gamma$$

$$_6C^{13} + {}_1H^1 \rightarrow {}_7N^{14}$$

$$_7N^{14} + {}_1H^1 \rightarrow {}_8O^{15}$$

$$_8O^{15} \rightarrow {}_7N^{15} + {}_1e^0 + \gamma$$

$$_7N^{15} + {}_1H^1 \rightarrow {}_6C^{12} + {}_2He^4$$

the whole reaction can be summarized as

$$4\,{}_1H^1 \rightarrow {}_2He^4 + 2\,{}_1e^0 + 26.7\text{MeV}$$

It is believed that both the above cycles are observed with equal probabailty in sun. However, in hotter star carbon-nitrogen cycle is responsible for the liberation of huge energy. In these reactions it is clear that hydrogen is consuming and coonverting into helium nuclues. It is being estimated that it would be $\approx 3 \times 10^{10}$ years before the hydrogen is all converted into helium in the sun. These reaction are not possible in the laboratory due to the high temperature requirement for the fusion lighter nuclei. However, in the laboratory these reactions are made possible when the matter is in plasma form. These plasma should be heated so that kinetic energy is imparted to the reacting nuclei for the fusion to occur. There may be variuos nuclear fusion reactions by using plasma.

6. INTERACTION OF RADIATION WITH MATTER

Nuclear radiations are detected only through their interaction with matter. It is therefore important to understand the different processes through which charged particles interact with matter. The diferent processes involved in the energy loss of the charged particle and photons are given below.

I. Interaction of Heavy Charged Particles with Matter

Energy loss of heavy charged particles such as protons, α-particles and heavy ions is primarily caused by coulomb interaction of the ions with the electrons bound to the nuclei of the absorbing medium. It results in the excitation and ionization of atoms and molecules of the medium. The heavy charged particles while passing through medium give up its kinetic energy to the atoms of the absorber. The electrons of atom either get excited to higher energy level or leave the atom, depending on kinetic energy gained by it. While considering the energy loss of protons and a - particles it is convenient to define stopping power, S (E).

Stopping power is defined as the amount of energy lost over per unit length of the path of the particle in a given absorber, *i.e.*,

$$S(E) = -\frac{dE}{dx}$$

II. Interaction of Light Charged Particles with Matter

Light charged particles like electrons lose their kinetic energy by inelastic radiative and elastic collision with atomic nucleus. The inelastic collision of electrons with matter is an important mechanism. Incident electrons passing through the field of nucleus experience a deflection or deceleration emittng radiation. This process is known as Bremstrahlung (breaking of radiation). The rate of energy loss by the interaction is proportional to Z^2, where Z is atomic number of target atom. The incident electron can have elastic collision with the nucleus, resulting into deflection of electrons without any radiation loss or excitation of nucleus. The total energy loss per unit path length of the electron is sum of ionization and radiation losses, *i.e.*,

$$\left(\frac{dE}{dx}\right)_{Total} = \left(\frac{dE}{dx}\right)_{Ionization} + \left(\frac{dE}{dx}\right)_{Radiation} \quad ...(10.6)$$

III. Interaction of Electromagnetic Radiation with Matter

When a monoenergetic beam of γ - rays passes through an absorbing material, its intensity gets reduced. If I is the intensity of gamma radiations incident on absorber of thickness x, then according to eqn.(10.3) the intensity of radiation absorbed is given by

$$dI \propto -Idx \quad \Rightarrow \quad dI = -\mu Idx$$

$$\Rightarrow \qquad \frac{dI}{I} = -\mu dx$$

$$\Rightarrow \qquad I = I_0 e^{-\mu x}.$$

It is known as exponential law of absorption of gamma rays.

7. ABSORPTION OF GAMMA (γ) RAYS THROUGH MATTER

The interaction of photons with the orbital electrons in the atom of the matter depends on the perpendicular distance between the direction of incident photon and the center of the nucleus. This distance is known as the interaction parameter, b and is shown in Fig. 10.4. Here we are considering the interacting force to be operating perpendicular to the incident photon direction. Depending on the relationship between b and the atomic radius, R, the three physical phenomena are taking place.

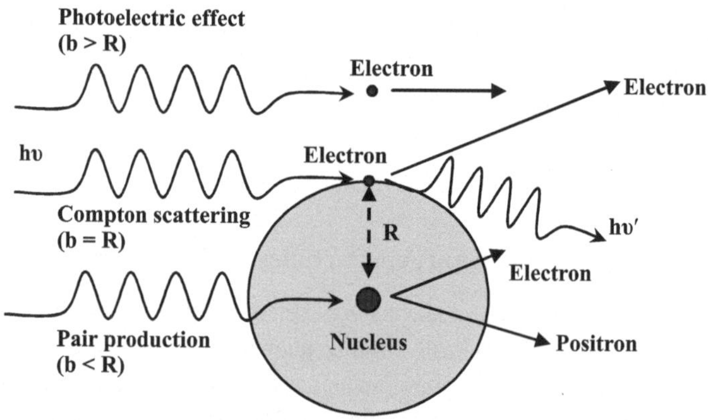

Fig. 10.4 *Photoelectric effect, Compton effect and Pair production in interactions of photons with matter*

If $b > R$, the electron is ejected by the photon and is called the Photoelectric Effect.

If b is comparable to R the incident photons are scattered by the electron of the atom and the electron itself also gets scattered. This is known as Compton Effect.

Lastly, if b is less than R so that the photon is incident in the vicinity of the field of nucleus, the photon is directly converted into an electron – positron pair. This is known as Pair Production.

We can represent the partial absorption coefficient for photoelectric effect, compton effect and pair production as μ_{pe}, μ_c, μ_{pp} respectively. The total absorption cross section, μ, is given by

$$\mu = \mu_{pe} + \mu_c + \mu_{pp}.$$

The above three processes are explained below in brief.

I. Photoelectric Effect

After the interaction of incident photon with the atomic electron (mainly outer K or L shell electron), the bound electron absorbs the photon energy and is ejected from the atom. The maximum kinetic energy, KE, imparted to the ejected electron is given by the Einstein relation,

$$KE = h\nu - BE.$$

Here υ and BE are the frequency of incident radiation and binding energy of the electron (also known as the work function of the material) respectively. About 75% of the photoelectrons are ejected by the incident photon from K-shell of the atom. However, lesser number of electrons are ejected from L- shell. When the energy of the incident photon is high enough, the maximum number of K-shell electrons are ejected than any other shell. The photoelectric effect is more predominant in heavy metals like lead that the light ones like aluminium.

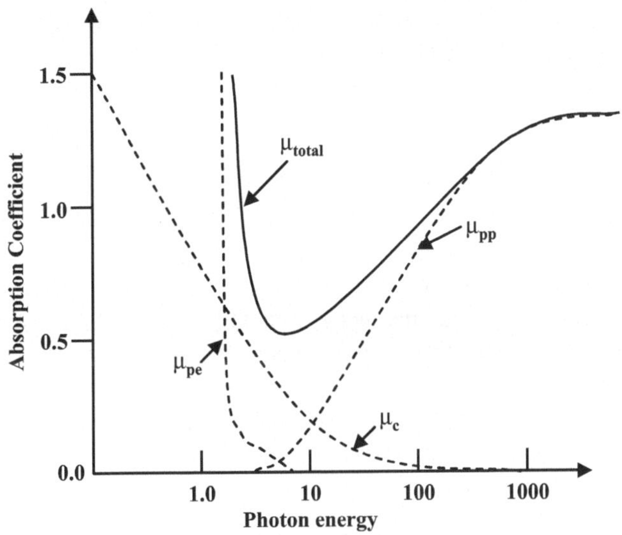

Fig. 10.5 *Variation of absorption coefficient versus photon energy for Photoelectric effect, Compton effect and Pair production*

At low photon energies, below 50 keV for aluminium and for 500 kV for lead, the photoelectric effect gives the main contribution to the absorption cross section and is shown in figure 10.5 for lead. Fig. 10.5 gives the variation of different absorption coefficients along with the total absorption coefficient for different incident photon energies.

II. Compton Effect

When the energy of the γ-radiations increases Compton scattering is taking place in place of photoelectric absorption as the chief means of removing photons from the initial beam. Here the incident γ-ray may interact with any one of the orbital electrons. These electrons are considered as free as the incident photon energy is larger to electrons binding energy and so after interaction the electron becomes detached from the atom. The photon energy is distributed between the recoil electron and the secondary photon. The secondary or scattered photon travels in different direction than the incident one.

The energy of this scattered photon is given by

$$h\nu' = \frac{h\nu}{1+\frac{h\nu}{m_0c^2}[1-\cos\phi]}.$$

Here, φ is the angle of scattering and $h\nu$ is the energy of the incident γ-ray photon. The variation of absorption coefficient of Compton Scattering with incident photon energy in lead is shown in Fig. 10.5. It indicates that μ_C decreases with the γ-ray energy from 500 keV to 5 MeV.

III. Pair Production

When γ-ray photon of high energy passes close to the atomic nucleus, it interacts with the field of the nucleus in such a way that the interaction parameter is less than R and the incident photon is converted to electron – positron pair. This process is possible only when the incident photon energy exceeds the rest pass energies of both electron and positron (*i.e.*, photon energy > (0. 51 eV + 0. 51 eV). The absorption coefficient for pair production μ_{pp} is zero below 1. 02 MeV and it increases slowly with energy and reaches saturation value for energies exceeding a few billion eV. This is also shown in Fig. 10.5.

8. NUCLEAR RADIATION DETECTORS

The dependence of interaction of charged particles and electromagnetic radiation on various factors is used to detect these particle and determine their properties. The various instruments used for the detection of nuclear particles and radiations are based on the principle of excitation or ionization of atoms of the medium in which the incident charged particles pass through. These detectors are classified as given in Fig. 10.6.

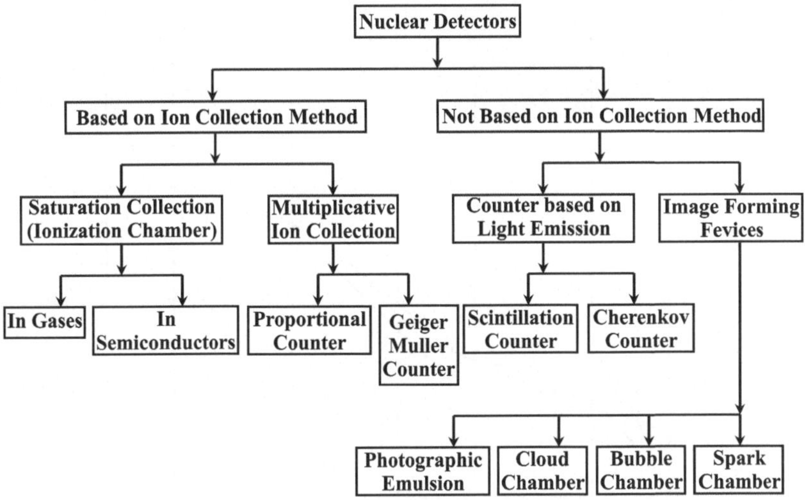

Fig. 10.6 *Classification of Nuclear Detectors*

I. Gas Filled Detectors

The ionization chamber, the proportional counter and Geiger Muller counter are the gas filled detectors operating in their specific regions. A gas filled detector usually consists of a gas filled chamber (whose conducting walls act as cathode) and central electrode (act as anode). A potential difference is applied between cathode and anode by high tension supply through an external registor *R*. This is shown in Fig. 10.7.

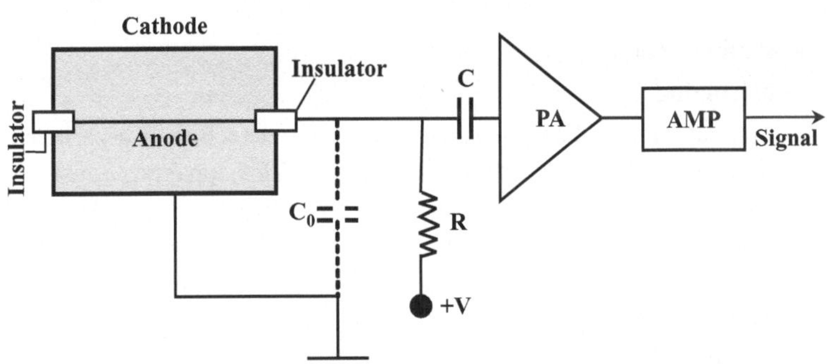

Fig. 10.7 *Schematic diagram of Gas filled detectors*

When high electric radiation or charged particles enter through such a chamber, the ionization of the gas starts resulting in the formation of ion pairs. In the absence of electric field, the ion pairs thus created will just recombine. As the potential difference is applied, the ion pairs are collected by the respective electrodes, thus, giving rise to current pulse.

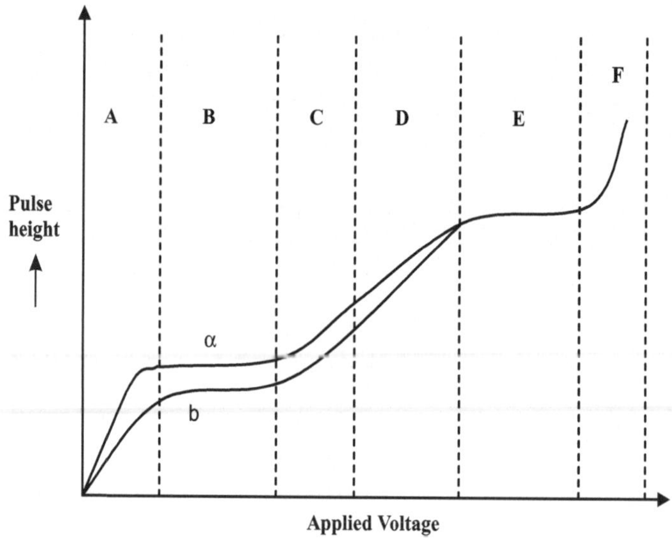

Fig. 10.8 *Variation of pulse height with applied voltage*

This current pulse passing through R gives rise to electrical signal which can be amplified and then detected by pulse counter. Fig. 10.8 represents the typical plot of log of pulse height versus applied potential difference V for two different types of events in the counter. One event is for the passage of α-paritcles and other for β- particles. In the graph, *B*-region shows operating region of ionization chamber. *C* and *D* regions shows operating region of proportional counter and E-region shows operating region of Geiger –Muller counter.

(*a*) Ionization Chamber

Construction

An ionization chamber is used mainly for the detection of a-paritcles, b-particles and other ions with discrete spectra. This apparatus consists of a chamber filled with a gas like air or argon at normal pressure. The wall of chamber acts as cathode and the central wire inside the chamber acts as anode. A potential difference is applied between cathode and anode through resistor *R*.

Working

When charged particle are passed through the chamber, the gas becomes ionized. If no potential difference is applied between cathod and anode, there will be no current through the gas as they recombine together due to electrostatic forces acting between ions. In the absense of potential difference, there is no directing force acting on ions. When the potential difference is applied between the electrodes current starts to increase as the ions are drifted by applied voltage. If the applied voltage is small most of the ions get recombine together and all ion pairs will not reach at electrodes. On increasing further, the applied voltage recombination phenomenon reduces, resulting into large amount of current. If the applied voltage becomes sufficiently high, then all ion pairs reach their respective electrodes without recombination. At this voltage, the pulse size becomes constant. Now on increasing more and more voltage current or pulse size remains constant. The ionization chamber works in the region of constant pulse size. In this region, the applied voltage is high enough to prevent recombination of ions and low enough to prevent multiplication.

(*b*) **Proportional Counter :** When a particle of low speccific ionization passes through an ionization chamber, the pulse produced is too small to detect it. This counter works in higher applied potential difference region than ionization counter. In proportional counter, the output signal is proportional to the number of ions formed by primary ionization process. It can be used to detect α-particle, β-particle, *X*-rays, neutrons, fission fragments etc. The chief disadvantage of the counter is that the amplification facctor depends on the applied voltage and therefore this voltage must be maintained within narrow limits.

Construction

It consists of cylindrical gas filled detector with a very thin central wire, which is insulated from the tube. The central wire (diameter of 0.1 mm) is always made positive with respect to the metallic cylindrical tube. A mixture of nobel gas Ar (90%) and some polyatomic gas (methane 10%) are usually used as filler gas in this counter.

Working

When the potential difference across two electrodes is increased beyond the saturation region, the ions gain sufficient energies to produce secondary ionization due to the collision with gas atoms in the chamber. In this region, electrons gain sufficient energy from primary ionization to cause secondary ionization while drifting towards central wire. If V_0 is applied voltage, a and b are radii of central wire and the cylindrical electrodes respectively, then the electric field strength E produced at a radial distance r from the central wire is given by

$$E = \frac{V_0}{r \ln(b/a)}$$

The ions so produced in secondary ionization again cause further ionization and the process continues. As a result, each original ion pair leads to the formation of a large number of secondary ion pairs. The total number of ion pairs so produced by a single primary ion is called as gas multiplication factor. Multiplication factor, M, depends on the anode diameter, the radius of cathode voltage, nature and pressure of the gas. Usually the gas multiplication factor is in the range of $10^2 - 10^4$. Let n be the total number of secondary electrons produces by a electron in primary ionization. The production of secondary electrons during the collision in the gas will also give rise to photons which in turn can produce photoelectrons in the volume of the counter. Let p be the probabiltiy that each secondary electron will give rise to a photoelectron. So, there will be np photoelectrons in all. If each photoelectron in turn produces n electrons by further collision, we shall get second generation avalanche of n^2p electrons which in turn will give rise to more photoelectrons and so on.

So, the multiplication factor will be given by

$$M = n + n^2p + n^3p^2 + \dots$$

In practice $np < 1$ and the series can be written as

$$M = \frac{np}{1 - np}$$

(*c*) **Geiger-Muller counter:** The Gieger Muller (*GM*) counter is the oldest type of gas filled detector which continue to be used to the present time very commonly as a radiation measurement device mainly due to its simple design and operation. Its operating voltage range is very high ($\approx$ 1000 volts).

Principle

If the applied voltage between the electrodes of a proportional counter exceeds a certain limit, the charge collected is no longer proprotional to the charge created. The gas multiplication factor becomes so large that the insulating properties of gas breakdown and the discharge passes inside the gas. This operation is called Geiger–Muller mode of operation. In this region, the total number of ions produce become independent of the initial number of ions formed by the entering particle.

Construction

It consists of cylindrical metallic normally copper cathode with a thin coaxial tungsten wire filled with monoatomic gas (Argon or Neon) at a pressure of 10 cm of mercury. The central wire (diameter of 0.1 mm) is always made positive with respect to the metallic cylindrical tube.

Working

When an ionising particle enters into the tube, it ionises the argon gas atoms. Since the applied voltage is very high, the electron accelerates towards anode. This highly energetic electron causes further ionisation while moving towards anode. The process is cumulative and large avalanche of electrons are produced. These energetic electron raise some gas atoms to its excited state followed by ultraviolet radiation, the absorption of which give rise to new avalanche. As ultaviolet radiation are neutral, it is unaffected by electric field and can go in any direction Hence, avalanche spreads rapidly in the entire volume of the tube (in contrast to localised action in proprotional counter). The total number of ions so produced is now independent of initial number of ions formed by incident ionising particle. The highly energetic positive ion strikes to cathode and detach secondary electron from the cathode and becomes neutral atom causing to excite some atoms. These atoms radiate high energy photons which on striking cathode eject photoelectrons to cause fresh avalanches. Thus, the entry of a single particle in the tube causes pulse of current, whose magnitude is independent of energy and nature of ionising particle. The current pulse causes the corresponding voltage pulse across resistor R. This voltage pulse is fed to a pulse amplifier through a capacitor *C*. The amplified pulse is finally passed to a counter which records each individual pulse separately. The combination of GM tube and the recorder is called GM counter. A plot of counting rate versus the applied voltage is shown in Fig.10. 9 (*a*).

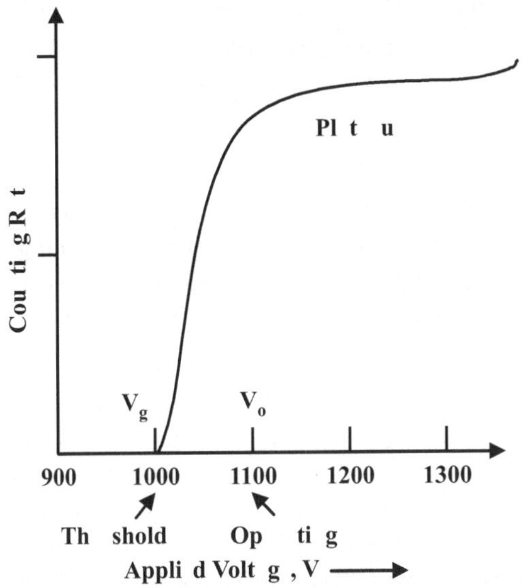

Fig. 10.9 (*a*) *Variation of counting rate with aplied voltage*

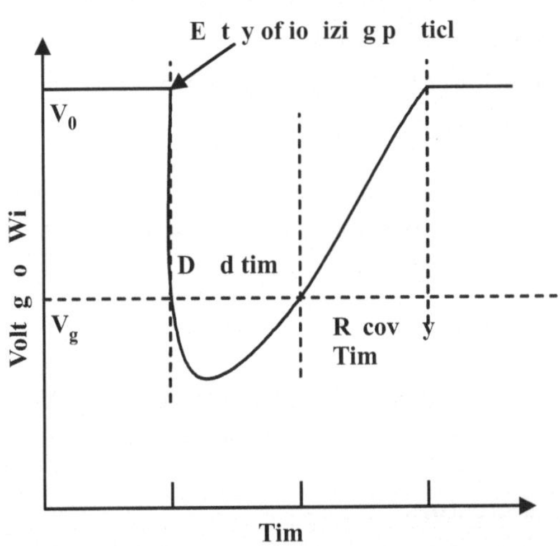

Fig. 10.9 (*b*) *Variation of Voltage on wire with time*

The counting starts at a voltage known as thrshold voltage V_g and then shows a sharp rise in counting with applied voltage. It then settle into plateu region where counting remains constant with applied voltage, the counter is preferred to operate in this region which is known as Geiger or Plateau region. This is shown in Fig.10.9 (*b*). A good *GM* counter has a plateau about 200 volts wide and has about 1% slope per 100 volts. Beyond the plateau, curve leads to continuous discharge. The main limitation of *GM* counter is its fairly long insensitive time following the entry of each particle. This is the dead time during which the counter voltage drops below the threshold voltage due to the presence of massive positive ion cloud around anode. This causes electric field to decrease and hence ionisation stops for a while. The *GM* counter remains dead for about 300 microsecond after each ionizing event.

Quenching

The positive ions striking the cathode liberate electrons. Since the field at wire is no longer reduced because of positive ions already reaching the cathode. A single ionizing event may cause continuous discharge. The process of prohibiting the undesirable secondary pulse is called as Quenching. There are two ways of quenching known as internal and external quenching.

(*i*) **Internal quenching:** In this method a polyatomic gas like ethyl alcohal vapor is added to the argon gas with ratio of 90% argon to 10% ployatomic vapor. This method is also known as self quenching. Since the ionisation potential of ethyl alcohal is lower (11.3 eV) than that of argon gas (15.7 eV), the moving inert gas ions towards cathode transfers its charge to alcohal ions. These alcohal ions gets neutralised with electrons emitting from cathode. After being neutralised these alcohal ions dissociate and are got exhausted from counter. Hence there is no multiple pulsing and the discharge is quenched soon after the initial ionization.

(*ii*) **External quenching:** In this method voltage across anode and cathode is lowered by applying suitable electronic circuit. Therefore a large series resistance is used in the counting circuit. When a large current pulse is produced due to the secondary emission of electrons, there occur a large voltage drop across series resistance. This lowereing the potential difference between electrodes of the counter. Hence further ionisation is avoided.

Applications

It is highly sensitive detector. It is used for detecting *X*-rays, γ-rays and β-particles. It is one of the cheapest nuclear radiation detector.

II. Scintillation Detectors

This counter was originally designed by Rutherford for counting α-particles.

Principle

When a screen coated with zinc sulphide or barium platinocyanide exposed to α-particles, material produce light flashes or scintillation. These substances are called scintillators or luminiscent materials. This phenomenon of emission of light due to energy incidence in the form of radiation, forms the basis of working of this detector.

Construction and Working

In this detector scintillating nuclear radiation to be detected, is made to incident on scitillating material, which produces flashes of light. These flashes of light were viewed through a low power microscope. In the modern detector photomultiplier tube is used to view the scintillation. In this detector flashes of light is allowed to fall on photocathode of phtomultiplier tube and a pulse is extracted out to signal the passage of nuclear radiation. Actually, when light flashes incident on photocathode, it causes photoelectric emission. A number of anodes called dynodes are arranged in tube at increasing positive potential. These dynodes have specially prepared surface to give multiple secondary emissions. The arrangement is shown in the Fig.10.10.

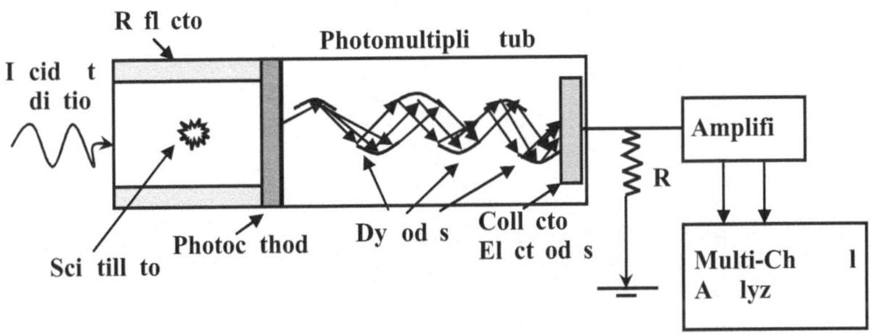

Fig. 10.10 *Schematic diagram of a Scintillation counter*

The elctrons emitted from the photocathode accelerate towards first dynodes. This causes the emission of secondary electrons. These electrons again then accelerate towards next dynodes causing higher secondary electron emission. This process goes on resulting into large multiplication of the electrons. The last dynode act as a collector. A wide variety of scintialltor are used to detect the nuclear radiation, some of them are described below:

(*a*) **Sodium iodide (Thallium activated):** This is the most varsatile of all the phosphors and is of outstanding importance for the study of γ radiation. But it is hygroscopic, so it must sealed in a aluminium can with reflecting of diffusing walls. The efficiency for γ radiation detection is many times greater than that of a GM counter.

(*b*) **Caesium iodide:** This material is not hygroscopic amd it is frequently used in preference to sodium iodide for the detection of low energy charged particles.

(*c*) **Zinc sulphide:** This is an excellent phosphor for particles of short range, but can't be used in thick layer since it rapidly becomes opaque to its own radiation.

(*d*) **Anthracene and stilbene:** Organic phosphors have a faster decay time than the inorganic phosphores but with a poorer efficiency, especially for heavy particles. It can detect β-particlesin the presence of γ-radiation.

(*e*) **Plastic and liquid scintillators:** In this scintillator energy is transferred from solvent to solute, which then re-emitts radiation in the wavelength range for which the solvent is transparent. Plastic scintillator are used in counter telescope which are generally used in high energy physics to define the path of particle.

(*f*) **Gases:** Xenon is particularly used when havy charged particle are to be counted in the presence of ã radiation. The main emmission of light takes place in the ultra violet region.

It has been obeserved that for sodium iodide and anthracene scintillators, the reponse is quite linear with electrons and protons as shown in Figure 10.11(*a*) and 10.11(*b*).

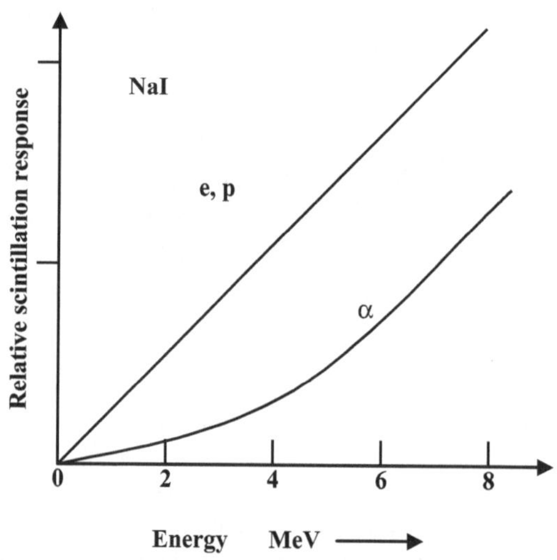

Fig. 10.11 (*a*) *Variation of scintillation response with energy for sodium iodide scintillator*

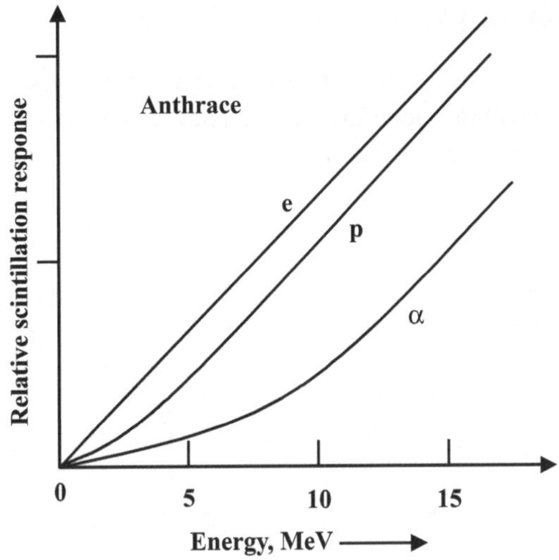

Fig. 10.11 *(b) Variation of scintillation response with energy anthracene scintillator*

But for heavy particle such as α-particles, it is nonlinear.

III. Solid State Detectors

The first solid state detector was designed by P. J Van Heerdan in 1945. He used crystal of silver chloride (intrinsic semiconductor) between two electrodes across which a potential difference is applied. When the crystal of silver chloride was exposed to nuclear radiation, electron hole pairs are generated. These charged particle move according to applied voltage and give current pulse. This simple solid state detector was used to limited extent only. Crystal counter have been improved with development of semiconducting materials like silicon and germanium crystal in recent years. These devices are now considered as efficient proportional conters for detection of heavy charged particles like protons, alpha particles fission fragments and can also detect electrons and γ radiation. With the introduction of impurities in these semicoductors intrinsic semiconductors converts into extrinsic semiconductors like *p*-type semicondutor and *n*-type semiconductor. Thus, impuritites in crytal changes charge carrier motion in applied electric field. Solid state detectore are of two types, *i.e.*, crystal detector and junction type detector.

(*a*) **Crytal detector**

Principle

When radiation of suitable energy is incident on semiconducting crystal electrons from valence band get excited to conduction band of the crytal and hence electron hole pair are generated. These charge carriers get drifted according to the applied voltage at the both ends of the crystal.

Constrution and working

A crytal counter consists of a semiconducting crystal made of silicon or gallium arsenide crytal with ohmic junctions at the ends as shown in Fig.10.12. The semiconducting material already contains an equilibrium population of current carrier which is given by

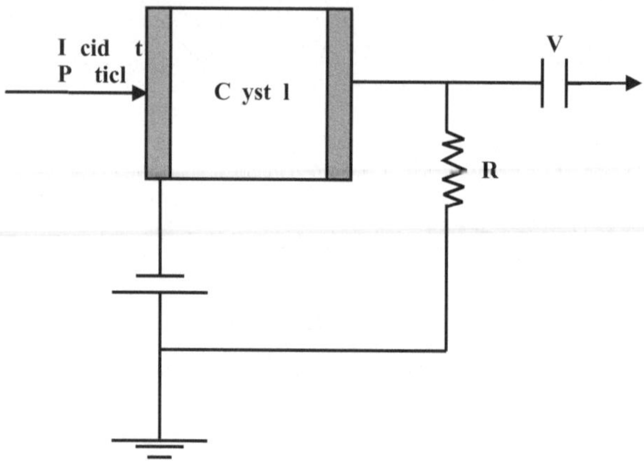

Fig. 10.12 *Schematic diagram of a crystal detector*

$$N = N_0 e^{-\frac{E_g}{kT}}$$

where E_g is the gap energy between the valence band and conduction band and T is the absolute temperature. When a particle of incident energy W is absorbed by the crystal, a large number of additional ion pairs are formed in it.

This number is given by

$$N = W/E_g$$

If an external electric potential is applied across two electrodes, these additional ion pairs will be collected by appropriate electrode within a short time across the resistor R in the external circuit. If all the carriers are collected efficiently without internal trapping or recombination, a proportional pulse corresponding to the energy loss of the particle in the material is observed. For this the initial equilibrium population N_e should be small and charge collection eficiency should be high.

(*b*) Diffused Junction Type Detector

Principle

In this detector the depletion region between *p-n* juntion diode plays an active role to detect the nuclear radiation. When nuclear radiation is passed through the detector, it produces electron hole pairs in the depletion region of revers biased *p-n* diode. The built in electric field of depletion region direct these charge carriers motion and corresponding current is measured in the electronic circuit.

Construction and Working

This type of detector consists of *p-n* junction diode. This type of diode is formed by diffusing *n*-type impurity into *p*-type silcon wafer. In the absence of external electric field electron and holes are majority charge carrier in *n*-type and *p*-type semi conductor and a depeltion region is formed between *p-n* juntion opposing charge carrier motion across the junction. Such a diode is shown in the Fig. 10.13. When diode is reverse biased, this depletionregion is further widened *i.e.*, width

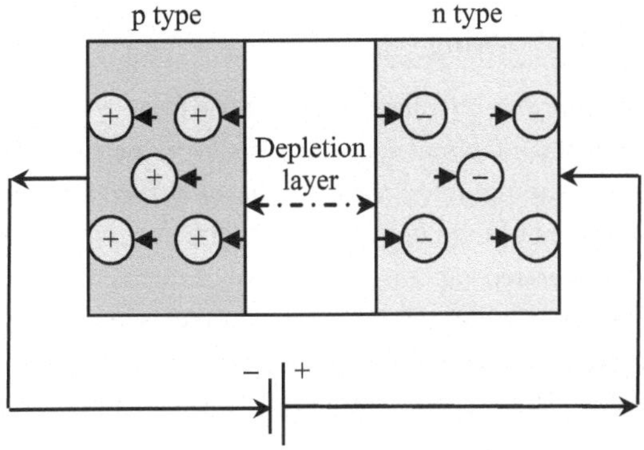

Fig. 10.13 *A reverse biased p-n diode*

of depletion region is increased with applied voltage. This is clear from the Fig. 10.14. Now when nuclear radiation is passed through it electron hole pairs are generated in this depletion region. These charge carrier swept across the juntion according to the built in electric field in depletion region. Hence, pulse representing the passage of radiation is measured in respective terminals.

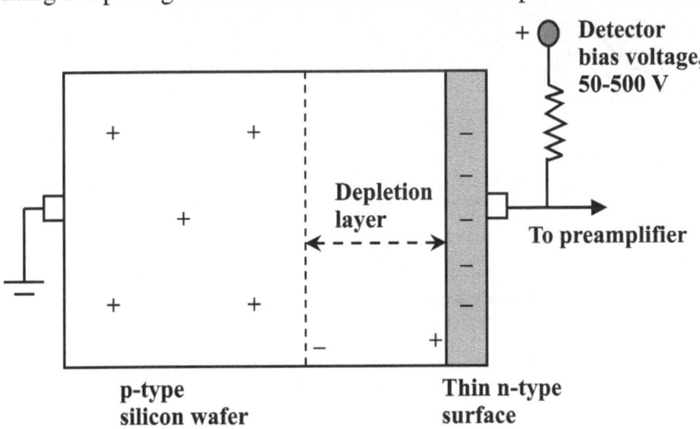

Fig. 10.14 *Schematic diagram of the semiconductor detector*

IV. Cloud Chamber

It was first invented by C. T. R Wilson, therefore, this chamber is named as Wilson Cloud Chamber. Over the years a wide variety of cloud chamber have been designed of various sizes and with different combination of gases and liquid mixtures at pressure upto about ten times of atmospheric pressure.

Principle

It is based on the principle that when a charged particles are passed through the supersaturated vapour, it produces ionization along the trajectory of the particle and droplets are formed on the ions due to the condensation of vapour.

Construction and Working

It is made of a cylindrical chamber with transoparent windows as shown in Fig. 10.15 The chamber contains air and water mixture which is cooled by adiabatic expansion to form the supesaturated vapour. The adiabatic expansion is produced by the controlled motion of a rubber diaphragm, this form one boundary of gas–vapour volume. If a saturated vapur is adiabatically expanded, it becomes unstable vapour. In this position when an ionizing particle is passes through such a supersaturated vapour, the vapour condenses on the ions which grow to form drops of visible size. This experiment shows as supersaturation increases, the negative ions first serve as centre of condensation. As the volume increases both positive and negative ions serve as the nuclei of droplets. A close array of fine droplets *i.e.*, a kind of linear cloud called a cloud track is formed. The trajectory of the particle is illuminated by a light beam incident through the walls of cylinder.

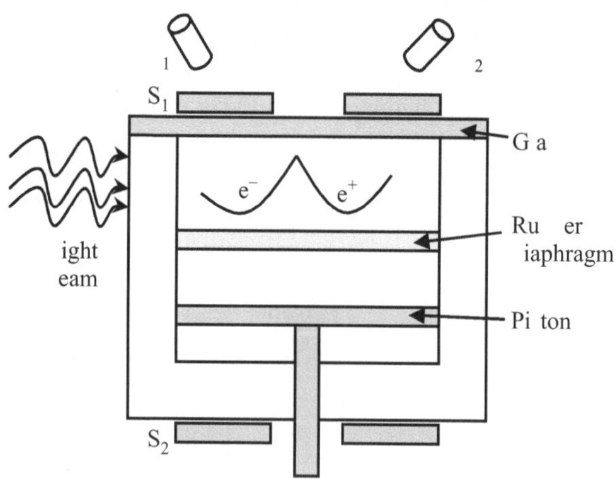

Fig. 10.15 *Schematic diagram of Wilson's cloud chamber*

If V_1, V_2 and T_1, T_2 are intitial and final volume and temperature of water vapours of the chamber respectively, then

$$T_2 = T_1 \left[\frac{V_1}{V_2}\right]^{\gamma - 1}, \qquad \text{where } \gamma = C_p/C_v.$$

Here, γ is the ratio of two specific heats of vapour. If initial and final pressure are p_1 and p_2,then

$$p_2 = p_1 \left[\frac{V_1}{V_2}\right]^{\gamma}$$

The ratio (p_2/p_1) is known as supersaturation. If this ratio exceeds the critical value (for water vapour it is 4.2 at 0°C) condensation occurs on drops carrying a charge and these grows. Externally applied electris field deflects the charged particle along the curved path. The trajectory of the particle is photgraphed by camera and recorded. The radius of curvature of the track is measured to estimate the momentum and energy of the incident particle and the nature of the particle can also be detected. Wilsons's cloud chamber had been earlier used for the study of cosmic radiation and had led to discovery of positron and electron, both particle have opposite curvature.

The cloud chamber suffers from following drawbacks.

(*i*) This device is too slow and the ionization tracks persists for quite some time and therefore a number of tracks are photographed.

(*ii*) The interaction of partcles with different materials can't be studied as clound chamber will not work with the introduction of additional material plates.

(*iii*) The chamber remains ineffective for considerable time.

Despite these drawbacks this chamber has made number of discoveries which are given below.

(*i*) Discovery of positrons, K^0 mesons, hyperons.

(*ii*) The muons were first discovered in secondary cosmic radiation.

(*iii*) The first accurate measurement of determination of the rest mass of the muon was made with its help.

V. The Bubble Chamber

The high energetic particle can't be studied in cloud chamber as these particles can not be completely observed in the cloud chamber. This has led to the development of Bubble Chamber. The bubble chamber was first designed by D. A. Glaser for the detection of charged particles. In the cloud chamber Wilson made use of the instability in supersaturated vapour while Glaser made use of instabilty in superheated liquid (at pressure and temperature such that it is below the boiling point).

Principle

A liquid normally boils with the evolution of bubble of vaopurs at the boiling point. If the liquid is heated under a high pressure to a temperature well above its normal point a sudden release of pressure would leave the liquid in a superheated state. If an ionizing particle is passed through the liquid within a few milliseconds after the pressure is released, the ions left in the track of a particle act as condensation centre for the formation of vapour bubbles.

Construction

Glaser first designed the apparatus using ether which was heated to 285°F under high presure. The schematic diagram of bubble chamber is shown in Fig. 10.16. An all glass chamber can not be used as it could not withstand high pressure changes. So, metal chamber chamber with glass windows was designed. It was found that rough spots on the metal surface and at the joints caused the liquid to start boiling. The rise in pressure due to the release of vapours by the boiling at the wall counteracted the reduction of pressure when the chamber was expanded. This made difficult to attain desired degree of superheating. This difficulty could be removed by expanding the chamber more quickly The first chamber built was two inch chamber of duralumin and glass with a diaphragm acturated by compressed air which could be fully expanded the chamber in five thousandth of seconds. The liquid remain sensitive for seven thousandth of seconds.

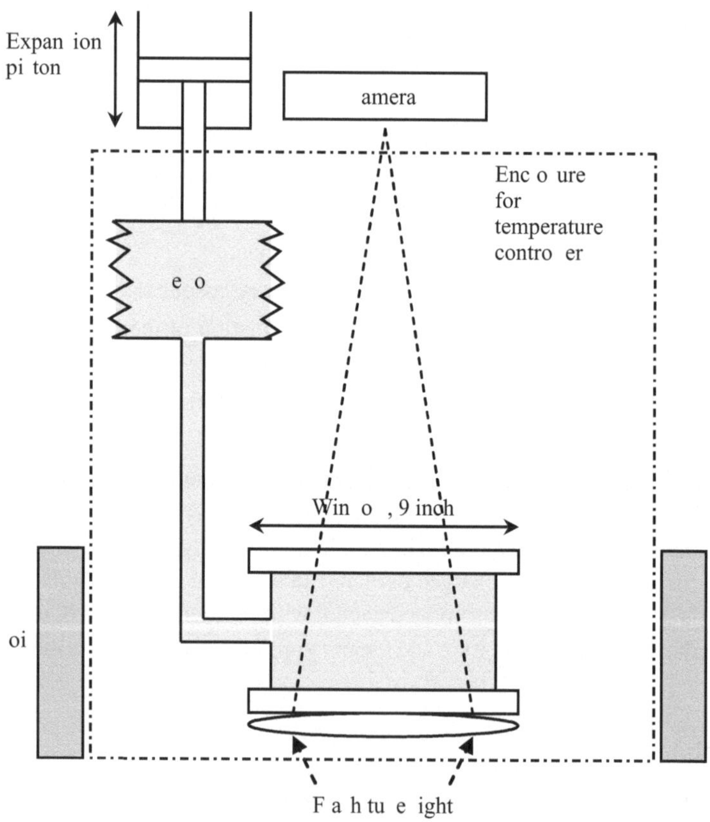

Fig. 10.16 *Schematic diagram of bubble chamber*

When charged particle are passed through the chamber the trajectory of particle is formed in the form of vapour bubbles whose size grow at a rapid rate so that it can attain a visible size in about 10 to 100 microseconds. The bubbles are visible under strong illumination and the cameras record the track of the particles.

The liquid used in the chamber should have following properties.

(*i*) It should be non-conducting so that ions would retain their charges.

(*ii*) It should have low surface tension so that the force tending to collapse a bubble would be weak.

(*iii*) It should have high vapour pressure which would tend to enlarge each bubble formed.

The greatest advantage of bubble chamber over cloud chamber is that it catches a large number of events withount any background effect. Another advantage is that it can be filled with any experimental liquid.

11

Superconductivity

The fascinating term of zero resistivity of superconductors make this the most wanted materials of the last century. Within a certain critical limits these materials practically shows zero resistivity along with its perfect diamagnetism. First property makes them a potential candidate for carrying electrical power of undiminished strength over very large distances, making very high power magnets etc. The anti-gravitation or levitation property of a diamagnetic material like superconductors becomes important for other use. Till now large amount of money had been spent in its research to have room temperature superconductor and research is still on. Here, we are going to discuss some important basic theories of superconductors and their uses.

1. INTRODUCTION

Superconductivity is one of the most fascinating physical phenomena having a vast potentiality in practical applications. In 1908 Dutch physicist Kammerlingh Onnes succeeded in his efforts to liquify Helium gas at 4.2 K at atmospheric pressure. Using liquid helium as coolant the variation of electrical resistance of metals at low temperature was studied by Onnes.

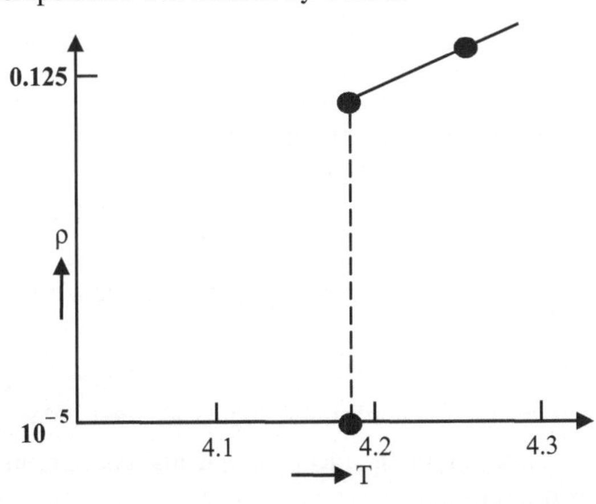

Fig. 11.1 *Variation of resistance of mercury with absolute temperature*

In 1911 Onnes discovered that the electrical resistivity temperature of highly purified mercury dropped abruptly to nearly zero at 4.15 K as shown in Fig.11.1 aside. This sudden drop of resistance was not in accordance with the expectations and was a completely new phenomenon **'Superconductivity'**. Later on this phenomenon was observed in Pb, Sn, Zn, Al and other metals.

- The sudden disappearance of electrical resistance in materials below a certain temperature is known as **'Superconductivity'**.
- The materials which exhibit superconductivity and which are in 'Superconducting' state are called **'Superconductors'**
- The temperature at which a normal materials turn into a superconductor is called the **'Critical temperature, T_C'**.
- Every superconductors have its own T_C at which it passes over into the superconducting state. A sudden fall in resistance for semiconductors indicates the transition to the superconducting state as shown in Fig. 11.2. Till 1973 the highest T_C was known for Nb_3Ge alloy as 23.2 K. In 1986 a T_C of 35 K was reported for La-Ba-Cu-O system. Till date many researchers have found $T_C > 100$ K in many superconducting systems. The big saving in the cost of these applications is due to the fact that liquid nitrogen (boiling point : 77 K) could be used to cool them rather than liquid helium (boiling point : 4.2 K), then the era of High T_C superconductor commenced.

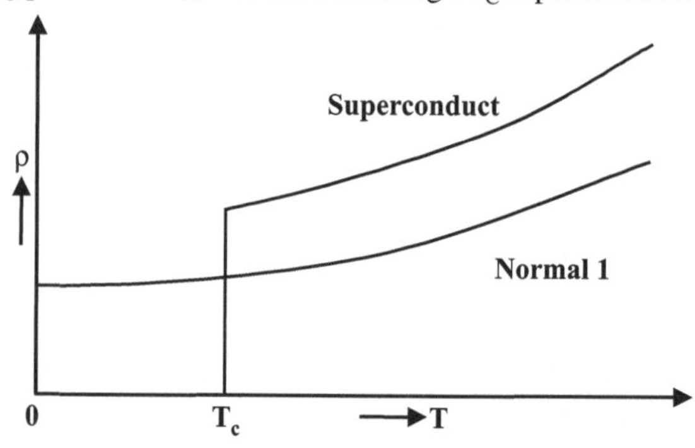

Fig. 11.2 *Variation of electrical resistivity with absolute temperature for normal metal and supeconductors*

Good electrical conductors are not superconductors. On the otherhand, the superconductors are very poor conductors at room temperatures. Berilium (Be), aluminium (Al), niobium (Nb), molybdenum (MO) and zinc (Zn) exhibit superconductivity. Semiconductors like Si, Ge, selenium (Se) and tellurium (Te) become metal at very high pressure and then at low temperature become superconductors.

- Those metals which are very good conductors at room temperatures, *i.e.*, Ag, Au, Cu, etc., are not at all superconductors even at 0.05 K.
- Some semiconductors (or chemical compounds) became superconductors at low temperatures.

- No general criteria for determining a material to become superconductors or not, has been established. The superconducting properties can change by varying (*i*), Temperature, (*ii*), Magnetic field, (*iii*), Stress, (*iv*) Impurity, (*v*) Atomic structure, (*vi*) Size, (*vii*) Frequency of excitation of applied electric field, (*viii*) Isotope mass.

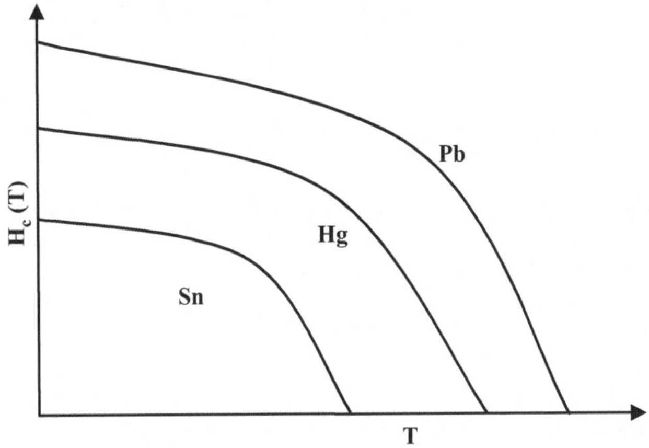

Fig. 11.3 *The variation of critical magnetic field with absolute temperature*

2. EFFECT OF MAGNETIC FIELD

If a magnetic field is applied parallel to the length of the superconducting wire, the resistance of the wire is suddenly restored at a finite value of the magnetic field strength and hence the superconductivity will disappear. This magnetic field is known as 'Critical Field' and depends on temperature. This is denoted by $H_C(T)$. This restoration of resistance is abrupt if the metal is perfectly pure and free from stains. Thereby at $T = T_C$, $H_C(T_C) = 0$. The variations of H_C with T are shown in Fig.11.3 for different materials. The curves are nearly parabolic and can be represented by $H_C = H_0\left[1 - \frac{T^2}{T_C^{\,2}}\right]$, where H_0 is the critical field at $T = 0\,K$.

3. PERSISTENT CURRENT

When we start a current to pass through a close loop of superconducting material, the current will continue to keep flowing on its own till the temperature of the loop kept at below the critical temperature. This current of constant magnitude is known as persistent current. It has been found theoretically that once the current flow is started it will continue for more than 10^5 years. This is very useful for superconducting magnets as it does not require any power supply to maintain its magnetic field.

4. HEAT CAPACITY

For any superconducting metal its transition below critical temperature to the superconducting state does not involve any crystallographic structural changes. This is only a thermodynamic phase change. Here, the specific heat (C_v) changes discontinuously at the critical temperature. This variation of C_v is given by $C_v = e^{-bT_c/T}$, where b is constant.

5. ISOTOPE EFFECT

Reynolds and Maxwell observed the variation of T_C with isotopic mass, M, as

$$T_C \propto M^{-1/2} \Rightarrow T_C M^{1/2} = \text{constant}.$$

Flux Exclusion : The Meissner Effect

In an ordinary conductor the electric field is given by $E = \frac{1}{\sigma} J = \rho J$, where J is the current density and ρ is resistivity of the conductor. So for a perfect conductor, $\rho = 0$. Hence, $E = 0$.

If such a perfect conductor is placed in a magnetic field and the magnetic field is withdrawn, the changing magnetic flux will not induce an e.m.f (e) in it.

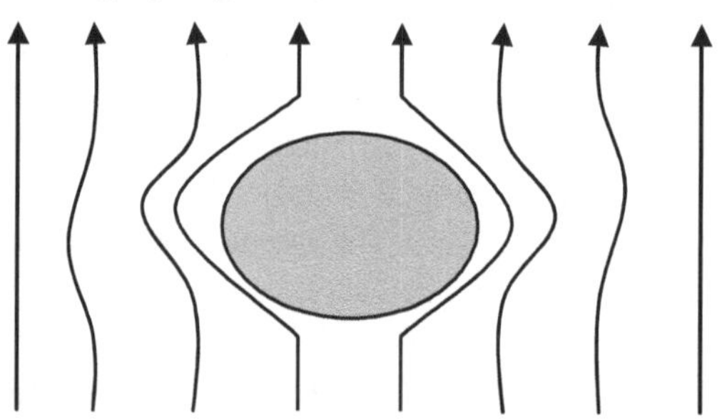

Fig. 11.4 *Exclusion of magnetic flux in Meissner Effect*

From Faraday's Law,

$$e = -\frac{d\phi}{dt}.$$

So when $e = 0$, ϕ becomes constant. So, the magnetic flux through the perfect conductor remains unchanged and the applied magnetic field is frozen or trapped in the perfect conductor. Also if a perfect conductor is taken into a magnetic field the magnetic field lines are pushed aside as shown in Fig. 11.4. A superconductor behaves in same way as with perfect conductor when placed into a magnetic field. It does not allow field lines to penetrate into it.

However, in 1933 Meissner found that a superconductor completely expels any magnetic field lines that were initially penetrating in the normal state. The magnetic flux was totally expelled from the sample when cooled beyond T_C.

This expulsion of magnetic flux during the transition from the normal to the superconducting state is called the **Meissner Effect**.

It shows that for superconductors both $\frac{dB}{dt} = 0$ and $B = 0$, independent of the path by which the state is reached.

Perfect Diamagnetism

A superconductor behaves as a perfect diamagnetic material as $B = 0$.

So, $$B = \mu_0 (H + M) = 0.$$

$\Rightarrow$ $$M = -H$$

Hence, magnetic susceptibility,

$$X = \frac{M}{H} = -1.$$

Because of diamagnetic nature all superconducting materials strongly repel external magnets.

$\Rightarrow$ **Levitation Effect,**

$\Rightarrow$ **Suspension Effect.**

Here, a piece of superconductor hangs beneath a magnet. This is a sure test for superconductor.

6. LONDON EQUATIONS

To explain Meissner effect in superconductors London (and London) in 1935 proposed a modification of the electromagnetic wave equation. The new equations besides explaining Meissner effect also explain observations on thin film superconductors. These equations are known as London equations. It was assumed that there are two type of electrons, namely 'normal' and 'super electron' in superconductors. Let n_n and n_s are the densities of normal and super electrons respectively, the conduction electron n becomes

$$n = n_n + n_s.$$

The normal current and super current are assumed to flow in parallel. Super current flows without resistance, while normal electron flow remains quite inert and is thus ignored in the superconducting state.

The force equation for super electrons is

$$m\frac{d\vec{v}}{dt} = -e\vec{E} \qquad ...(11.1)$$

Here, m and v are the mass and velocity of electron respectively, $-e$ is electronic charge and E is the applied electric field.

Now, current density in the sample is

$$\vec{J} = -ne\vec{v} \qquad ...(11.2)$$

Here, n is no. of electrons per unit volume.

Hence, on differentiation with respect to time from eqn. (11.2),

$$\frac{d\vec{J}}{dt} = -ne\frac{d\vec{v}}{dt}$$

$$\Rightarrow \qquad \frac{d\vec{J}}{dt} = -ne\left(-\frac{e\vec{E}}{m}\right)$$

$$\Rightarrow \qquad \frac{d\vec{J}}{dt} = \frac{ne^2}{m}\vec{E} \qquad ...(11.3)$$

For $E = 0$, J becomes constant. This is the basic relation describing the absence of resistance and is known as the 1st London equation. It states that no electric field is necessary unless the current changes, steady currents may be set up in the absence of electric field. This is the phenomenon of 'Superconductivity'. Superconductors consist of both normal and superconducting electrons, we are considering only the superconducting electrons here.

From Maxwell's equations,

$$\vec{\nabla}\times\vec{E} = -\frac{d\vec{B}}{dt} \qquad ...(11.4)$$

$$\text{and,} \qquad \vec{\nabla}\times\vec{H} = \vec{J} \qquad ...(11.5)$$

Taking curl of both sides of eqn.(11.3),

$$\Rightarrow \qquad \vec{\nabla}\times\frac{d\vec{J}}{dt} = \frac{ne^2}{m}\left(\vec{\nabla}\times\vec{E}\right) = \frac{ne^2}{m}\left(-\frac{d\vec{B}}{dt}\right) \quad \text{(using equation (11.4))}$$

$$\Rightarrow \qquad \vec{\nabla}\times\left(A\frac{d\vec{J}}{dt}\right) = \left(-\frac{d\vec{B}}{dt}\right), \left[\text{where } A = \frac{m}{ne^2}\right] \qquad ...(11.6)$$

On differentiation of eqn.(11.5) with respect to t,

$$\frac{d\vec{J}}{dt} = \vec{\nabla}\times\frac{d\vec{H}}{dt} \qquad ...(11.7)$$

Substituting eqn.(11.7) into eqn.(11.6),

$$\vec{\nabla} \times \left(A . \vec{\nabla} \times \frac{d\vec{H}}{dt} \right) = -\mu_0 \frac{d\vec{H}}{dt} \qquad \left(\text{as } \vec{B} = \mu_0 \vec{H}\right)$$

$$\Rightarrow A . \vec{\nabla} \times \left(\vec{\nabla} \times \frac{d\vec{H}}{dt} \right) = -\mu_0 \frac{d\vec{H}}{dt}$$

$$\Rightarrow \quad A\vec{\nabla} . \left(\vec{\nabla} . \frac{d\vec{H}}{dt} \right) - A\nabla^2 \left(\frac{d\vec{H}}{dt} \right) = -\mu_0 \frac{d\vec{H}}{dt}$$

$$\Rightarrow \quad A\vec{\nabla} . \left(\frac{d\left(\vec{\nabla} . \vec{H}\right)}{dt} \right) - A\nabla^2 \left(\frac{d\vec{H}}{dt} \right) = -\mu_0 \frac{d\vec{H}}{dt}$$

$$\Rightarrow \quad 0 - A\nabla^2 \left(\frac{d\vec{H}}{dt} \right) = -\mu_0 \frac{d\vec{H}}{dt} \qquad \left(\text{as } \vec{\nabla} . \vec{H} = 0\right)$$

$$\Rightarrow \quad A\nabla^2 \left(\frac{d\vec{H}}{dt} \right) = \mu_0 \frac{d\vec{H}}{dt} \qquad ...(11.8)$$

Integrating with respect to time,

$$\nabla^2 A\left(\vec{H} - \vec{H_0}\right) = \mu_0 \left(\vec{H} - \vec{H_0}\right) \qquad ...(11.9)$$

Here, H_0 is the constant of integration and also the field at $t = 0$. Equation (11.9) is a direct consequence of Maxwell's equation. However, from Meissner Effect, at T_C we have zero magnetic field, *i.e.*, at $t = 0$, $H_0 = 0$. So, according to Londons we should eliminate $\vec{H_0}$.

Hence, instead of eqn. (6), the correct equation will be by omitting time derivative, as,

$$\vec{\nabla} \times \left(A . \vec{J}\right) = -\vec{B} = -\mu_0 \vec{H} \qquad ...(11.10)$$

This is the fundamental equation in a superconductor as it gets rid of $\vec{H_0}$ term. This is also called the (second) **London equation**. Sometimes, equation (10) along with equation (3) together is known as **London equations.**

Explanation of Meissner Effect and Flux Penetration from London Equations

From Maxwell's Eqn. as eqn.(5),

$$\vec{\nabla} \times \vec{H} = \vec{J} \quad \Rightarrow \quad \vec{\nabla} \times \vec{B} = \mu_0 \vec{J}$$

Taking curl of both sides,

$$\vec{\nabla} \times \left(\vec{\nabla} \times \vec{B}\right) = \mu_0 \left(\vec{\nabla} \times \vec{J}\right)$$

$$\Rightarrow \vec{\nabla}.\left(\vec{\nabla}.\vec{B}\right)-\nabla^2\vec{B} = \mu_0\left(\vec{\nabla}\times\vec{J}\right)$$

$$\Rightarrow \qquad -\nabla^2\vec{B} = \mu_0\left[-\frac{ne^2}{m}\vec{B}\right] \qquad \left(\text{from } \vec{\nabla}.\vec{B}=0 \text{ and eqn. } (11.10)\right)$$

$$\Rightarrow \qquad \nabla^2\vec{B} = \frac{ne^2\mu_0}{m}\vec{B}$$

$$\Rightarrow \qquad \nabla^2\vec{B} = \frac{1}{\lambda^2}\vec{B} \qquad \text{...(11.11)},$$

where, $$\lambda = \frac{1}{\sqrt{\dfrac{ne^2\mu_0}{m}}}.$$

Equation (11.11) is a simple differential equation with a solution

$$B = B_0 e^{-\frac{x}{\lambda}} \qquad \text{...(11.12)}$$

in one dimension.

Here, B_0 is magnetic flux density at surface and x is the distance from the surface into the specimen. The solution in equation (11.12) agrees quite well with the experimental data. Experimental decrease of the magnetic field inside a superconductor is also shown in Fig. 11.5.

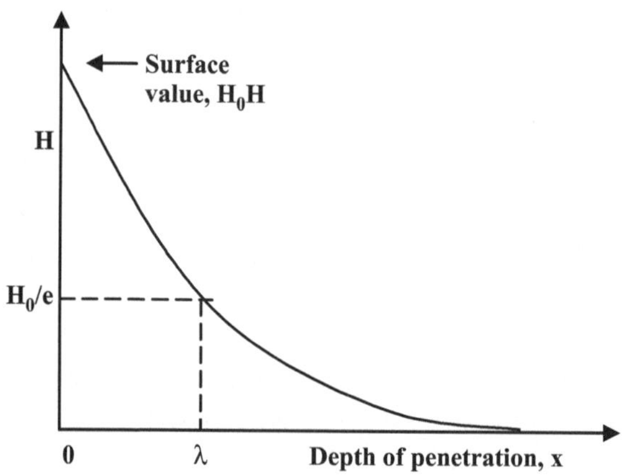

Fig. 11.5 *Penetration of magnetic flux lines on application of external magnetic field*

- Above equation shows that the magnetic field does not drop abruptly to zero into the specimen, but decays inside the superconductor through a distance (characteristic of the material) λ, called **London Penetration Depth.**
- The flux density decays exponentially in a superconductor falling (1/e)th of its value at the surface at a distance λ, called **London Penetration Depth.**

7. JOSEPHSON EFFECT

The electronic wave function has always a finite probability for quantum mechanical tunneling through a barrier. Normally, significant tunneling is possible for those electrons whose de-broglie wavelengths are comparable to or greater than the barrier thickness (normally, 50 $A°$ or less). In 1962 Josephson showed that tunneling of superconducting electrons (*i.e.*, Cooper pairs) was possible in the same way of the tunneling of normal electrons. Esaki, Giaver and Josephson received the Nobel prize in 1973 for their work on electron tunneling.

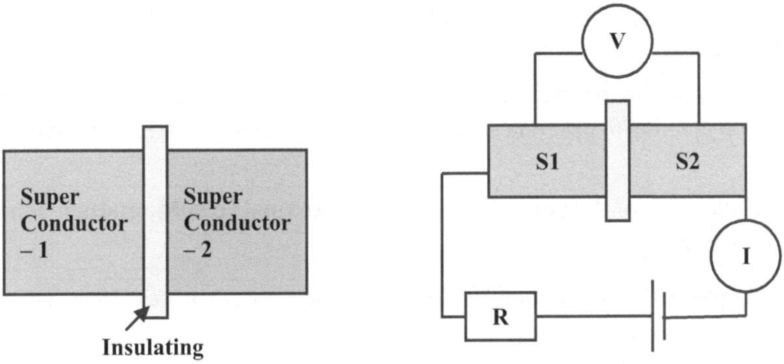

Fig. 11.6 (*a*) *Schematic diagram of a Josephson junction*
(*b*) *Explanation of Josephson effect in the junction*

Figure 11.6(*a*) showed a schematic arrangement of Josephson junction. It consists of two superconductors separated by a thin (10 – 20 $A°$) barrier insulating layer. As shown in Figure 11.6(*b*) when a dc current flows through this junction, the potential difference across its ends is zero. The current flows both ways, one through the superconductor and other across the junction. This is known as the D.C. Joshephson Effect. The current flows by tunneling of Cooper pairs from one superconducting film into the other while penetrating the barrier.

The super current passing through the junction is given by

$$I_S = I_C \sin \theta$$

Here, θ is the phase difference between the wave functions which describe the Cooper pairs on the both sides of the barrier interface, I_C is the critical current supported by the junction and is dependent on thickness, width of the interface. If I_S exceeds I_C large amount of current passes through the junction and yields a potential difference, V, between the two superconducting films (normally, then the barrier potential between the Cooper pairs of both sides is ≈ 2 eV). This causes a frequency difference, ν, and is given by

$$\nu = \frac{2eV}{h} \qquad ...(11.13)$$

and, phase difference is given by,

$$\theta = 2\pi\nu t = 2\pi t\left(\frac{2eV}{h}\right) \quad ...(11.14)$$

Hence using the equations (13) and (14), we have

$$I_S = I_C \sin\left(\frac{4\pi eV}{h}t\right).$$

This current is an alternating current and it exists only when a dc voltage exists across the junction. This is known as the AC Josephson Effect.

When $V = 1\ \mu V$, an AC current of frequency 483.6 MHz is developed across the junction. Due to AC Josephson Effect, we have now a new definition of volt.

One volt is now defined as the potential difference when applied across a Josephson junction to generate an electromagnetic radiation of 483597.9 GHz.

Classifications of Superconductors (Based on their Magnetic Field behavior)

(a) Type-I : They obey complete Meissner Effect, i.e., they are perfect diamagnetic as shown in figures below. The variation of resistivity and magnetization with externally applied magnetic field are shown in Fig.11.7. Pure specimen of many materials shows this behavior. Below H_C the specimen is superconducting abruptly and the magnetic field fully absent. They are often called soft because of their tendency to give away to low magnetic fields.

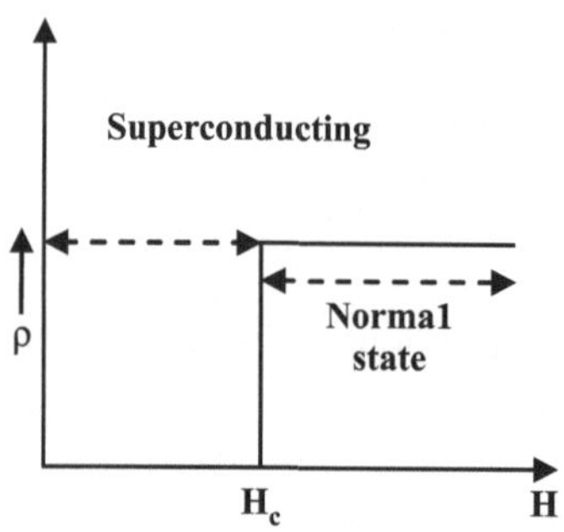

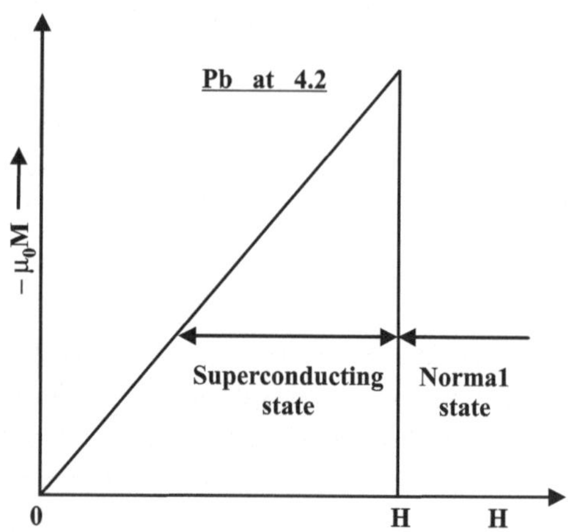

Fig. 11.7 *(a) Variation of resistivity with applied magnetic field*
(b) Variation of magnetization with applied magnetic field

(b) Type-II : These materials exhibit a magnetization as shown in figures below and are also known as hard superconductors. The variation of resistivity and magnetization with externally applied magnetic field are shown in Fig.11.8.

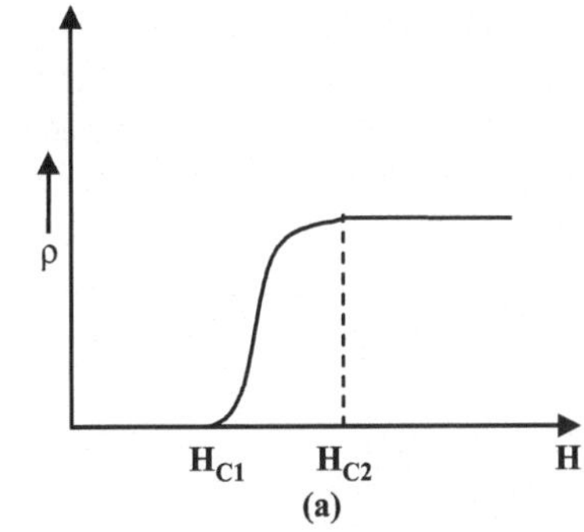

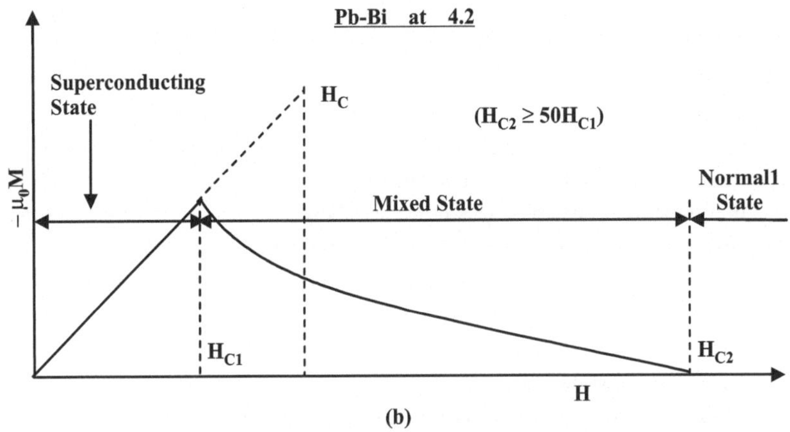

Fig. 11.8 *(a) Variation of resistivity with applied magnetic field*
(b) Variation of magnetization with applied magnetic field

For $H < H_{C1}$, the material is diamagnetic and $B = 0$. H_{C1} is known as the Lower Critical Field. At H_{C1} the flux begins to penetrate the specimen and penetration increases until H_{C2} is reached. Here, $M = 0$ and specimen becomes a normal material. This H_{C2} is known as the Upper Critical Field. Between H_{C1} and H_{C2}, $B \neq 0$ and the Meissner effect is said to be complete. There is a gradual fall in M as H increases in Type-II. In this region, between H_{C1} and H_{C2} the superconductivity is partially destroyed and is known as 'Mixed State'.

8. ELEMENTS OF BCS THEORY

The modern theory of superconductivity was put forward by Bardeen, Cooper and Schrieffer in 1957 and hence name as 'BCS theory'. This theory successfully explained all observable effects such as zero resistivity, the Meissner effect, isotope effect and some other phenomenon. Important aspects of this theory are discussed below in brief and qualitative manner.

I. Electron – Electron Interaction Via Lattice Deformation

In order to understand the direct interaction, let us consider an electron passing through the packing of positive ions. It is attracted by the neighbouring positive ions and gets screened by them. This screening reduces effective charge of electron. At the same time due to attraction between the electron and ion core, the lattice gets deformed on local scale. This deformation is greater for smaller mass of positive ions core. Second electron passing through this assembly see deformed lattice and gets attracted towards the assembly. This interaction is called a electron-electron interaction via lattice deformation. In the language of field theory, the above interaction is said to be due to exchange of virtual phonon "q" between two electrons. In terms of the wave vector k of the two electrons the process can be written as

$$k_1 - q = k_1' \text{ and } k_2 + q = k_2'$$

which gives net wave vector ($k_1 + k_2 = k_1' + k_2'$) as zero.

II. Cooper Pair

At absolute zero temperature all quantum states with energies $E \leq E_f$ are filled by electrons. In this situation ,cooper showed that if there is an attraction between the two electrons they are able to form a bound state so that their total energy is less than $2E_f$.These electron are paired to form a single system and their motion are correlated. These two electrons together form a Cooper pair and is known as a Cooper electron. These pair of electrons is in fact super electrons which are responsible for superconductivity.

III. BCS Ground state

In superconducting materials excitation is different from normal metals. In this case when a pair of electron lying just below the Fermi surface is taken just above it , they form a Cooper pair and their total energy is reduced , it can be done for other pairs until the system can gain no additional energy by pair formation and hence the total energy is further reduced.

9. APPLICATIONS

(*i*) Generation and transmission of electricity (saves lots of power loss)

(*ii*) Medical diagnostics, *e.g.*, MRI, NMR

(*iii*) Enormously powerful electromagnets

(*iv*) Supercomputers and information processing

(*v*) Magnetically levitating the world's fastest trains

(*vi*) Magnetic energy storage

(*vii*) Electromagnetic shielding

(*viii*) Superconducting transformers.

Appendix

1. INTERFERENCE IN WEDGE SHAPED THIN FILM

Let us consider a wedge shaped thin film of refractive index μ enclosed by two plane surfaces *OP* and *OQ* inclined at an angle θ. The thickness of film increases from *O* to *Q*. When a ray RA incident at an angle ion film, it produces two reflected ray *AS* and *DF* from wedge shaped film shown in Fig.1.

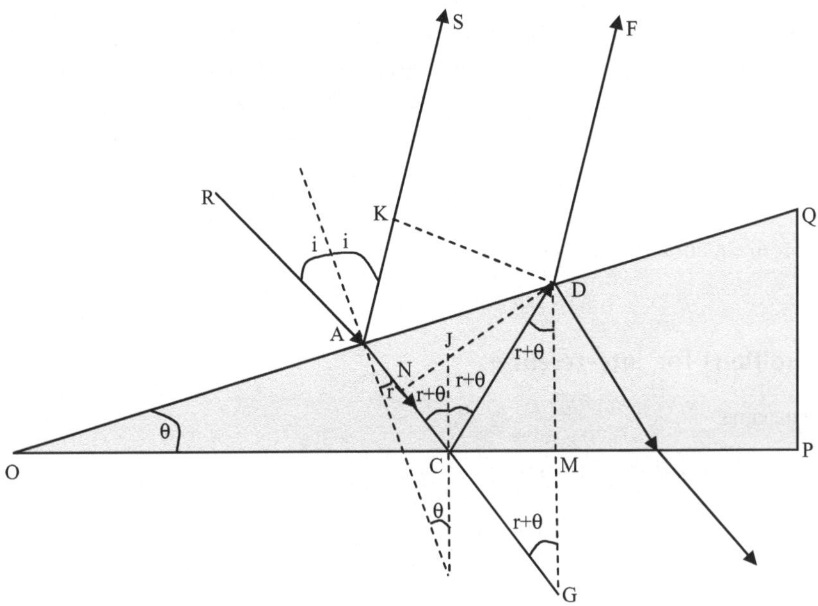

Fig.1 *Interference by reflected light from wedge shaped thin film*

The path difference between these two rays *AS* and *DF* is given by

$$\Delta = (AC + CD) \text{ in film} - \text{AK in air}$$
$$= \mu\,(AC + CD) - AK$$
$$= \mu\,(AN + NC + CD) - AK \qquad ...(1)$$

From geometry, $\angle ADK = i$, $\angle ADN = r$

$$\mu = \frac{\sin i}{\sin r} = \frac{AK/AD}{AN/AD} = \frac{AK}{AN}$$

$$\Rightarrow \qquad AK = \mu AN \qquad ...(2)$$

From eqns.(1) and (2)

$$\Delta = \mu\,(AN + NC + CD) - \mu\, AN$$
$$= \mu\,(NC + CD) \quad ...(3)$$

DM is perpendicular to OP producing AC such that it meets at G

In ΔCDM and ΔCMG

$$\angle DMC = \angle CMG = 90°,$$
$$\angle CDM = \angle MGC = r + \theta$$

and CM is common, so ΔCDM and ΔCMG are congruent.

Thus, $DM = MG = t$ and $CD = CG.$

Substituting value of CD in equation (3) we have

$$\Delta = \mu NC + \mu CG = \mu(NC + CG)$$
$$= \mu NG \quad ...(4)$$

As angle between OP and OQ is θ, then angle at normal to C for ray AC should be $(r + \theta)$. From the Fig.1 ACG cuts the parallel lines CJ and DG, so we have

$$\angle ACJ = \angle CDG = r + \theta$$

In ΔNDG

$$NG = DG \cos(r+\theta) = 2t\cos(r+\theta) \quad ...(5)$$

From equation (5) and (4)

$$\Delta = 2\mu t\cos(r+\theta) - \lambda/2 \quad ...(6)$$

Conditions for interference:

For maxima:

$$2\mu t\cos(r+\theta) - \lambda/2 = n\lambda$$

$$2\mu t\cos(r+\theta) = \frac{(2n+1)}{2}\lambda, \text{ where } n = 0, 1, 2, 3, 4 \ldots$$

For minima:

$$2\mu t\cos(r+\theta) - \frac{\lambda}{2} = \frac{(2n-1)}{2}\lambda$$

$$2\mu t\cos(r+\theta) = n\lambda, \text{ where } n = 0, 1, 2, 3, 4 \ldots$$

Nature of Interference pattern:

When a parallel beam of monochromatic light is incident on wedge shaped film μ, θ, r remain constant at each point of the film, but thickness of the film varies for interfering reflected light from the edge of the film. Since the locus of all points of constant thickness is straight line in inclined plane parallel to the edge of the film, therefore, the interference pattern comes out in the form of straight line.

From equation (6) the film will appear bright if

$$t = \frac{(2n+1)\lambda}{4\mu\cos(r+\theta)} \quad \text{where } n = 0, 1, 2, 3, 4 \ldots$$

i.e.,

$$t = \frac{\lambda}{4\mu\cos(r+\theta)}, \frac{3\lambda}{4\mu\cos(r+\theta)}, \frac{5\lambda}{4\mu\cos(r+\theta)}, \ldots$$

Film will appear dark if

$$t = \frac{n\lambda}{2\mu\cos(r+\theta)} \quad \text{where } n = 0, 1, 2, 3, 4 \ldots$$

i.e.,

$$t = 0, \frac{\lambda}{2\mu\cos(r+\theta)}, \frac{2\lambda}{2\mu\cos(r+\theta)}, \ldots$$

Since at the edge of the wedge $t = 0$, therefore, film appear dark as it satisfies the condition of destructive interference. Afterwards as the thickness of the film increases, pattern appears in the form of alternative dark and bright fringes parallel to the edge of the film.

2. DISPERSION IN AN OPTICAL FIBER

When a light pulse is launched into a fiber, its amplitude decreases while traveling along the fiber. Thus the pulse becomes wider at output end. Such a distortion in pulse occurs due to dispersion effect. It is measured in nanoseconds per kilometer. This is shown in Fig.2.

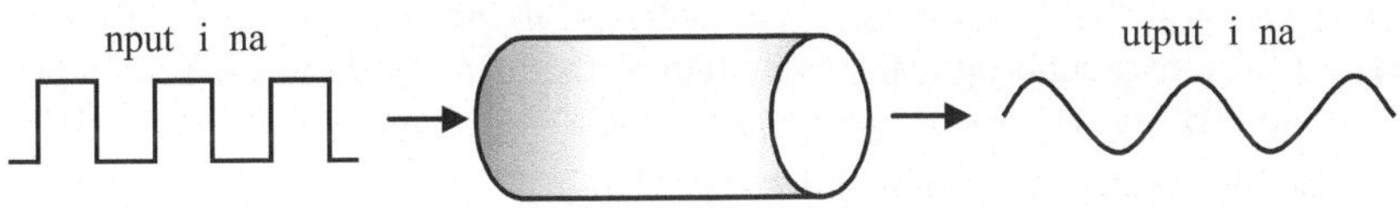

Fig. 2 *Signal broadening in a fiber*

Following mechanisms are responsible for distortion in light pulse in a fiber.

(i) Material Dispersion

Here the light moves of different wavelength travel at different speeds in a medium. Short wavelength pulse travels slower than larger wavelength pulse. This is known as material dispersion. The amount of material dispersion (D_m) can be calculated by following relation

$$D_m = \frac{\lambda\Delta\lambda}{c} L \frac{d^2n}{d\lambda^2} \qquad \ldots(7)$$

Here λ, $\Delta\lambda$, L, n are peak wavelength, spectral width, length of fiber core and refractive index of core respectively.

(ii) Waveguide dispersion

This dispersion arises due to variation of refractive index for any mode with wavelength. The amount of waveguide dispersion is governed by an equation similar to equation (7) by replacing refractive index with effective refractive index.

(iii) Inter-modal dispersion

When a large number of modes are propagating in a fiber, they travel with different velocities with respect to fiber axis. Part of the wave arrives at the output before other parts leading to a spread of the input pulse. This is known as inter-modal dispersion. It does not depend upon spectral width ($\Delta\lambda$) of source.

In MMF all three dispersion occur while in SMF only material and waveguide dispersion occur. Low Numerical Aperture (NA) fibers exhibit smaller dispersion while high NA fibers show longer dispersion. Therefore GRIN fibers are preferred for low dispersion property. But GRIN fiber fabrication is very expensive. This dispersion limits the bandwidth of fiber. The information flow i.e., pulse rate, must be slow enough that dispersion does not cause adjacent pulse to overlap.

3. ATTENUATION IN THE OPTICAL FIBER

Signal attenuation (also known as fiber loss or signal loss) is one of the most important properties of fiber for determining repeater less separation between a transmitter and a receiver. Since repeaters are very expensive to fabricate, maintain and install therefore low attenuation is always desirable to reduce the system cost. Signal attenuation is usually expressed in terms decibel. The decibel may be defined for a particular optical wavelength as the logarithmic ratio of the input (transmitted) optical power P_i into a fiber to the output (receiver) optical power P_o from the fiber which is given by following relation,

$$\text{Number of decibel} = 10 \log_{10} \left(P_i / P_o\right)$$

A signal propagating through a fiber gets progressively attenuated. The signal attenuation can be expressed in terms of decibel per kilo meter length of a fiber (dB/Km) and it is defined as the logarithmic ratio of the output optical power from a fiber of length L to the input optical power. Thus if α is the signal attenuation per unit length in decibel then

$$\alpha = \frac{10}{L} \log_{10} \left(P_i / P_o\right)$$

Where P_i is the power of optical signal launched at the one end of fiber and P_o is the power of optical signal emerging from the other end of fiber. In case of ideal fiber we should have,

$$P_o = P_i,$$

so $\alpha = 0$ dB/Km for ideal fiber.

In normal practice typical attenuation is 3dB/Km. Attenuation is dependent on a wavelength of light signal propagating in fiber.

There are following three mechanisms responsible for attenuation in optical signal.

(*i*) Absorption by atomic defect in glass composition.

(*ii*) Extrinsic absorption by impurity atoms in the glass material.

(*iii*) Intrinsic absorption by the basic constituent atoms of fiber material.

Usually absorption losses by atomic defects in fiber are negligible as compared to extrinsic and intrinsic absorption loss.

4. ULTRASONIC WAVES

Sound waves are mechanical longitudinal waves which can be generated in all forms of the matter. These waves can be divided into three classes according to their frequencies.

(*i*) Infrasonic waves:- These wave frequency lies below 20 Hertz. Our ear is not sensitive to these waves and is produced by large vibrating bodies such as during an earthquake.

(*ii*) Audible waves:- These longitudinal waves frequency lies in the range from 20 Hertz to 20,000 Hertz and produce sensation in our ears. These waves are produced by vibration bodies such as vocal cords (human and animal voice) stretched string (violin, guitar, etc) air columns and also by irregular motions of the bodies like doors, tree leaves, etc. Regular motion of vibrating body produces pleasant effect while irregular motion of vibrating body produce disturbance to our ears.

(*iii*) Ultrasonic waves:- These longitudinal waves frequency lies above 20,000 Hertz and do not produce sensation to our ears. These waves can be produced by high frequency vibration of quartz crystal under an electric field (piezo-electric effect) or by vibrations of ferromagnetic rod under an alternating magnetic field (magneto-striction effect).

Apart from above classes, the longitudinal waves having frequency more than 10^8 Hertz are called hypersonic waves.

These longitudinal waves having frequency above the upper limit of audibility are called ultrasonic or supersonic waves. Their wavelength is H″1 cm in air.

Production of ultrasonic waves.

There are two methods of production of ultrasonic waves.

I. **Piezoelectric method:** This method is based on piezoelectric effect. The word piezoelectric means pressure + electricity. There are different kinds of piezoelectric effect. In one kind of effect if certain crystal like quartz, rochelle salt, tourmalline, etc. are stretched or compressed along certain axis (known as mechanical axis), an electric potential difference is produced along a perpendicular axis (known as electric axis) *i.e.*, in this effect mechanical vibration (pressure) are converted into electric signal.

The converse of this effect is also true *i.e.*, when a crystal is subjected to alternating potential difference, it generate mechanical vibrations perpendicular o electric axis. When a crystal such as quartz, rochelle salt or tourmaline is subjected to an alternating potential difference, it is set into mechanical vibrations. If the frequency of electrical oscillations coincides with the natural frequency of the crystal, a large amplitude of vibrations results. This phenomenon is utilized to generate ultrasonic waves

The experimental arrangement for generating this effect is shown in figure.

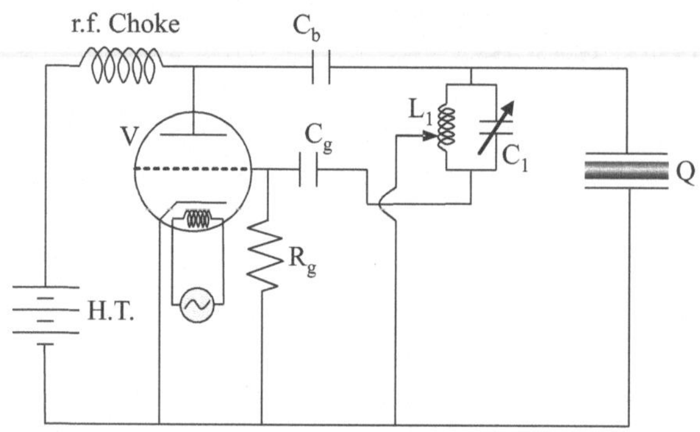

Fig. 3 *The Experimental arrangement in a Piezoelectric Crystal Oscillator*

The alternating voltage to crystal is applied by hartley oscillatory circuit. This circuit consists of a tuned oscillatory circuit (inductance L_1 and variable condenser C_1 in parallel). This tuned circuit is connected the triode value V. the coil L_1 is tapped and connected to the cathode. The quartz crystal is connected in parallel of variable condenser C_1.

The proper grid bias is obtained by grid leak resistance R_g and grid condenser C_g. The dc voltage is applied to the plate through radio frequency choke. C_b is a blocking capacitor which prevents the direct current to pass through the tank circuit and by pass radiofrequency currents. The capacity of variable condenser C_1 is adjusted so that the frequency of the oscillatory circuit is tuned to the natural frequency of quartz crystal. A large amplitude vibration of crystal takes place and ultrasonic waves are produced. This method can produce ultrasonic waved upto 500k Hz with a moderate size crystal.

II. Magnetostriction effect: This method is based on the phenomenon of magnetostriction. According to this phenomenon when a rod of ferromagnetic material (such as nickel) placed in a magnetic field parallel to its length, change in the form contraction or extension of rod takes place. This change is independent of the sign of the field and depends on the magnitude of field and nature of material. It the applied magnetic field is alternating and has a frequency equal to natural frequency of rod, a large amplitude longitudinal vibration set up due to resonance. The experimental set up for generating ultrasonic waves is shown in figure 2.

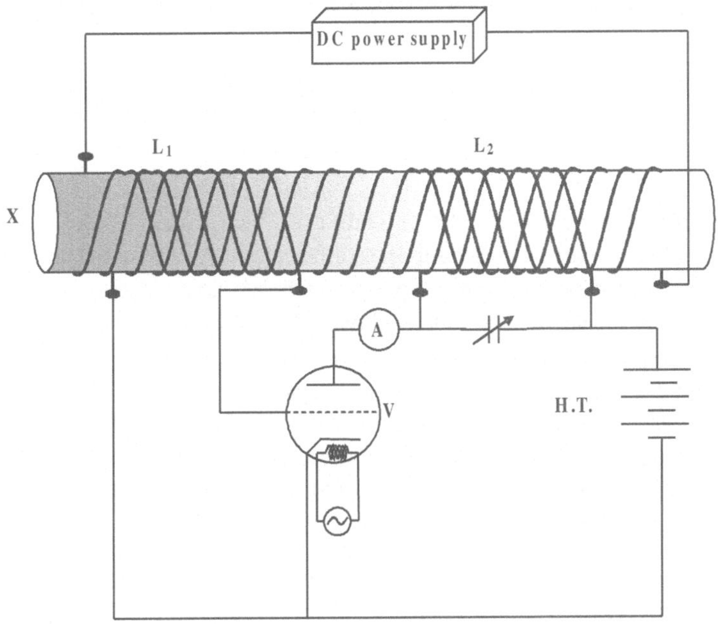

Fig. 4 *The experimental set up for generating ultrasonic waves by using Magnetostriction effect.*

A rod XY of nickel is placed in a solenoid, fed by D.C supply, which produces a steady polarizing magnetic field. Two other coils L_1 and L_2 are included in the grid and anode circuits of triode value. The frequency of anode circuit is adjusted with the variable chapter C. When the frequency equals the natural frequency of the rod, longitudinal oscillations are produced due to resonance. Thus ultrasonic waves are generated in surrounding medium. The procedure of generating ultrasonic waves is as follows. The rod is magnetized in the beginning by passing a direct current in the coil which is wrapped round the rod. When the frequency of plate circuit is adjusted to the natural frequency of rod, resonate vibrations are produced in the surrounding medium. The direct current millimeter (mA) reads the plate current. The vibrations are maintained due to the coupling provided by coil $L_{1.}$ During resonance millimeter records maximum current variation in the anode current passing through the coil L,which causes a corresponding change in the magnetization of the rod. Thus there is a change in the length of rod due to magnetostriction effect. This change in length of the rod obviously changes magnetic flux through coil L_1 which in turn changes the e.m.f developed across $L_{1.}$ This e.m.f amplifies current in the plate circuit i.e. in the coil, L_1. In this way plate current build up to a large amplitude with a frequency determined by the frequency of longitudinal vibration of the rod. When the frequency of the circuit becomes equal to the frequency of the rod, resonance occurs and the sound waves of maximum amplitude are generated. By adjusting the length of the rod and the capacity of condenser, high frequency oscillation of different frequencies are obtained.

Detection of Ultrasonic Waves: We can not directly detect the ultrasonic although some animals specially the bat can do so. It can be defected by following methods.

(i) **Kundt's tube method:** A Kundt's tube can be used to detect ultrasonic waves of relatively large wavelength when ultrasonic waves are passed through the tube, the lycopodium powder sprinked in the tube and collects in the form of heaps at the nodal points and is blown off at the antinodal points.

(ii) **Acoustic Grating method:** When ultrasonic waves are passed through a liquid, a diffraction pattern similar to those a observed in ruled grating formed. The complete arrangement is shown in Fig.1. The ultrasonic waves passing through a liquid(Kerosene oil) are elastic waves and forms compression and rarefactions in the liquid. These compression and rarefactions after reflection from sides of a cell form a stationery wave pattern, in which nodes and anti-nodes are spaced periodically. If light is passed perpendicular to direction of propagation of ultrasonic waves, liquid behave as diffraction grating, where grating element (a+b) become equal to wavelength of ultrasonic waves, λ_u Such a grating is known as acoustic grating.

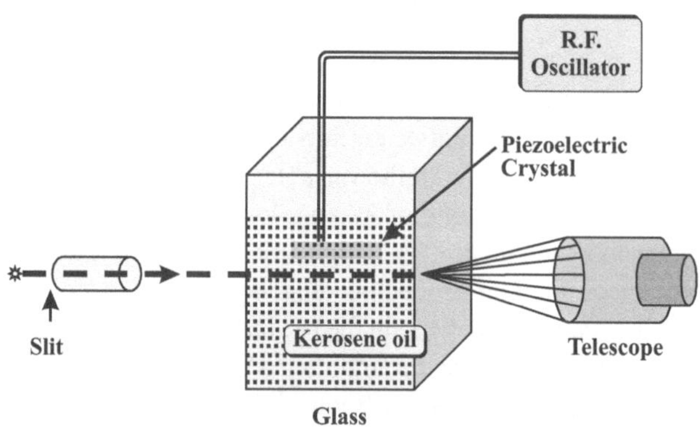

Fig. 5 *The experimental arrangement to determine the frequency of the ultrasonic wave*

If λ_u be the wavelength of ultrasonic wave in liquid, θ_n be the wavelength of light used and θ_n be the angle of diffraction for nth order, then,

$$(a+b)\sin\theta_n = n\lambda$$

$$As\ (a+b)=\lambda_u,\ then, \lambda_u,\ sin\theta_n = n\lambda \Rightarrow \lambda_u = \frac{n\lambda}{\sin\theta_n}$$

Procedure

(*a*) Rotate the telescope and focus it on some distant object. Make clear image of distant object. The telescope is in a position to focus parallel beam of rays coming from distant object at focus. Do not disturb this position throughout the observations.

(*b*) Bring telescope in line with the collimator. Illuminate the slit of collimator with light source (sodium light) and adjust its width with the help of adjusting screw provided on the slit. Look through the telescope and adjust collimator till a clear image of slit is observed. In this position the collimator is providing parallel beam of rays.

(*c*) Mount the glass / PVC cell on the prism table and fill it with kerosene oil up to nearly 3/4th of it height. Turn the prim table till the front and back faces of cell are perpendicular to the incident light.

(*d*) Immerse the crystal in the liquid and rotate it such that the ultrasonic waves generated by it travel in a direction perpendicular to the direction of propagation of light. Connect the crystal to the output terminal of the r.f. oscillator.

(*e*) Switch on the r.f. oscillator and adjust it frequency to match with the natural frequency of the crystal. At this instant resonance takes place and diffraction pattern will be observed in the spectrometer telescope. Usually five lines including slit are seen in the telescope on either side as shown in Fig.2.

(*f*) Measure angle $2\theta_1$ between first order spectral line. Similarly $2\theta_1$, $2\theta_1$, and so on are between the higher orders spectral lines on either side of central maxima.

(g) Note down the frequency of r.f. oscillator.

Fig. 6 *The diffraction spectra obtained in the experiment*

Application of ultrasonic waves: some of the important application of ultrasonic waves are given below.

(*i*) As the frequencies of ultrasonic are comparable to the frequency of vibration and those of rotation of the molecules of matter, they can be used to investigate molecular energies.

(*ii*) It is used to detect flaws or cracks in metals.

(*iii*) These waves are used for directional signaling.

(*iv*) These waves are used for determining the depth of sea, lakes etc.

(*v*) Ultrasonic waves can be used to detect reflecting objects like a submerged submarine, hidden iceberg etc.

(*vi*) These waves can be used for drawing utensils, washing clothes, removing dust and soot from the chimney.

(*vii*) Ultrasonic waves are used for soldering and drilling purpose.

(*viii*) It can be used for producing different alloys.

(*ix*) It can be used to detect abnormal growth in the brain, certain tumors.

(*x*) It can be used in the treatment of neuralogic pain.

(*xi*) Ultrasonic waves are used incoagulation of suspended liquid particles and crystallization of method.

(*xii*) It has got wide range of application in biological and medical uses like billing smaller animals rats, fist, frogs etc. and also bacteria. It can destroy red blood corpuscles.

Concepts of holography:- Fundamentally holography is a method of recording optical image. The word holography originates from the Greek word hole meaning the whole and 'graphy' meaning the writing i.e. holography means complete writing. A photographs represents a two dimensional recording of a three dimensional scene. In photograph the intensity distribution is recorded that prevailed on the plane of photograph. Since the intensity pattern has been recorded in photograph, the three dimensional character of the object scene is lost. Holography is a method invented by Gabor in 1947, in which one not only records the amplitude but also the phase of light waves. This is done by using interferometer techniques, the image produced by holography has a true three dimensional form. The capability to produce images as true as the object itself is the factor behind gaining wide popularity of holography technique. In the recording of hologram, superimposition of the object wave and another wave called the reference wave takes place.

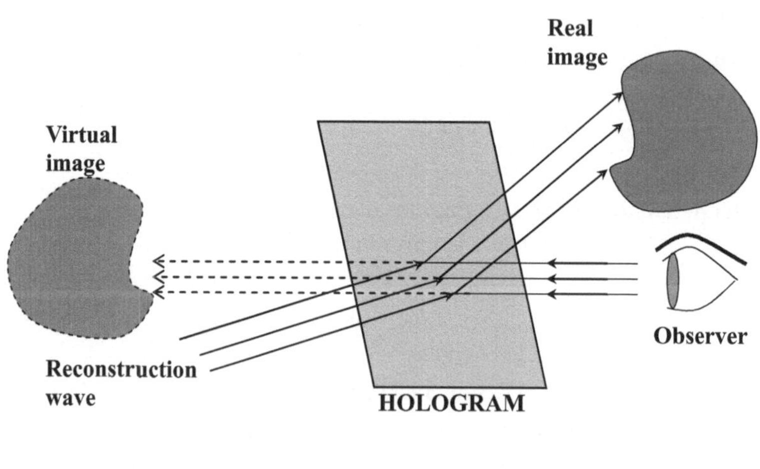

Fig. 5 *Reconstruction process to create a Hologram*

The photographic plate is made to record the resulting interference pattern. The reference wave is usually a plane wave. The recorded interference pattern is called hologram and it containing the information not only about the amplitude but also about the phase of the object wave. This hologram has no resemblance with the object but it contains all the information about the object in kind of optical code. To view the image, the hologram, is illuminated by another wave called as reconstruction wave(in most cases identical to reference wave used during the formation of hologram). The process of image formation from hologram is known as reconstruction process. Figure depicts the reconstruction process. This process forms a virtual and a real image of the object scene. Thus one can move the position of eye and look behind the object or one can focus at different distances.

Numerical Problems with Solutions

Chapter 1. Interference

Problem 1. Two parallel slits 0.5×10^{-3} mt apart are illuminated by a monochromatic light of $\lambda = 5890$ Å. Calculate the width of the fringes which are obtained on a screen a distance of 0.5 mt from the slit.

Solution: The fringe width in Young's double slit experiment is given by

$$\beta = \frac{D\lambda}{2d}$$

Given data are $D = 0.5\text{m}$, $2d = 0.5\text{m}$, $\lambda = 5890$ Å, then,

$$\beta = \frac{0.5 \times 5890 \times 10^{-10}}{0.5 \times 10^{-03}} = 0.5890 \times 10^{-3} m$$

Problem 2. Green light of $\lambda = 5100$ Å from a narrow slit is incident on a double slit. If the overall separation of 10 fringes on a screen 2 mt away is 0.02 mt. Find the double slit separation.

Solution: The fringe width in Young's double slit experiment is given by

$$\beta = \frac{D\lambda}{2d}$$

The overall separation of 10 fringes *i.e.*, $x = 10\beta = 0.02 = 10 \times \frac{D\lambda}{2d}$

Given data are, $D = 2$ mt., $\lambda = 5100$ Å Hence,

$$0.02 = \frac{10 \times 2 \times 5100 \times 10^{-10}}{2d} \Rightarrow 2d = 51000 \times 10^{-10} m = 5.1 \mu m$$

Problem 3. In Young's double slit experiment. The separation of the slit is 1.9×10^{-3} mt and the fringe separation is 0.31×10^{-3} mt at a distance of 1 mt from the slits. Calculate wavelength.

Solution: Fringe width in Young's double slit experiment is given by

$$\beta = \frac{D\lambda}{2d}$$

Hence, $\lambda = \frac{\beta \times 2d}{D}$,

$$\lambda = \frac{0.31 \times 10^{-3} \times 1.9 \times 10^{-3}}{1} = 0.5890 \times 10^{-6} m$$

Problem 4. Sodium light of $\lambda = 5890 \times 10^{-10}$ mt, falls on a double slit of separation 2 mm. The distance between the slits and screen is 0.04 mt. Locate the position of 10th bright fringe.

Solution: In Young's double slit experiment, the position of the bright fringe is given by

$$x_n = \frac{nD\lambda}{2d}$$

Given data are $\lambda = 5890 \times 10^{-10}$ m, $2d = 2$ mm, $D = 0.04$m hence, the position of 10th bright fringe x_{10} is given by

$$x_{10} = \frac{10 \times 0.04 \times 5890 \times 10^{-10}}{2 \times 10^{-3}} = 1.178 \times 10^{-4}\,\text{m}$$

Problem 5. A Fresnel's biprism (BP) is located at 0.05 mt from a slit illuminated by Sodium light of $\lambda = 5890 \times 10^{-10}$ mt, . The width of the fringes obtained on a screen 0.75 mt away from the biprism is 9.424×10^{-4} mt. What is the distance between the coherent sources?

Solution: In the Fresnel's Biprism experiment, the fringe width β of the observed interference pattern is given by

$$\beta = \frac{D\lambda}{2d}$$

Given data are $D = 0.8\text{m},\ \lambda = 5890 \times 10^{-10}\text{m},\ \beta = 9.424 \times 10^{-4}\,\text{m}$.
The distance between two coherent sources 2d is given as

$$2d = \frac{D\lambda}{\beta} = \frac{0.8 \times 5890 \times 10^{-10}}{9.424 \times 10^{-4}}\,\text{m} = 5 \times 10^{-4}\,\text{m}$$

Problem 6. The inclined faces of a glass Fresnel's bi-prism (μ = 1.5) makes an angle of 1 degree with the base of the prism. The slit is 0.1 mt from the BP and is illuminated by a light of $\lambda = 5900 \times 10^{-10}$ mt, find the fringe width observed at a distance of 1 mt from the *BP*.

Solution: The fringe width β of the observed interference pattern is given by $\beta = \frac{D\lambda}{2d}$

It α is the angle of inclined faces with faces with the base of prism and μ is the refractive index, then separation $2d$ is given as

$$2d = 2a(\mu - 1)\alpha$$

Here, $\alpha = 1^0 = \pi/180$ radian, $\mu = 1.5$, $a = 0.1\text{m}$

So, $2d = 2\times 0.1(1.5-1)\times\dfrac{3.14}{180} = 0.00174\,\text{m}$

Since $D = 1.1\text{m}$, $\lambda = 5900A^o$, we have

$$\beta = \frac{1.1\times 5900\times 10^{-10}}{0.00174}m = 37.2\times 10^{-5}\,\text{m}$$

Problem 7. In a Fresnel's biprism experiment fringes for light $\lambda = 5000\times 10^{-10}$ are observed 0.2×10^{-3} mt apart at a distance of 1.75 mt from the prism. The prism is made of glass of refractive index 1.5 and it is at distance of 0.25 mt from the illuminated slit. Calculate the angle at the vertex of the Fresnel's biprism.

Solution: The distance between two coherent sources $2d$ is given as

$2d = 2a(\mu-1)\alpha$.

The angle of inclined faces of biprism with base α, is given by

$$\alpha = \frac{2d}{2a(\mu-1)} \Rightarrow \alpha = \frac{D\lambda/\beta}{2a(\mu-1)}$$

Given data are, D=2m, $\lambda = 5890A^o$, $a = 0.25m$, $\mu = 1.5$,

$\beta = 0.2\times 10^{-3}\text{m}$

$$\text{So, } \alpha = \frac{2\times 5890\times 10^{-10}}{2\times 0.25(1.5-1)\times 0.2\times 10^{-3}} = \frac{5890\times 10^{-10}}{0.25\times 0.5\times 0.2\times 10^{-3}}$$

$$= \frac{589}{0.125}\times 10^{-6} = 235.6\times 10^{-4}\,\text{radian}$$

$$= \frac{235.6\times 10^{-4}\times 180}{3.14}\,\text{degree} = 1.351^\circ$$

Angle of prism at the vertex is

$A = 180 - 2(1.351) = 177.2980 = 177^o 17.8'$.

Problem 8. The distance between the slit and the Fresnel's biprism and screen are 50 cm each. Angle of Fresnel's biprism and refractive index are 179° & 1.5 respectively. Calculate the wavelength if the distance between successive fringes is 0.0135 cm.

Solution: If A is the angle of Fresnel's Biprism at vertex and α is the inclined face angle with the base of prism, then

$$\alpha = \frac{180-A}{2} = \frac{180-179^o}{2} = 0.5^o = \frac{0.5\times\pi}{180} = 0.0087\,\text{radian}$$

The separation '2d' between coherent sources is given as

$2d = 2a(\mu - 1)\alpha$

The fringe width 'β' is given by

$$\beta = \frac{D\lambda}{2d} = \frac{D\lambda}{2a(\mu - 1)\alpha} \Rightarrow \lambda = \frac{\beta \times 2a(\mu - 1)\alpha}{D}$$

Given data are D=100cm =1m, α=0.0087 radian, $\mu = 1.5$,

$\alpha = 0.5^\circ$, $\beta = 0.0135 \times 10^{-2} m$

$$\lambda = \frac{0.0135 \times 10^{-2} \times 2 \times 0.5(1.5 - 1) \times 0.0087}{1}$$

$= 5872.5 \times 10^{-10} m = 5872.5 A^o$

Problem 9. In a Fresnel's biprism experiment the fringe width observed is 0.087 mm. What will it become if the slit to BP distance is reduced to 0.75 times the original distance?

Solution: If a is the distance between slit to biprism and μ & α are refractive index and the inclined face angle of prism, then the distance between two coherent sources $2d$ is given by $2d = 2a(\mu - 1)\,\alpha$

For the actual distance a, we have,

$(2d)_1 = 2a(\mu - 1)\,\alpha$ ...(1)

When the distance is reduced to 0.75a, we have

$(2d)_2 = 2(0.75a)\,(\mu - 1)\,\alpha$ ...(2)

Using eqns.(1) & (2), we have,

$$\frac{(2d)_1}{(2d)_2} = \frac{2a\,(\mu - 1)\,\alpha}{2a\,(0.75a)\,(\mu - 1)\,\alpha} = \frac{1}{0.75}$$

$\Rightarrow (2d)_2 = 0.75 \times (2d)_1$

The fringe width β for original distance a is given as

$$\beta = \frac{D\lambda}{(2d)_1} = 0.087 \times 15^3 \text{m}$$

Here $D = a + b$

Since in the biprism experiment, D>>a, therefore on reducing a upto 0.75a, there is no observed change in D. Thus after reducing a into 0.75a, the fringe width β_2 is given by

$$\beta_2 = \frac{D\lambda}{(2d)_2} = \frac{D\lambda}{0.75\,(2d)_1} = \frac{D\lambda \times 0.087 \times 10^{-3}}{0.75 \times D\lambda} = 0.116 \text{ mm}$$

Problem 10. In a Fresnel's biprism experiment the eyepiece is placed at a distance of 1.2 mt from the source the distance between the virtual sources was found to be 7.5×10^{-4} mt. Find the wavelength if the eyepiece is to be moved transversely through a distance of 1.888×10^{-2} mt for 20 fringes.

Solution: If the eyepiece is moved to 1.88×10^{-2} m for 20 fringes then the fringe width b is given as $\beta = \dfrac{1.88 \times 10^{-2}}{20} = 0.094 \times 10^{-2}$ m

As we know $\beta = \dfrac{D\lambda}{2d} \Rightarrow \lambda = \dfrac{\beta \times 2d}{D}$

Given data are, D=1.2m, 2d = 7.5×10^{-4} m, β =0.094×10^{-2} m

So, $\lambda = \dfrac{0.094 \times 10^{-2} \times 7.5 \times 10^{-4}}{1.2} = 0.5875 \times 10^{-6}\text{ m} = 5875 A^o$

Problem 11. Interference fringes are produced by Fresnel's biprism in the focal plane of a reading microscope which is 1 mt from the slit. A lens interposed between the BP and the microscope gives two images of the slit in two positions. If the images of the slits are 4.5 mm in one position and 2.90 mm in the other position for $\lambda = 5870 \times 10^{-10}$ mt, find the distance between the consecutive interference bands.

Solution: Let us consider the case when lens L & L′ are introduced in Fresnel's Biprism experiment, then respective image of an object of $2d$ distance are d_1 & d_2 respectively.

The distance between two coherent sources is given by

$2d = \sqrt{d_1 d_2} = \sqrt{4.5 \times 2.9} \times 10^{-3}\text{ m} = 3.6125\text{mm}$

The distance between the consecutive interference bands *i.e.*, fringe width 'β' is given by

$$\beta = \frac{D\lambda}{2d}$$

Now given data are, $D = 1\text{ m}, \lambda = 5870 \times 10^{-10}\text{ m}$;

and, $2d = 3.6125 \times 10^{-3}\text{ m}$,

Hence, $\beta = \dfrac{1 \times 5870 \times 10^{-10}}{3.6125 \times 10^{-3}} = 1624.91 \times 10^{-7}\text{ m} = 1.625 \times 10^{-4}\text{ m}$

Problem 12. In a Fresnel's biprism experiment with sodium light bands of width 0.0195 cm are observed at 100 cm from the slit. On introducing a convex lens 30 cm away from the slit, two images of the lit are seen 0.7 cm apart at 100 cm distance from the slit. Calculate the wave-length.

Solution: Given data are β =0.0195 cm, $v = 30$ cm , $u = 70$ cm,

$I = 0.7$ cm, $O = 2d$.

For convex lens,

$$\frac{I}{O} = \frac{u}{v} \Rightarrow \frac{0.7}{2d} = \frac{70}{30} \Rightarrow 2d = \frac{30\times 07}{70} = 0.3 \text{ cm}$$

The fringe width is $\beta = \dfrac{D\lambda}{2d}$

And given $D = 100 \text{ cm} = 1 \text{ m}, 2d = 0.3 \text{ cm} = 0.3\times 10^{-2}$ m ;

Hence, $$\lambda = \frac{\beta\times 2d}{D} = \frac{0.0195\times 10^{-2}\times 0.3\times 10^{-2}}{1} = 5850\times 10^{-10} \text{ m}$$

Problem 13. A Fresnel's biprism arrangement is set with sodium light of $\lambda = 5893\times 10^{-10}$ mt, and in the field of view of the eyepiece 62 fringes are observed. How many fringes shall we get in the same field of view if we replace the source by mercury lamp using green filter $\lambda = 5461\times 10^{-10}$ m·

Solution: The fringe width in Fresnel's biprism experiment is given by

$$\beta = \frac{2d\lambda}{D} = \frac{fov}{\gamma} \qquad ...(3)$$

Here $2d, D, \lambda, fov \,\&\, \gamma$ are distance between two coherent sources, distance between slit to the screen wavelength of light, wavelengt field of view and number of fringes respectively. We need to determine the number of fringes in the same field of view if the wavelength of incident light is changed.

Let n_1, λ_1 are number fringes & wavelength of light for *fov* and n_2, λ_2 are changed number of fringes & wavelength of light for the same *fov* . Then from eqn.(3)

$$n_1\lambda_1 = n_2\lambda_2$$

Given data are $n_1 = 62, \lambda_1 = 5893\times 10^{-10}$m and

$\lambda_2 = 5461\times 10^{-10}$m , hence,

$$62\times 5893\times 10^{-10} = n_2\times 5461\times 10^{-10}$$

$$\Rightarrow n_2 = 66.9 \approx 67$$

Problem 14. When a thin sheet of transparent material of thickness 6.3×10^{-6} mt is introduced in the path of one of the interfering beams the central fringe shifts to a position occupied by the 6th bright fringe. If $\lambda = 5.46\times10^{-7}$ mt, find refractive index of the sheet.

Solution: If β is the fringe width, then on introducing thin sheet of thickness t in the path of one of the interfering beam, the central fringe shifts by a distance

$$s = 6\beta \qquad ...(4)$$

The displacements of fringes due to introduction of thin sheet is given by

$$s = \frac{\beta(\mu-1)t}{\lambda} \qquad ...(5)$$

Here λ, μ are wavelengths of light & refractive index respectively.

From eqns.(4) & (5), $6\beta = \frac{\beta(\mu-1)t}{\lambda} \Rightarrow 6\lambda = (\mu-1)t$ ...(6)

Given data are $t = 6.3\times10^{-6}m, \lambda = 5.46\times10^{-7}m$; hence, from eqn.(6),

$$6\times5.46\times10^{-7} = (\mu-1)\times6.3\times10^{-6} \Rightarrow (\mu-1) = 0.52 \Rightarrow \mu = 1.52$$

Problem 15. In a Fresnel's biprism experiment fringes were first observed with sodium light of $\lambda = 5893\times10^{-10}$ mt, and fringe width was measured to be 0.347 mm. Sodium light was then replaced with white light and the central fringe was located. On introducing a thin glass sheet in half of the beam the central fringe was shifted by 2.143 mm, calculate the thickness of the glass sheet if $\mu = 1.542$.

Solution: The fringe width β is given by $\beta = \frac{D\lambda}{2d}$

Given data are, $\beta = 0.347\times10^{-3}\text{m}, \lambda = 5893\times10^{-10}\text{m}$;

hence, $$0.347\times10^{-3} = \frac{D}{2d}\times5893\times10^{-10} \Rightarrow D/2d = 588.83 \qquad ...(7)$$

The displacement of things s is given by

$$s = \frac{\beta(\mu-1)t}{\lambda}$$

Given data are, $s = 2.143\times10^{-3}m, \mu = 1.542$; hence,

$$s = \left(\frac{D\lambda}{2d}\right) \times \frac{(\mu - 1)t}{\lambda} = \frac{D}{2d}(\mu - 1)t$$

Then from eqn.(7),

$$2.143 \times 10^{-3} = 588.83 \times (1.542 - 1) \times t$$

$$\Rightarrow 319.73 \times t = 2.143 \times 10^{-3} \Rightarrow t = 6.71 \mu m$$

Problem 16. White light is reflected from an oil film of thickness 0.01 mm and refractive index (RI) 1.4 at an angle of 45° to the vertical. If the reflected light falls on the slit of a spectrometer, calculate the no. of dark bands seen between λ =4000 Å and λ = 5000 Å.

Solution: The condition of minima for the reflected light from thin film is given by

$$2\mu t \cos r = n\lambda$$

It n_1 & n_2 are orders of minima corresponding to wavelength λ_1 & λ_2 respectively, then,

$$2\mu t \cos r = n_1 \lambda_1 \qquad ...(8)$$

$$2\mu t \cos r = n_2 \lambda_2 \qquad ...(9)$$

According to Snell's law

$$\mu = \frac{\sin i}{\sin r} \Rightarrow \frac{Sin 45^o}{\sin r} = 1.4$$

$$\Rightarrow \sin r = \frac{1}{1.4 \times 1.4142} = \frac{1}{1.9799} = 0.5051$$

So, $\cos r = \sqrt{1 - \sin^2 r} = \sqrt{1 - (0.505112)^2}$

$= \sqrt{1 - 0.2551} = \sqrt{0.7449} = 0.8631$

Given data are: $\lambda_1 = 4000A^o, \lambda_2 = 5000A^o, t = 0.01 \times 10^{-3}$ m; hence,

$$n_1 = \frac{2 \times 1.4 \times 0.01 \times 10^{-3} \times 0.8631}{4000 \times 10^{-10}} = 0.6042 \times 10^2 = 60.42 \approx 60$$

$$n_2 = \frac{2 \times 1.4 \times 0.01 \times 10^{-3} \times 0.8631}{5000 \times 10^{-10}} = 48.33 \approx 48$$

The no. of dark bands seen between λ_1 & λ_2 are

$\Delta n = n_1 - n_2 = 60 - 48 = 12$.

Problem 17. White light is incident on a soap film at an angle $\sin^{-1}(4/5)$ and the reflected light shows dark bands. The consecutive dark bands corresponds to $\lambda = 6.1 \times 10^{-5}$ cm & $\lambda = 6.0 \times 10^{-5}$ cm. If $\mu = 1.33$ is for the film calculate its thickness.

Solution: It 'μ' & 't' are refractive index and thickness of thin film respectively, the dark band of wavelength λ of nth order is given by

$$2\mu t \cos r = n\lambda_1 \qquad ...(10)$$

For the next order dark band of wavelength λ_2 we have,

$$2\mu t \cos r = (n+1)\lambda_2 \qquad ...(11)$$

From eqns.(10) & (11), we have,

$$n\lambda_1 = (n+1)\lambda_2 = n\lambda_2 + \lambda_2 \Rightarrow n = \frac{\lambda_2}{\lambda_1 - \lambda_2} \qquad ...(12)$$

Thus thickness of soap film is $t = \dfrac{n\lambda_1}{2\mu \cos r} = \dfrac{\lambda_1\lambda_2}{2\mu(\lambda_1 - \lambda_2)\cos r}$

(using (12))

Given data are $\lambda_1 = 6.1\times10^{-5} cm, \lambda_2 = 6\times10^{-5} cm, \mu = 1.33,$

angle of incidence $= i = \sin^{-1}(4/5)$.

Using Snell's law,

$$\frac{1}{\sin r} = \frac{\mu}{\sin i} \Rightarrow \sin r = \frac{\sin i}{\mu} = \frac{4}{5\times1.33} = 0.6015$$

$$\text{So, } \cos r = \sqrt{1-(0.6015)^2} = \sqrt{0.6382} = 0.7989$$

$$\text{Hence, } t = \frac{6.1\times10^{-5}\times6\times10^{-5}}{2\times1.33\times0.7989\times(6.1-6)\times10^{-5}}$$

$$= 172.22\times10^{-5} cm = 0.0017 cm$$

Problem 18. A parallel beam of sodium light of $\lambda = 5893 \times 10^{-10}$ mt, strikes the film of oil floating on water. When viewed at an angle of 30° from the normal, 8th dark band is seen. Find film thickness.

Solution: Let' μ, t ' are the refractive index & thickness of oil film respectively, then for nth order dark band of wavelength λ , we have,

$$2\mu t \cos r = n\lambda$$

Given data: $\lambda = 5893\times10^{-10} m, \sin i = \sin 30^\circ = \frac{1}{2}, \mu = 1.5$

So, $\cos r = \sqrt{1-\sin^2 r} = \sqrt{1-\left(\frac{\sin i}{\mu}\right)^2}$ (using Snell's law)

$$= \sqrt{1-\left(\frac{1}{2\times\frac{3}{2}}\right)^2} = \sqrt{1-\frac{1}{9}} = \frac{2\sqrt{2}}{3}$$

Hence, thickness of the film is

$$t = \frac{n\lambda}{2\mu\cos r} = \frac{8\times5893\times10^{-10}\times3}{2\times1.5\times2\sqrt{2}} = 1.665\,\mu m$$

Problem 19. A soap film of RI 4/3 and thickness $\lambda = 1.5\times10^{-6}$ mt is illuminated by white light incident at an angle of 60°. The light reflected by it is examined by a microscope in which a dark band is found corresponding to $\lambda = 5\times10^{-7}$ mt. Calculate the order of interference of the dark band.

Solution: Given data $\mu = \frac{4}{3}, t = 1.5\times10^{-6} m, i = 60^\circ, \lambda = 5\times10^{-7} m$

Then the nth order dark band of wavelength 1 at angle of refraction 'r' is given by

$2\mu t\cos r = n\lambda$

Now, $\cos r = \sqrt{1-\sin^2 r} = \sqrt{1-\left(\frac{\sin i}{\mu}\right)^2}$

(using Snell's law)

$$= \sqrt{1-\left(\frac{\sin 60^\circ}{\frac{4}{3}}\right)^2} = \sqrt{0.5781} = 0.7603$$

The order of interference band is

$$n = \frac{2\mu t\cos r}{\lambda} = \frac{2\times\frac{4}{3}\times1.5\times10^{-6}\times0.7603}{5\times10^{-7}} = 6.082 \approx 6$$

Problem 20. White light falls normally on a film of soapy water whose thickness is $\lambda = 5\times10^{-7}$ & $\mu = 1.33$. Which visible wavelength will be reflected most strongly?

Solution: Condition of reflected light maxima from a thin film of refractive index 'μ' & thickness 't' at angle of refraction 'r' is given as

$$2\mu t\cos r=\left(\frac{2n+1}{2}\right)\lambda$$

Where n & λ are order & wavelength of light respectively.
Since light falls normally, therefore,

$$\cos r=1\ \&\ \lambda=\frac{4\mu t}{(2n+1)}$$

Given data: $\mu=1.33, t=5\times10^{-7}m$, then,

$$\lambda=\frac{4\times1.33\times5\times10^{-7}}{(2n+1)}=26000A^{\circ}, 8666A^{\circ}, 5200A^{\circ}\ \text{respectively}$$

$n=0, n=1, n=2.$

Out of these only 5200 A° lies in the visible range for n=2.

Problem 21. In Newton's ring experiment the diameter of 25th ring was found to be 0.8 cm and that of 5th ring was 0.3 cm. If the radius of curvature of plane-convex lens is 1.0 mt, calculate the wavelength.

Solution: In Newton's ring experiment, the wavelength λ is given by

$$\lambda=\frac{D_{n+p}^{2}-D_{n}^{2}}{4pR}$$

D_{n+p} & D_{n+p} are diameter of (n+p)th & nth order rings, p is difference amongst order of rings, R is the radius of curvature of plano-convex lens.

Given data: $D_{n+p}=D_{25}=0.8\,\text{cm},\ D_{n}=D_{5}=0.3\,\text{cm},$

$R=1\,\text{m}, p=20.$

$$\text{So}\ \lambda=\frac{D_{25}^{2}-D_{5}^{2}}{4\times20\times100}=\frac{(0.8)^{2}-(0.3)^{2}}{80\times10^{2}}=\frac{0.64-0.09}{80\times10^{2}}\,\text{cm}=6875A^{\circ}$$

Problem 22. Newton's rings are observed in reflected light of $\lambda=5893\times10^{-10}$ mt, The diameter of 10th dark ring is 0.5 cm. Find the radius of curvature of the lens and the thickness of the air film for that ring.

Solution: The diameter of nth order dark ring in Newton's ring experiment is given by

$$D_{n}^{2}=4Rn\lambda$$

Given data: $D_{n}=D_{10}=0.5$ cm, $x=5893\times10^{-10}$ m
The radius of wavelength of plano-convex lens R is given by

$$R = \frac{D_n^2}{4n\lambda} = \frac{(0.5)^2}{4\times10\times5893\times10^{-8}} \text{ cm} = 106.2cm$$

The thickness of air film 't' is given as

$$t = \frac{D_n^2}{8R} = \frac{D_{10}^2}{8R} = \frac{(0.5)^2}{8\times106.2} cm = 2.946 \text{ ìm}$$

Problem 23. In the Newton's ring experiment a source emits two wavelengths $\lambda_1 = 6\times10^{-7}$ mt, & $\lambda_2 = 5.9\times10^{-7}$ mt, it is found that nth dark ring due to one wavelength coincides with $(n + 1)^{th}$ dark ring due to the other. Find the diameter of the nth dark ring if radius of curvature of the plano-convex lens (R) is 0.9 mt.

Solution: According to given condition if nth dark ring of wavelength λ_1 coincides with (n+1)th dark ring of wavelength λ_2, then,

$$n\lambda_1 = (n+1)\lambda_2 \quad \text{...(13)}$$

Thus diameter of rings holds the given relation as $D_n^2 = D_{n+p}^2$

Given data: $R = 0.9\text{m}; \lambda_1 = 6\times10^{-7}\text{ m}; \lambda = 5.9\times10^{-7}$; hence,

$$n\times6\times10^{-7} = (n+1)\times5.9\times10^{-7}$$

$$\Rightarrow 0.1n = 5.9 \Rightarrow n = 59$$

If R is radius of curvature of plano-convex lens, then diameter of 59th order dark ring is

$$D_n^2 = D_{59}^2 = 4Rn\lambda_1 = 4R\times59\times6\times10^{-7}$$

$$\Rightarrow D_{59} = \sqrt{4\times0.9\times59\times6\times10^{-7}} = 11.28\times10^{-3}m = 1.128 \text{ cm}$$

Problem 24. In Newton's ring experiment the diameter of 4th and 12th rings are 0.4 cm and 0.7 cm respectively. Find the diameter of the 20th dark ring.

Solution: In Newton's ring experiment, the wavelength λ is given by

$$\lambda = \frac{D_{n+p}^2 - D_n^2}{4pR}$$

D_{n+p} & D_n are diameter of (n+p)th & nth order rings, p is difference amongst order of rings, R is the radius of curvature of plano-convex lens.

Given data: $D_{n+p} = D_{12} = 0.7cm, D_n = D_4 = 0.4cm, p = 8$

$$\lambda = \frac{D_{12}^2 - D_4^2}{4\times8\times R} \Rightarrow \lambda R = \frac{(0.7)^2 - (0.4)^2}{32} \Rightarrow \lambda R = 0.0103 \text{ cm}^2$$

The diameter of 20th dark ring D_{20} is given by

$$D_n^2 = D_{20}^2 = 4Rn\lambda = 4\times 20(\lambda R)$$

$$\Rightarrow D_{20} = \sqrt{4\times 20\times 0.0103cm} = 0.905\text{ cm}$$

Problem 25. Newton's rings are formed between a plane surface of glass and lens. The diameter of the 3rd dark ring is 10^{-2} mt. When a light of $\lambda = 5890\times 10^{-10}$ mt, is used at such an angle such that light passes through the air film at an angle of 30° to the normal. Find *R*.

Solution: The condition for nth order dark ring in reflected light from air film of thickness of t is given as $2\mu t\cos r = n\lambda$

For air film, $\mu = 1$, so, $2t\cos r = n\lambda$ ...(14)

In Newton ring experiment, thickness of air film for nth order dark ring is given as

$$t = \frac{D_n^2}{8R} \qquad ...(15)$$

Given data: $D_n = D_3 = 10^{-2}m, \lambda = 5890A^\circ, r = 30^\circ$

Using eqns.(14) & (15),

$$2\times\frac{D_n^2}{8R}\cos r = n\lambda \Rightarrow \frac{2\times D_3^2}{48R}\cos 30^\circ = 3\times 5890\times 10^{-10}$$

$$\Rightarrow R = \frac{10^{-4}\times\sqrt{3/2}}{12\times 5890\times 10^{-10}} = \frac{72200}{5890} = 12.25m$$

Problem 26. A plano-convex lens of radius 3 mt is placed on an optically flat glass plate and is illuminated by monochromatic light. The diameter of 8th bright ring in the reflected system is 0.72×10^{-2} mt. calculate the wavelength.

Solution: The diameter of nth order bright ring D_n in Newton's ring experiment is given by $D_n^2 = 2(2n-1)R\lambda$

Given data : $n = 8, D_8 = 0.72\times 10^{-2}m, R = 3m$

$$\lambda = \frac{D_n^2}{2(2n-1)R} == \frac{(.072)^2 10^{-4}}{2(16-1)\times 3} = 5760A^\circ$$

Problem 27. If the diameter of the nth dark ring in an arrangement giving Newton's ring changes from 0.30 cm to 0.25 cm as a liquid is introduced between the lens & the plate. Calculate the velocity of light in the liquid when the velocity of light in vacuum is 3×10^{10} mt/s.

Solution: In Newton's ring experiment the nth order dark ring diameter for reflected light from liquid thin film of refractive Index 'ì' is given by

$$D_n^2\Big|_{liquid} = \frac{4Rn\lambda}{\mu}$$

For air film, $D_n^2\Big|_{air} = 4Rn\lambda$

Hence, $$\frac{D_n^2\Big|_{air}}{D_n^2\Big|_{liquid}} = \frac{4Rn\lambda}{4Rn\lambda/\mu} = \mu$$

Green Data: $D_n^2\Big|_{liquid} = 0.25$ cm, $D_n^2\Big|_{air} = 0.30$ cm; so,

$$\mu = \frac{(0.3)^2}{(0.25)^2} = 1.44$$

Velocity of light in vacuum is $3\times10^{10}\ m/s,\ i.e$ so velocity of light in liquid is given by

$$u = \frac{c}{\mu} = \frac{3\times10^{10}}{1.44} = 2.08\times10^{10}\ m/s$$

Problem 27. Newton's rings by reflection are formed between two bi-convex lens having equal radii of curvatures of 100 cm. Calculate the distance between the 5th and 15th dark ring using $\lambda = 5400\times10^{-10}$ mt.

Solution: In Newton's rings experiment two convex lens having radius of curvatures R_1 & R_2, are used. The radius of nth order dark ring 'r_n' is given by

$$r_n^2 = \frac{n\lambda}{\left(\frac{1}{R_1}+\frac{1}{R_2}\right)}$$

Given $R_1 = R_2 = R = 100cm, \lambda = 5400A^\circ, n = 5$; hence,

$$r_n^2 = \frac{n\lambda}{2/R} = \frac{Rn\lambda}{2} \Rightarrow r_5^2 = \frac{10^{-2}\times100\times5\times5400\times10^{-10}}{2} = 0.0116$$

Similarly when n=15, we have,

$$r_{15}^2 = \frac{10^{-2}\times100\times15\times5400\times10^{-10}}{2} = 0.0201\text{m}$$

The distance between 5th & 15th dark ring is

$$d = r_{15} - r_5 = 0.0201 - 0.0116 = 0.0085m$$

Problem 29. A Michelson Interferometer is set for the white light fringes (straight). When a mica sheet of thickness 0.005 cm is put in front of the fixed mirror, then in order to bring back the coloured fringes to their original position, the movable mirror is moved by 0.0025 cm, calculate refractive index of mica.

Solution: In Michelson Interferometer, when a glass plate of thickness t and refractive index μ is introduced in the path of any light ray divergent from inclined glass plate, the fringe pattern is changed by l and this is given by

$$l = (\mu - 1)t$$

Given data: $l = 0.0025cm, t = 0.005cm$; hence,

$$0.0025 = (\mu - 1)\times 0.005\mu \Rightarrow \mu = 1.5$$

Problem 30. In a Michelson Interferometer 100 fringes cross the field of view when the movable mirror is displaced through 0.02948 mm calculate the wavelength.

Solution: In Michelson Interferometer, when a movable mirror is displaced by a distance l and the crosswire moves n number of fringes in the field of view, then,

$2l = n\lambda$; Here λ is the wavelength of monochromatic light.

Using the given data: $\lambda = \frac{2l}{n} = \frac{2\times 0.02948\times 10^{-3}}{100} = 5896A^{\circ}$

Problem 31. The wavelength of two components of *D* lines of sodium are λ = 5890 Å and λ = 5896 Å. By how much distance one of the mirrors of Michelson Interferometer is moved so as to obtain consecutive positions of maximum distinctness.

Solution: In interferometer, the difference $\Delta\lambda$ between two wavelength, & λ_2 is given by

$$\lambda_1 - \lambda_2 = \Delta\lambda = \frac{\lambda_1\lambda_2}{2l}$$

Here l is the distance moved by movable mirror to obtain the crosswire at the position of maximum indistinctness.

Given data: $\lambda_1 = 5896A^{\circ}, \lambda_2 = 5890A^{\circ}$; hence,

$$\Delta\lambda = (5896 - 5890)\times 10^{-10} = \frac{5896\times 5890\times (10^{-10})^2}{2\times l}$$

$$\Rightarrow l = 2893953.00\times 10^{-10} m \Rightarrow l = 0.02894$$

Problem 32. In a Michelson Interferometer experiment the distance traveled by the mirror for two successive positions of maximum distinctness was 0.2945 mm. If the mean wavelength of sodium *D*-lines is λ = 5893 Å, calculate the difference between the two wavelengths.

Solution: In Michelson Interferometer, the difference $\Delta\lambda$ between two wavelength, is given by

$$\lambda_1 - \lambda_2 = \Delta\lambda = \frac{\lambda_1\lambda_2}{2l}$$

Here l is the distance moved by movable mirror to obtain the crosswire at the position of maximum indistinctness.

Given data: $l = 0.2945mm, \lambda_1 = \lambda_2 = \lambda = 5893A^\circ$; hence,

$$\Delta\lambda = \frac{\lambda_1\lambda_2}{2l} = \frac{\lambda^2}{2l} = \frac{(58931^2\times(10^{-10})^2}{2\times 0.2945\times 10^{-3}} = 5.9A^\circ \approx 6A^\circ$$

Chapter 2. Diffraction

Problem 1. A light of wavelength 6000 Å falls normally on a straight slit of width 0.10 mm. Calculate the total angular width of the central maximum and also the linear width as observed on a screen placed 1 mt away.

Solution: When a light of wavelength $\lambda = 6000A^\circ$ falls normally on a straight slit of width of 0.1mm, the angular width θ and linear width y of central maxima has a relation and

given by $\theta = {}^{y}\!/_{d}$

The position of nth order minima in single slit diffraction experiment is given by

$$e\sin\theta = n\lambda$$

For central maxima, n=1; hence,

$$e\sin\theta = \lambda \Rightarrow 1\times 10^{-3}\times\theta = 6000\times 10^{-10}$$

(as θ is very small, $\sin\theta \approx \theta$)

$$\Rightarrow \theta = 6\times 10^{-3} \text{ radian}$$

Total angular width $= 2\theta = 2\times 6\times 10^{-3} = 1.2\times 10^{-2}$ radian

The total linear width of central maxima, $2y = 2\theta D$

$$\Rightarrow 2y = 2\times 6\times 10^{-3}\times 1 = 1.2\times 10^{-2} m = 1.2cm$$

Problem 2. A single slit of width 0.14 mm is illuminated normally by monochromatic light and diffraction bands are observed on a screen mt away. If the center of the 2nd dark band is 1.6 cm from the middle of the central bright band, deduce the wavelength of light.

Solution: The position of nth order minima in single slit diffraction is given by

$$e\sin\theta = n\lambda \qquad ...(15)$$

Where e is the width of slit, λ is the wavelength of monochromatic light. If screen is at distance D from the single slit, then the relation between angular width θ and linear widh of maxima is given by

$$\theta = y/d$$

Given data; $e = 0.144mm, y = 1.6cm$ for $n = 2, D = 2m$

We have, $\sin\theta = \frac{n\lambda}{e}$

As θ is very small, $\sin\theta \approx \theta$, hence,

$$\theta = \frac{n\lambda}{e} = y/d \Rightarrow \frac{2\times\lambda}{0.14\times10^{-3}} = \frac{1.6\times10^{-2}}{2} \Rightarrow \lambda = 5600A^\circ$$

Problem 3. Light of wavelength 5000 Å is incident normally on a single slit. The central maximum spreads out at a 30° on both sides of the direction of the incident light. Calculate the width of the slit.

Solution: Direction of first minima for normally incident light of wavelength λ is given by

$$e\sin\theta = \lambda$$

Given data: $\theta = 30°, \lambda = 5000A^\circ$

Thus the width of single slit,

$$e = \frac{\lambda}{\sin\theta} = \frac{5000\times10^{-8}}{\sin 30°} cm = 10^{-4} cm$$

Problem 4. In Fraunhofer diffraction due to a a single slit the screen is placed 2 mt away from the slit. If the slit width is 0.2 mm and the first minima lies 5 mm on either side of the central maxima, find the wavelength of the incident light.

Solution: In Fraunhofer diffraction through a single slit, the direction of first minima θ is given by

$$e\sin\theta = \lambda \Rightarrow \sin\theta = \lambda/e$$

However, $\sin\theta = y/d$

So, $\lambda/_e = y/_d$

Given data; $y = 5mm, D = 2m, e = 0.2mm$; hence,

$$\lambda = {ey}/_b = \frac{0.2\times10^{-3}\times5\times10^{-3}}{2} = 5000A^\circ$$

Problem 5. A single slit is illuminated by light of two wavelengths, λ_1 and λ_2. One observes that due to Fraunhofer diffraction, the 1st minima obtained for λ_1 coincides with the 2nd diffraction minima of λ_2. What is the relation between λ_1 & λ_2.

Solution: In Fraunhofer diffraction through single slit the direction minima of nth order is given by

$e\sin\theta = n\lambda$

According to the given condition, direction of 1st minima for λ_1 coincides with direction of 2nd minima for λ_2, *i.e.*,

$e\sin\theta_1 = \lambda_1$

$e\sin\theta_1 = 2\lambda_2$

Thus relation between λ_1 & λ_2 is $\lambda_1 = 2\lambda_2$

Problem 6. A lens of focal length 1 mt forms Fraunhofer diffraction pattern of a single slit of width 0.05 cm in its focal plane. The incident light contains two wavelengths λ_1 & λ_2. It is found that the fourth minimum corresponding to λ_1 and the fifth minimum corresponding to λ_2 occur at the same point 0.5 cm from the central maximum. Find the values of λ_1 & λ_2.

Solution: In Fraunhofer diffraction through single slit, the direction of nth order minima is given by

$e\sin\theta = n\lambda$

The direction of 4th minima corresponding to λ_1 is given as

$$e\sin\theta_1 = 4\lambda_1 \quad ...(16)$$

Similarly direction of 5th minima corresponding to λ_2 is given as

$$e\sin\theta_2 = 5\lambda 2 \quad ....(17)$$

According to given condition $\theta_1 = \theta_2 = \theta$ and if y & D are the linear position of θ on the screen and distance of slit from screen respectively, then,

$$\theta_1 = \theta_2 = \theta = y/_d \quad ...(18)$$

Given data: $D = 1m, e = 0.05cm, y = 0.5cm$

From eqns.(16) & (18), we have,

$$e\sin\left(\frac{y}{d}\right) = 4\lambda_1$$

$$\Rightarrow 0.5\times10^{-2}\times\sin\left(\frac{0.5\times10^{-2}}{1}\right) = 4\lambda_1$$

$$\Rightarrow \lambda_1 = 0.05\times10^{-2}\times\sin(0.005)/4$$

$$= 0.05\times10-2\times0.005/4 = 6250A^\circ$$

Similarly from eqns.(17) & (18),

$$e\sin\left(\frac{y}{d}\right) = 5\lambda_2 \Rightarrow \lambda_2 = \frac{0.05\times10^{-2}\times\sin(0.005)}{5} = 5000A^\circ$$

Problem 7. A parallel beam of monochromatic light is allowed to be incident normally on a plane diffraction grating having 5000 lines per cm. The second order spectral line is observed to be deviated through 30°. Calculate the wavelength of the spectral line.

Solution: In Fraunhofer diffraction through diffraction grading, the direction θ of nth order principle maxima is given by

$$(e+d)\sin\theta = n\lambda$$

Here $(e+d)$ is diffraction element.

Given data: $\theta = 30, n = 2, \frac{1}{(e+d)} = 5000$ lines per cm; hence,

$$\lambda = \frac{(e+d)\sin\theta}{n} = \frac{\sin\theta}{2\times\frac{1}{(e+d)}} = \frac{\sin 30^\circ}{2(5000)}\text{ cm} = 5000A^o$$

Problem 8. Light of wavelength 5000 Å is incident normally on a plane transmission grating. Find the difference in deviations in the 1st and 3rd order spectra. The number of lines per cm on the grating surface is 6000.

Solution: In Fraunhofer diffraction through diffraction grading, the direction q of nth order principle maxima is given by

$$(e+d)\sin\theta = n\lambda$$

Here $(e+d)$ in diffraction element.

So direction of first order separation is given by

$$(e+d)\sin\theta_1 = 1\times\lambda \quad ...(19)$$

And third order spectra as

$$(e+d)\sin\theta_3 = 3\lambda \quad ...(20)$$

Given data: $\lambda = 5000^\circ, \frac{1}{e+d} = 6000$ lines per cm; hence,

$$\sin\theta_1 = \frac{\lambda}{(e+d)} = 5000\times10^{-8}\times6000 = 30\times10^{-2} = 0.3$$

$$\Rightarrow \theta_1 = \sin^{-1}(0.3) \Rightarrow \theta_1 = 17.458°$$

$$\sin\theta_3 = \frac{3\lambda}{(e+d)} = 3\times5000\times10^{-8}\times6000 = 90\times10^{-2} = 0.9$$

$$\Rightarrow \theta_3 = \sin^{-1}(0.9) \Rightarrow \theta_3 = 64.1581°$$

The angular difference $\Delta\theta$ for 1st & 3rd order spectra is

$$\Delta\theta = \theta_3 - \theta_1 = 64.158° - 17.458° - 46.7°$$

Problem 9. How many orders will be visible if the wavelength of the incident light is 5000 Å and the numbers of line per inch on the grating is 2620?

Solution: In Fraunhofer diffraction through diffraction grading, the direction θ of nth order principle maxima is given by

$$(e+d)\sin\theta = n\lambda$$

Here $(e+d)$ is diffraction element.

Maximum order will be visible at diffraction angle $\theta = 90°$, hence,

$$(e+d)\sin 90° = n\lambda \Rightarrow n = \frac{(e+d)}{\lambda}$$

Given data: $\lambda = 5000 A°, \frac{1}{(e+d)} = 2620$ lines per inch;

so, $\frac{1}{e+d} = \frac{2620}{2.54}$ lines per cm.

Thus maximum possible order, 'n' is given as

$$n = \frac{1}{\lambda\times\left(\frac{1}{e+d}\right)} = \frac{1}{5000\times10^{-8}\times\frac{2620}{2.54}}$$

$$= \frac{2.54}{13.1\times10^{-2}} = 0.194\times10^2 \approx 19$$

Problem 10. How many orders will be observed by a grating having 4000 lines per cm, if it is illuminated by a visible light of wavelength range between 4000 Å & 7000 Å ?

Solution: In Fraunhofer diffraction through diffraction grading, the direction θ of nth order principle maxima is given by

$$(e+d)\sin\theta = n\lambda$$

Here $(e+d)$ in diffraction element.

For wavelength λ_1, $$n_1 = \frac{(e+d)}{\lambda_1}$$

For wavelength λ_2, $$n_2 = \frac{(e+d)}{\lambda_2}$$

Given data: $\lambda_1 = 4000A°, \lambda_2 = 7000A°$, $\frac{1}{e+d} = 4000$ lines per cm;

So,

$$n_1 = \frac{1}{\lambda_1 \times \left(\frac{1}{e+d}\right)} = \frac{1}{4000 \times 10^{-8} \times 4000} = 6.25 \approx 6$$

$$n_2 = \frac{1}{\lambda_2 \left(\frac{1}{e+d}\right)} = \frac{1}{7000 \times 10^{-8} \times 4000} = 3.57 \approx 3$$

Thus 3rd to 6th order spectrum will be visible if $4000A°$ & $7000A°$ wavelengths are used.

Problem 11. Two spectral lines have wavelengths λ & $\lambda + d\lambda$ respectively. Show that if $d\lambda << \lambda$, their angular separation '$d\theta$' of a grating spectrum is given by $d\theta = \frac{d\lambda}{\sqrt{\left(\frac{e+b}{n}\right)^2 - \lambda^2}}$.

Solution: In Fraunhofer diffraction through diffraction grating, the direction θ of nth order principle maxima is given by

$$(e+d)\sin\theta = n\lambda$$

Here $(e+d)$ in diffraction element.

Let $(\theta + d\theta)$ is the direction of maxima for wavelength $(\lambda + d\lambda)$, then,

$$(e+d)\sin(\theta + d\theta) = n(\lambda + d\lambda)$$

$$\Rightarrow \frac{n\lambda}{\sin\theta}(\sin\theta \cos d\theta + \cos\theta \sin d\theta) = n(\lambda + d\lambda)$$

$$\Rightarrow (\cos d\theta + \cot\theta \sin d\theta) = \frac{1 + d\lambda}{\lambda}$$

As $d\theta$ is small, so, $\cos d\theta \approx 1$ & $\sin d\theta \approx \tan d\theta$

$$1+\cot\theta.\tan\theta = 1+\frac{d\lambda}{\lambda} \Rightarrow \cot\theta.\tan\theta = \frac{d\lambda}{\lambda} \Rightarrow \tan\theta = \frac{d\lambda}{\lambda}\times\frac{1}{\cot\theta}$$

$$\frac{(\sin\theta)d\lambda}{(\cos\theta)\lambda} = \frac{\theta\lambda}{\lambda}.\frac{n\lambda}{(e+d)}.\frac{1}{\sqrt{1-\sin^2\theta}}$$

Hance, from above

$$\Rightarrow \tan d\theta = \frac{n d\lambda}{(e+d)\sqrt{1-\left(\frac{n\lambda}{e+d}\right)^2}}$$

$$= \frac{d\lambda}{\sqrt{(e+d)^2/n^2 - \lambda^2}}$$

$$\Rightarrow \tan d\theta \approx d\theta = \frac{d\lambda}{\sqrt{\frac{(e+d)^2}{n^2} - \lambda^2}}$$

Problem 12. In a plane transmission grating the angle of diffraction for the 2nd order principal maximum for the wavelength 5000 Å is 30°. Calculate the number of lines per cm in the grating.

Solution: Direction of principal maximum of nth order for Fraunhofer diffraction through a plane transmission grating is given by

$(e+d)\sin\theta = n\lambda$

Here $(e+d)$ is diffraction element.

For second order diffraction $(e+d)\sin\theta = 2\lambda$

Given data: $\lambda = 5000A^\circ, \theta = 30^\circ$; we have,

$(e+d)\sin 30^\circ = 2\times 5000\times 10^{-8}\,cm$

$\Rightarrow (e+d) = 4\times 5000\times 10^{-8}\,cm$

$$\Rightarrow \frac{1}{e+d} = \frac{10^4}{2} = 5000$$

The number of line cm in grating is 5000.

Problem 13. A diffraction grating used at normal incidence gives yellow line (6000 Å) in a certain spectral order is superimposed on a blue line (4800 Å) of next higher order. If the angle of diffraction is $\sin^{-1}(3.4)$, calculate the grating element.

Solution: Direction of principal maximum of nth order for Fraunhofer diffraction through a plane transmission grating is given by

$$(e+d)\sin\theta = n\lambda \qquad ...(21)$$

Here $(e+d)$ is diffraction element.

It λ_2 of next order, *i.e.*, $(n+1)$th order, is superimposed on nth order of λ_1 then,

$$(e+d)\sin\theta = (n+1)\lambda_2 \qquad ...(22)$$

From eqns.(21) & (22), we have,

$$(n+1)\lambda_2 = n\lambda_1$$

Given data : $\lambda_1 = 6000A^\circ, \lambda_2 = 4800A^\circ$; hence,

$$(n+1)\times 4800 = 6000\times n$$

$$\Rightarrow n = 4$$

Thus, for, we have from eqn.(21), $(e+d) = \dfrac{n\lambda_1}{\sin\theta}$

If $\theta = \sin^{-1}\left(3/4\right)$, then,

$$(e+d) = \frac{4\times 6000\times 10^{-8}}{3/4} = 32\times 10^{-5} = 3.2\times 10^{-4}\, cm$$

Problem 14. A grating is made of 200 wires per cm placed at eual distances apart. The diameter of each wire is 0.025 mm. calculate the angle of diffraction for the third order spectrum and also find the absent spectra, if any. The wavelength of light used is 6000 Å.

Solution: Direction of principal maximum of nth order for Fraunhofer diffraction through a plane transmission grating is given by

$$(e+d)\sin\theta = n\lambda$$

Here $(e+d)$ is diffraction element.

Given data: $\dfrac{1}{e+d} = 200$ wires per cm, $n = 3, \lambda = 6000A^\circ$; hence, from eqn.(21),

$$\sin\theta = \frac{n\lambda}{(e+d)} = 3\times 6000\times 10^{-8}\times 200 = 0.036 \Rightarrow \theta = 2^\circ 4'$$

Direction of mth order minima in Fraunhoffer diffraction through a single slit is given by

$$e\sin\theta = m\lambda \qquad ...(23)$$

Hence, for the condition of absent spectra by using eqns. (21) & (22),

$$\frac{(e+d)\sin\theta}{e\sin\theta} = \frac{n\lambda}{m\lambda}$$

$$\Rightarrow \frac{(e+d)}{e} = \frac{n}{m} \qquad ...(24)$$

At this angle θ, corresponding to the nth order principle maxima of grating spectra will remain absent for any mth order minima of single slit and is given by,

$$\frac{1}{e+d} = 200 \Rightarrow e+d = \frac{1}{200}cm = 0.005cm \text{ , } e = 0.025mm$$

Using eqn.(24),

$$n = m\frac{(e+d)}{e} = \frac{(0.005)}{0.025\times10^{-1}}m = 2m$$

For $m = 1$, the $2^{nd}, 4^{th}, 6^{th}, ...$ order principal maximas will remain absent.
For $m = 2$, the $4^{th}, 8^{th}, 12^{th}, ...$ order principal maximas will remain absent.

Problem 15. A diffraction grating produces an angular separation of 0.01 radian between two wavelengths observed at an angle of 30°. If the men value of the wavelength is 5000 Å and the spectrum is observed in the second order, calculate the difference in the two wavelengths.

Solution: Direction of principal maximum of nth order for Fraunhofer diffraction through a plane transmission grating is given by

$$(e+d)\sin\theta = n\lambda$$

Here $(e+d)$ is diffraction element.
Taking the derivative of eqn.(21), we get,

$$d\theta = \frac{nd\lambda}{\left(\dfrac{n\lambda}{\sin\theta}\right)\cos\theta} \qquad ..(25)$$

Given data: $\lambda = 5000A°, \theta = 30°, d\theta = 0.01$ radian; using these into eqn.(25),

$$d\lambda = \left(\frac{\lambda\cos\theta}{\sin\theta}\right)d\theta = \frac{5000\times10^{-10}\times\cos 30°}{\sin 30°}\times0.01$$

$$= 5000\times1.732\times0.01\times10^{-10} = 86.6A°$$

Problem 16. A plane transmission grating has $N = 40{,}000$ lines and $(e+b) = 12.5 \times 10^{-5}$ cm. Calculate the maximum resolving power. The wavelength of light used is 5000 Å.

Solution: The resolving power of diffraction grating is given by resolving power $= nN$.

It will be maximum for maximum 'n' value.

So for maximum resolving power, $n = n_{\max}$,diffraction grating equation is given by

$(e+d)\sin\theta = n\lambda$

For $n = n_{\max}$, $\theta = 90°$ and so from eqn.(21),

$$n_{\max} = \frac{(e+d)}{\lambda}$$

Hence, maximum resolving power is $\frac{(e+d)}{\lambda}N$

Given data: $(e+d) = 12.5\times10-5cm, N = 40,000, \lambda = 5000A°$;

$$\text{So, } RP_{\max} = \frac{12.5\times10^{-5}\times40000}{50000\times10^{-8}} = 10^{5}$$

Problem 17. Calculate the minimum number of lines in a grating which will just resolve the lines of wavelengths 5890 Å & 5896 Å in the 2nd order.

Solution: According to Rayleigh's criteria of just resolution, the resolving power of diffraction grating for wavelength λ & $\lambda + d\lambda$ is given

$$\text{as } RP = \frac{\lambda}{d\lambda} = nN \qquad ...(26)$$

Given data: $\lambda = 5890A°, d\lambda = 5896 - 5890 = 6A°, n = 2$; hence using eqn.(26),

$$RP = \frac{5890}{6} = 2N$$

$$\Rightarrow N = 490.8 = 491$$

Problem 18. A grating has 6000 lines per cm drawn on it. If its width is 10 cm, calculate (*a*). the resolving power in the 2nd order, (*b*). the smallest wavelength that can be resolved in the 3rd order in 5896 Å wave-length region.

Solution: Resolving power of diffraction grating is given as

$$RP = \frac{\lambda}{d\lambda} = nN$$

Given data: $\frac{1}{e+d} = 6000$ lines per cm.

Hence in 10 cm, total number of lines is $N = 6000 \times 10 = 60000$
For 2nd order spectra, we have,

$RP = 2 \times 60,000 = 1,20,000$

The smallest wavelength $d\lambda$ that can be resolved in 3rd order is given by eqn.(26) as

$$\frac{\lambda}{d\lambda} = 3 \times 60,000 \Rightarrow d\lambda = \frac{5896}{180,000} = 0.0328A^\circ$$

Chapter 3. Polarization

Problem 1. A glass plate is used as a polarizer. Find the angle of polarization. Also find the angle of refraction, given μ = 1.54.

Solution: If 'i_p' is the polarization angle, 'μ' is the refractive index of glass plate, then,

$$\mu = \tan i_p \qquad ...(27)$$

Given data: $\mu = 1.54$ for glass plate; hence using eqn.(27),

$$i_p = \tan^{-1}(1.54) = 57.0023^\circ$$

The polarization angle holds following relation with angle of refraction '*r*',

$$i_p + r = 90^\circ \Rightarrow r = 90 - 57.0023^\circ = 33^\circ$$

Problem 2. A ray of light strikes a glass plate at an angle of incidence of 60°. If the reflected and refracted rays are perpendicular to each other, find the refractive index of glass.

Solution: If at any angle of incidence, the refracted & reflected rays are perpendicular to each other, then, the light reflected is polarization light and incident angle is polarizing angle. If 'i_p' is the polarization angle, 'μ' is the refractive index of glass plate, then,

$$\mu = \tan i_p \Rightarrow \mu = \tan 60^\circ = 1.7321$$

Problem 3. At a certain temperature the critical angle of water for total internal reflection is 60° for a certain wavelength. What is the polarizing angle and the angle of refraction for light incident on water at an polarizing angle.

Solution: Let the light ray is traveling from air to water and there is total internal reflection at critical angle 'C', 60°. Then according to Snell's law,

$$\mu_{air} \times \sin C = \mu_{water} \times \sin 90^\circ$$

$$\Rightarrow \mu_{water} = \frac{1}{\sin 90^\circ}.\sin 60^\circ = 0.8660$$

Thus polarization angle, 'i_p' for water with respect to air is given by

$$\mu_{water} = \tan i_p \Rightarrow i_p = \tan^{-1}(\mu_{water}) = \tan^{-1}(0.8660) = 40.892^\circ$$

The angle of refraction 'r' is given by $i_p + r = 90^\circ$

$$\Rightarrow r = 90^\circ - 40.89^\circ = 49.11^\circ$$

Problem 4. Calculate the thickness of a doubly refracting crystal required to introduce a path difference of $\lambda/2$ between ordinary (O) and extra-ordinary (E) rays for $\lambda = 6000$ Å, $\mu_0 = 1.55$, $\mu_E = 1.54$.

Solution: In case of half wave plate, the thickness 't' of plate is given by

$$t = \frac{\lambda}{2(\mu_0 - \mu_e)} \qquad ...(28)$$

Given data: $\lambda = 6000A^\circ, \mu_0 = 1.55, \mu_e = 1.54$; hence using eqn.(28),

$$t = \frac{6000 \times 10^{-10}}{2(1.55 - 1.54)} = \frac{6000 \times 10^{-10}}{0.02} = 3 \times 10^{-5} m = 0.003cm$$

Problem 5. Calculate the thickness of a doubly refracting crystal required to introduce a path difference of $\lambda/4$ between O-ray and E-ray for $\lambda = 5893$ Å, $\mu_0 = 1.53$, $\mu_E = 1.54$.

Solution: In case of quarter waveplate, when $u_e > u_o$, the thickness 't' of plate is given by

$$t = \frac{\lambda}{4(\mu_e - \mu_0)} \qquad ...(29)$$

Given data: $\mu_e = 1.54, \mu_0 = 1.53, \lambda = 5893A^\circ$; hence using eqn.(29),

$$t = \frac{5893 \times 10^{-10}}{4(1.54 - 1.53)} = 1.47 \times 10^{-3} cm$$

Problem 6. Calculate the thickness of half wave plate for sodium light, given $\mu = 1.54$ and ratio of velocity of O-ray to E-ray is 1.007. Is the crystal is positive?

Solution: The velocity v of light traveling in a medium of refractive index μ is given by

$$v = \frac{c}{\mu}$$

Here, c is the velocity of light in free space.
For ordinary and extra-ordinary light rays, we have,

$$v_o = \frac{c}{\mu_o} \;\&\; v_e = \frac{c}{\mu_e}$$

Thus, $$\frac{v_o}{v_e} = \frac{\mu_e}{\mu_o}$$

Given data: $\frac{\mu_0}{\mu_e} = 1.007, \mu_0 = 1.54, \lambda = 5893 A°$; hence,

$$\frac{\mu_e}{\mu_0} = 1.007 \Rightarrow \mu_e = 1.007 \times 1.54 = 1.551$$

Since $\mu_o > \mu_o$, therefore crystal is positive with thickness t of half waveplate and is given by

$$t = \frac{\lambda}{2(\mu_e - \mu_0)} = \frac{5893 \times 10^{-8}}{2(1.551 - 1.54)} = 267.86 \times 10^{-5} = 0.00268 \text{ cm}$$

Problem 7. For quartz *RI* for right and left handed circularly polarized vibrations are 1.55810 and 1.55821 respectively for $\lambda = 4000$ Å. Find the amount of optical rotation produced at $\lambda = 4000$ Å by a plate of quartz 2 mm thick and with its faces perpendicular to the optic axis.

Solution: If μ_L & μ_R are the refractive indexes for right & left landed circularly polarized vibration and t is thickness of quartz plate, then optical path difference Δx is given by

$$\Delta x = (\mu_L - \mu_R)$$

Thus the total amount of optical rotation δ due to the optical path difference Δx is given by

$$\delta = \frac{2\pi}{\lambda} \times \Delta x = \frac{2\pi}{\lambda} \times t(\mu_L - \mu_R) \qquad ...(30)$$

Given data: $\mu_L = 1.55821, \mu_R = 1.55810, t = 2mm, \lambda = 4000 A°$; hence using eqn.(30),

$$\delta = \frac{2 \times 3.14}{4000 \times 10^{-10}} \times 2 \times 10^{-3} (1.555821 - 1.55810)$$

$$\Rightarrow \delta = \frac{1256 \times 11}{400} rad = 3.4540 rad = 198°$$

Problem 8. The values of *RI* for *E*-ray and *O*-ray for quartz are 1.5508 and 1.5418 respectively. Calculate the phase retardation for λ = 5000 Å when plate thickness is 0.032 mm.

Solution: If μ_e & μ_o are the refractive indexes for extraordinary & ordinary light for plate thickness 't' respectively, then optical path difference 'Δx' is given by,

$$\Delta x = t(\mu_e - \mu_o) \qquad ...(31)$$

The phase difference 'δ' due to path difference 'Δx' is given by,

$$\delta = \frac{2\pi}{\lambda} \times \Delta x = \frac{2\pi}{\lambda} \times t(\mu_e - \mu_o)$$

Given data: $\lambda = 5000 A°, \mu_e = 1.5508, \mu_o = 1.5418, t = 0.032$ mm; so using eqn.(31),

$$\delta = \frac{2\pi}{5000 \times 10^{-10}} \times 0.032 \times 10^{-3}(1.5508 - 1.5418) = 1.152\pi \text{ radian}$$

Problem 9. The plane of polarization of plane polarized light is rotated through 6.5° in pasing through a length of 20 cm of sugar solution of 5% concentration. Calculate the specific rotation of the sugar solution.

Solution: The specific rotation 's' for any optically active substance is given by,

$$s = \frac{\theta}{l \times c} \qquad ...(32)$$

Here,'θ' is optical rotation, 'l' is the length in decimeter and concentration 'c' in gm/cc

Given data: $\theta = 6.5°, l = 20cm = 2dm, c = 5\% = 0.05$ gm/cc, hence using eqn.(32),

$$s = \frac{6.5}{20 \times 0.05} = 6.5 \text{ degree/(cm/(gm/cc))}$$

$$= \frac{6.5}{2 \times 0.05} = 65 \text{ degree/(dm/(gm/cc))}$$

Problem 10. A 20 cm ling tube containing sugar solution is placed between crossed Nicols and illuminated by light of $\lambda = 6 \times 10^{-5}$ cm. If specific rotation is 60° and optical rotation produced is 12° , what is the strength of the solution?

Solution: The specific rotation 's' for any optically active substance is given by,

$$s = \frac{\theta}{l \times c}$$

Here, θ is optical rotation, 'l' is the length in decimeter and concentration 'c' in gm/cc

Given data: $s = 60$ degree/(dm/(gm/cc)), $\theta = 12^\circ, l = 20cm = 2dm$; hence, using eqn.(32),

$$c = \frac{\theta}{ls} = \frac{12}{60 \times 2} = \frac{1}{10} = 0.1 gm/cc = 10\%$$

Problem 11. A mass of 80 gm of sugar is dissolved in 1 liter of water. The solution contained in a tube of length 20 cm causes an optical rotation of 12° for $\lambda = 5000$ Å at 17°C. What mass of sugar should be dissolved in 2 liter of water so that the solution placed in a tube of length 30 cm would cause an optical rotation of 24° with wavelength and at same temperature?

Solution: The specific rotation 's' for any optically active substance is given by,

$$s = \frac{\theta}{l \times c}$$

Here, 'θ' is optical rotation, 'l' is the length in decimeter and concentration 'c' in gm/cc

When a mass of 80gm is dissolved in liter of water then concentration is

$$c_1 = \frac{80}{10 \times 10 \times 10} \text{ gm/cc} = 0.08 \text{ gm/cc}$$

Given data: Tube length $l_1 = 20cm = 2dm, \theta_1 = 12°$ for

$\lambda = 5000 A°$ at temperature $17°C$

$$\text{So, } s = \frac{\theta_1}{l_1 c_1} = \frac{0.08}{2 \times 12} = \frac{12}{2 \times 0.08} = 75 \text{ degree/(dm/(gm/cc))}$$

Now, if $l_2 = 30cm = 3dm, \theta_2 = 24°$, $s = 75$ degree/(dm/(gm/cc)), at same λ & temperature, then, concentration, c_2, is given by

$$c_2 = \frac{\theta_2}{s \times l_2} = \frac{24}{3 \times 75} = 0.1067 gm/cc$$

Mars of sugar, 'm', in one liter is

$= 10 \times 10 \times 10 \times 0.1067 = 106.7 gm$

Thus mass of sugar dissolved in 2 liter of water for optical rotation 24° is given by

$= 106.7 \times 2 = 213.4 gm$

Problem 12. The specific rotation of quartz at 5086 Å is 29.73°/mm. Calculate the difference in *RI*.

Solution: The phase retardation or total optical rotation is given by

$$\delta = \frac{2\pi}{\lambda}(\mu_L - \mu_R)t$$

For an average optical rotation we have,

$$\theta = \delta/2 = \pi/\lambda\,(\mu_L - \mu_R)t$$

Here μ_L, μ_R are refractive indexes for left handed and right handed quartz, t is the thickness of quartz plate.
Now specific rotation is given by

$$s = \theta/t \Rightarrow 29.73° = \pi/\lambda\,(\mu_L - \mu_R) \quad ...(33)$$

Given data: $\lambda = 5086A°$, $\Delta\mu = (\mu_L - \mu_R)$; using eqn.(33),

$$\text{So, } \Delta\mu = \frac{\lambda \times 29.73°}{\pi} = \frac{29.73 \times \pi}{180} \times \frac{5080 \times 10^{-7}}{\pi} = 8.4 \times 10^{-5}$$

Problem 13. A sugar solution in a tube of 20 cm produces optical rotation of 13°. The solution is then diluted to one-third of its previous concentration. Find optical rotation produced by 30 cm long tube containing the diluted solution.

Solution: The specific rotation s for any optically active substance is given by,

$$s = \frac{\theta}{l \times c}$$

Here, θ is optical rotation, l is the length in decimeter and concentration c in gm/cc
Given data: $l = 20cm = 2dm$, $\theta = 13°$; hence,

$$s = \frac{13}{2 \times c} = \frac{6.5}{c} \quad ...(34)$$

It concentration is diluted to $c/3$ and length of optical active substance is $l = 30cm = 3dm$, then using eqn.(34), optical rotation is given by

$$\theta = s \times l \times \left(c/3\right) = \frac{6.5}{C} \times 3 \times \frac{c}{3} = 6.5°$$

Problem 14. A 5% solution of cane sugar placed in a tube of length 40 cm, causes the optical rotation of 20°. How much length of 10% solution of the same substance will cause 35° rotation?

Solution: The specific rotation s for any optically active substance is given by,

$$s = \frac{\theta}{l \times c}$$

Here, 'θ ' is optical rotation, 'l' is the length in decimeter and concentration 'c' in gm/cc

It $l_1 = 40cm = 4dm, c_1 = 0.05gm/cc, \theta_1 = 20°$ then

$$s = \frac{\theta_1}{l_1 c_1} = \frac{20°}{4 \times 0.05} = 100 \text{ degree/(dm/(gm/cc))}$$

For the same substance, specific rotation will remain the same, then length of optical substance is given by

$$l = \frac{\theta}{s \times c}$$

Given data: $c = 0.1$ gm/cc, $\theta = 35°$; hence,

$$l = \frac{100 \times 0.1}{35} \frac{35}{0.1 \times 100} = \frac{35}{10} = 3.5 \text{ dm=35cm}$$

Problem 15. 80 gm of impure sugar is dissolved in a liter of water. The solution gives an optical rotation of 9.9° when placed in a tube of length 20 cm. If the specific rotation of pure sugar solution is 66° dm^{-1} $(gm/cc)^{-1}$ find the percentage purity of the sugar sample.

Solution: The concentration of sugar solution is given by

$$c = \frac{80}{10 \times 10 \times 10} = 0.08 \text{ gm/cc}$$

The specific rotation 's' for any optically active substance is given by,

$$s = \frac{\theta}{l \times c}$$

Here, θ is optical rotation, l is the length in decimeter.

Given data: $c = 0.08$ gm/cc, $\theta = 9.9°, l = 20cm = 2dm$; hence,

$$s = \frac{9.9}{2 \times 0.08} = 61.875 \text{ degree/(dm/(gm/cc))}$$

The standard value of specific rotation of sugar solution $s = 66$ degree/(dm/(gm/cc)).

The percentage of impurity of sample

$$= \frac{66 - 61.875}{66} \times 100 = 6.25\%$$

Hence, the percentage of purity of sample

$$= 100 - 6.25 = 93.75\%$$

Chapter 4 & 5 Fiber Optics and Laser

Problem 1. A light ray enters from air to a fiber. The RI of air is 1. The fiber has *RI* of core and claddings are 1.5 and 1.48 respectively. Find the critical angle, the fractional *RI*, the acceptance cone and numerical aperture.

Solution: According to Snell's law, the critical angle θ_c at the interface of core and cladding layer is given by

$$\sin\theta_c = \frac{\mu_{clad}}{\mu_{core}}$$

Given data: $\mu_{clad} = 1.48$, $\mu_{core} = 1.5$; hence,

$$\sin\theta_c = \frac{1.48}{1.5} \Rightarrow \theta_c = 80.63°$$

Fraction refractive index Δ is given by

$$\Delta = \frac{\mu_{core} - \mu_{clad}}{\mu_{core}} = \frac{1.5 - 1.48}{1.5} = 0.013$$

Acceptance angle θ_a is given by

$$\sin\theta_a = \sqrt{\mu_{core}^2 - \mu_{clad}^2} = \sqrt{(1.5)^2 - (1.48)^2}$$

$$= 0.244 \Rightarrow \theta_a = 14.13°$$

Numerical aperture is given by $NA = \sin\theta_a = \sin 14.13° = 0.244$

Problem 2. A graded fiber has a core diameter of 50 μm and an *NA* of 0.22 at λ = 0.850 μm. What are the *V* number and the number of modes guided in the core ?

Solution: V number is given by $V = \frac{2\pi a}{\lambda} NA$

Here a is core radius., λ is wavelength, NA is the Numerical Apeture.

Given data: $a = \frac{50}{2} = 25mm, \lambda = 0.85\times10^{-6}\,m, NA = 0.22$; hence,

$$V = \frac{2\times3.14\times25\times10^{-6}\times0.22}{0.85\times10^{-6}}\times0.22 = 40.66$$

Number of modes is given by

$$N = \frac{V^2}{4} = \frac{(40.66)^2}{4} = 413.31$$

Problem 3. Calculate the coherence length of a laser beam for which the bandwidth is $\Delta\upsilon = 3000$ Hz.

Solution: The coherence length 'l_c' for a laser beam is given by

$$l_c = \frac{c}{\Delta f}$$

Given data: $c = 3\times10^8 m/\sec, \Delta f = 3000Hz$; hence,

$$l_c = \frac{3\times10^8}{3000} = 10^5 m$$

Problem 4. If we consider the sun as the source, then we have the angle subtended by the sun at the slit, $\theta = 32'$, determine the transverse coherence length, if $\lambda = 5\times10^{-5}\,cm.$

Solution: The transverse coherence length 'l' is given by
$l = \lambda / \theta$

Given data: $\lambda = 5\times10^{-5}\,cm, \theta = 32' = \frac{32}{60}\times\frac{3.14}{180}\,radian$; hence,

$$l = \frac{5\times10^{-5}\times60\times180}{32\times3.14} = 537.42\times10^{-5} m = 0.005cm$$

Problem 5. For an ordinary source, coherence time $\tau_C = 10^{-10}$ sec. Obtain the degree of monochromaticity for $\lambda_C = 5400$ Å

Solution: The bandwidth 'Δf' and coherence time 't_c' is given by $\Delta f = \frac{1}{t_c}$

$$\text{The degree of monochromaticity} = \frac{\Delta f}{f_o} = \frac{\Delta f}{c}\times\lambda_0 = \frac{\lambda_0}{ct_c}$$

Given data: $\lambda_c = 5400A°, t_c = 10^{-10}s, c = 3\times10^{-8}m/s$; hence, the degree of monochromaticity is given by

$$= \frac{5400\times10^{-10}}{10^{-10}\times3\times10^8} = 18\times10^{-6}.$$

Chapter 6. Simple Harmonic Motion

Problem 1. A particle vibrates with SHM of amplitude 0.05 mt and a period of 6 sec. How long does it take to move from one end of its path to a position 0.025 m from the mean position on the same side ?

Solution: During SHM the displacement of particle at any time 't' is given by

$$y = A\sin(\omega t + \varphi) \quad \text{...(35)}$$

Here, $A \& \varphi$ are amplitude and phase change during motion respectively.

The angular frequency, $\omega = \frac{2\pi}{T} = \frac{2\pi}{6} = \pi/_3 \ radian / \sec$

Now at $t = 0, y = A$. So for the given condition the phase change φ is given as

$y = A\sin(\omega t + \varphi) \Rightarrow A = A\sin(0 + \varphi) \Rightarrow \sin\varphi = 1 \Rightarrow \varphi = \pi / 2$

Thus in the given condition using eqn.(35), the equation of motion is given as

$$y = A\sin\left(\omega t + \pi / 2\right) = A\cos\omega t \qquad ...(36)$$

So, using eqn.(36), time required 't' to travel a distance of 0.025m from the mean position is given by

$$0.025 = 0.05\cos(\omega t) \Rightarrow \cos\omega t - 1/_2$$

$$\Rightarrow \omega t = \pi/_3 \Rightarrow t = \pi/_3 \times 3/_\pi = 1 \ \sec$$

Problem 2. A particle vibrates with SHM of amplitude 0.06 mt and time period 31.4 sec. Calculate the maximum velocity.

Solution: During SHM the displacement of particle at any time 't' is given by eqn.(35) as

$y = A\sin(\omega t + \varphi)$

So velocity of particle is given by

$$v = \frac{dy}{dt} = A\omega\cos(\omega t + \varphi) \qquad ...(37)$$

Now, $v = v_{\max}$ only when $\cos(wt + \varphi) = 1$

Hence using eqn.(37),

$$v_{\max} = \omega A \qquad ...(39)$$

However, $\omega = \frac{2\pi}{T} = \frac{2\pi}{31.4} = \frac{2 \times 3.14}{31.4} = 0.2$ rad/sec and we have

$A = 0.06$ m, then using eqn.(38),

$v_{\max} = 0.06 \times 0.2 = 0.012$ m/sec

Problem 3. A mass of 8 kg stretches a spring 0.32 mt from its equilibrium position. The mass is removed and another body of mass 0.5 kg is suspended from the source spring. If the spring is now stretched and released, what is its period of motion ?

Solution: For a spring, $F = kx$ ...(39)

Here, 'F' is the force on the spring, 'k' is the spring constant and 'x' is the displacement of the spring. However, the gravitational force on mass 'm' is

$$F = mg \qquad ....(40)$$

Given data: $m = 8kg, g = 9.8m/s^2, x = 0.32m$; hence using

eqns.(39) & (40), $k = \frac{mg}{x} = \frac{8\times 9.8}{0.32} = 245$ N/m

The time period of oscillations is for a mass of $m = 0.5kg$ is given by

$$T = 2\pi.\sqrt{\frac{m}{k}} = = 2\pi \times \sqrt{\frac{0.5}{245}} = 0.28 \sec$$

Problem 4. For a damped harmonic oscillator the relaxation time is 50 sec. Find the time in which the amplitude and the energy of the oscillator falls to 1/e times of its initial value.

Solution: Amplitude of the damped oscillation at a time t is given by

$$A = ae^{-bt} \qquad ...(41)$$

Here, a is initial amplitude and b is the damping constant.

Now. the relaxation time is, $\tau = \frac{1}{2b} = 50$ sec, given, hence,

$$\Rightarrow b = \frac{1}{100} /\text{sec}$$

Now, after a time t, amplitude becomes a/e , hence using eqn.(41)

with $b = 1/100 / \sec$

we have, $\frac{a}{e} = a.e^{-bt} = a.e^{-t/100}$

$$\Rightarrow \frac{t}{100} = 1 \Rightarrow t = 100 \ \sec$$

Problem 5. The quality factor of a tuning fork is 10^4. Calculate the time interval after which it energy becomes 1/10 of its initial value. The frequency of the fork is 250 Hz.

Solution: The quality factor of the tuning fork is given by $Q = \omega\tau$

$$\Rightarrow \tau = \frac{Q}{\omega} = \frac{Q}{2\pi f}$$

Given data: $Q = 10^4, f = 250$ Hz; hence,

$$\tau = \frac{10^4}{2\pi \times 250} = \frac{20}{\pi} \ \sec$$

The energy of the fork after time t is $E = E_0 e^{-2bt}$ ...(42)

Here E_0 is the initial energy of the fork, then, eqn.(42) becomes,

$$E = E_0 e^{-t/\tau} \qquad (\text{as } \tau = \frac{1}{2b}) \qquad ...(43)$$

Let after a time t, the energy is reduced to $(E_0/10)$, then using eqn.(43), we have

$$\frac{E_0}{10} = E_0 e^{-t/\tau} \Rightarrow 10^{-1} = e^{-t/(20/\pi)} \Rightarrow t = 14.65 \text{ sec}$$

Problem 6. Quality factor of sonometer wire of frequency 240 Hz is 2000. Find the time in which the amplitude decreases to $1/e^2$ to its initial value.

Solution: The quality factor of the tuning fork is given by $Q = \omega\tau$

$$\Rightarrow \tau = \frac{Q}{\omega} = \frac{Q}{2\pi f}$$

Given data: $Q = 2000, f = 240$ Hz; hence,

$$\tau = \frac{2000}{2\pi \times 240} = \frac{25}{6\pi} \text{ sec}$$

After time t the initial amplitude is changed to a/e^2, hence,

$$\frac{a}{e^2} = a.e^{-bt} = ae^{-t/2\tau} \qquad (\text{as } \tau = \frac{1}{2b} \Rightarrow b = \frac{1}{2\tau})$$

$$\Rightarrow e^{-2} = e^{-t/2\tau} \Rightarrow 2 = t/2\tau \Rightarrow t = 4\tau \Rightarrow t = 4 \times 1.32 = 5.3 \text{ sec}$$

Problem 7. If the amplitude of forced vibration is $A = \dfrac{f}{\sqrt{\left(\omega^2 - p^2\right)^2 + 4b^2 p^2}}$

and Q-factor of the oscillator is 50, calculate the value of A/A_{MAX} when $p/\omega = 0.99$.

Solution: Given, $A = \dfrac{f}{\sqrt{(\omega^2 - p^2)^2 + 4b^2 p^2}}$ ...(44)

Here, at resonance, $\omega = p$, then $A = A_{\max}$. Hence using eqn.(44),

$$A_{\max} = \frac{f}{2bp} = \frac{f}{2b\omega} \qquad ...(45)$$

Now,

$$Q = \omega\tau = \frac{\omega}{2b} \Rightarrow \frac{\omega}{2b} = 50 \Rightarrow 2b = \frac{\omega}{50} \quad ...(46)$$

Using eqn.(46) into eqn.(46), we have,

$$A_{\max} = \frac{f}{2b\omega} = \frac{f}{\left(\frac{\omega}{50}\right)\omega} = \frac{50f}{\omega^2} \quad ...(47)$$

Using eqn.(44), we have,

$$A = \frac{f}{\sqrt{(\omega^2 - p^2)^2 + 4b^2 p^2}} = \frac{f}{\sqrt{(\omega^2 - p^2)^2 + \left(\frac{\omega}{50}\cdot\right)^2 p^2}}$$

$$= \frac{f}{(\omega^2)\sqrt{\left[1 - \frac{p^2}{\omega^2}\right]^2 + \left(\frac{\omega}{50}\right)^2 \cdot \frac{p^2}{\omega^4}}} = \frac{f}{(\omega^2)\sqrt{\left[1 - \frac{p^2}{\omega^2}\right]^2 + \frac{p^2}{2500\omega^2}}}$$

$$= \frac{f}{\omega^2\sqrt{[1-(0.99)^2]^2 + \frac{1}{2500}\times(0.99)^2}} \quad \left(\text{given, } \frac{p}{\omega} = 0.99\right)$$

$$= \frac{f}{\omega^2} \times \frac{50}{1.4} = \frac{50f}{1.4\omega^2} \quad ...(48)$$

Using eqns.(47) & (48), we have,

$$\frac{A}{A_{\max}} = \frac{\frac{50f}{1.4\omega^2}}{\frac{50f}{\omega^2}} = \frac{1}{1.4} = 0.71$$

Chapter 7. Dielectric

Problem 1. If the magnitude of h in plane e.m.wave is 1 A/mt, calculate the magnitude of E for plane wave in free space.

Solution: For free space, $\frac{E}{H} = \sqrt{\frac{\mu_0}{\varepsilon_0}}$ ...(49)

Given data: $\varepsilon_0 = 8.86\times10^{-12}$ F/mt, $\mu_0 = 4\pi\times10^{-7}$ H/mt, $H = 1$ A/m; hence, using eqn.(49)

$$E = H\sqrt{\frac{\mu_0}{\varepsilon_0}} = 1\times\sqrt{\frac{4\pi\times10^{-7}}{8.86\times10^{-12}}} = 376.6 \text{ V/m}$$

Problem 2. An e.m.wave traveling in *Z*-direction in an unbounded loss-free dielectric medium with $\mu_r = 1, \varepsilon_r = 2$. It has peak electric field strength $E_0 = 5$ V/mt.

Find (*i*) The impedence of the medium, (*ii*) Peak magnetic field intensity, (*iii*) Velocity of the wave.

Solution:

(*i*) The relative permittivity and permeability of the medium are given by ε_r and μ_r .

$$\text{We know, } \varepsilon_r = \frac{\varepsilon}{\varepsilon_0} \Rightarrow \varepsilon = \varepsilon_0 \varepsilon_r \text{ and } \mu_r = \frac{\mu}{\mu_0} \Rightarrow \mu = \mu_0 \mu_r$$

Given data: $\varepsilon_r = 2$ and $\mu_r = 1$; $\varepsilon_0 = 8.86 \times 10^{-12}$ F/mt,

$\mu_0 = 4\pi \times 10^{-7}$ H/mt; hence,

Impedance of the medium is given by

$$Z = \frac{E}{H} = \sqrt{\frac{\mu}{\varepsilon}} = \sqrt{\frac{\mu_0 \mu_r}{\varepsilon_0 \varepsilon_r}} = 266.8 \ \Omega$$

(*ii*) Given data,: $E_0 = 5$ V/m:

$$\text{We know, } Z = \frac{E}{H} = \frac{E_0}{H_0} \Rightarrow H_0 = \frac{E_0}{Z} = \frac{5}{266.8} = 0.187 \text{ A/m}$$

(*iii*) Velocity of the e.m.wave is given by

$$v = \frac{1}{\sqrt{\mu\varepsilon}} = \frac{1}{\sqrt{(\mu_0 \mu_r)(\varepsilon_0 \varepsilon_r)}} = \frac{c}{\sqrt{(\mu_r)(\varepsilon_r)}} = \frac{3 \times 10^8}{\sqrt{(1)(2)}}$$

$$= 2.12 \times 10^8 \text{ m/s}$$

Problem 3. In a plane e.m.wave the electric field oscillate at a frequency 3×10^{11} Hz and amplitude 50 V/mt. What is the wavelength of the wave? What is the amplitude of the oscillating magnetic field?

Solution:

For the e.m. wave of frequency f and wavelength λ we have

$$c = f\lambda \Rightarrow \lambda = \frac{c}{f}$$

Given data: $f = 3 \times 10^{11}$ Hz; $E_0 = 50$ V/m; hence,

$$\lambda = \frac{c}{f} = \frac{3 \times 10^8}{3 \times 10^{11}} = 10^{-3} \text{ m}$$

Also, we know,

$$\frac{E_0}{H_0} = \sqrt{\frac{\mu_0}{\varepsilon_0}} \Rightarrow \frac{E_0}{\mu_0 H_0} = \frac{1}{\mu_0} \times \sqrt{\frac{\mu_0}{\varepsilon_0}} = \sqrt{\frac{1}{\mu_0 \varepsilon_0}}$$

$$\Rightarrow \frac{E_0}{B_0} = \sqrt{\frac{1}{\mu_0 \varepsilon_0}} = c \Rightarrow B_0 = \frac{E_0}{c} \Rightarrow B_0 = \frac{50}{3\times 10^8} = 1.67\times 10^{-7} \text{ T}$$

Problem 4. If the average distance between the sun and the earth is 1.5×10^{11} mt and power radiated by the sun is 3.8×10^{26} *W*, show that the average solar energy incident on the earth is 2 cal/cm^2-min.

Solution: Given data: Average distance between the sun and the earth is $R = 1.5\times 10^{11}$ mt,

power radiated by the sun $P = 3.8\times 10^{26}$ W; hence,
average solar energy incident on the earth is given by

$$S = \frac{Energy}{area \times time} = \frac{Power}{area} = \frac{P}{4\pi R^2}$$

$$\Rightarrow S = \frac{3.8\times 10^{26}}{4\pi \left(1.5\times 10^{11}\right)^2} \text{ W/m}^2$$

But, $1 \text{ W} = \frac{1}{4.2}$ cal/sec; hence, $1 \text{ W/m}^2 = \frac{1}{4.2} \times \frac{1}{(100)^2}$ cal/cm^2/sec

$$\text{So,} \quad S = \frac{3.8\times 10^{26}}{4\pi \left(1.5\times 10^{11}\right)^2} \times \frac{1}{4.2} \times \frac{1}{(100)^2} \text{ cal/cm}^2\text{/sec}$$

$$= \frac{3.8\times 10^{26}}{4\pi \left(1.5\times 10^{11}\right)^2} \times \frac{1}{4.2} \times \frac{1}{(10)^2} \times 60 \text{ cal/cm}^2\text{/min}$$

$= 1.92$ cal/cm^2/min

Problem 5. If earth receives 2 cal/min/cm^2 solar energy what are the amplitudes of electric and magnetic field of radiation?

Solution: **Given data:** $S = 2$ cal/cm^2/min;
We have here,

$$S = \left|\vec{S}\right| = \left|\vec{E} \times \vec{H}\right| = EH \sin 90^\circ = EH$$

$$\Rightarrow EH = S = 2\times \frac{4.2\times 10^4}{60} = 1400 \text{ J/m}^2\text{/sec} \qquad ...(50)$$

Also, impedance is given by $Z = \frac{E}{H} = \sqrt{\frac{\mu_0}{\varepsilon_0}} = 376.6$ ...(51)

Using eqns.(50) & (51),

$$EH \times \frac{E}{H} = 1400 \times 376.6 \Rightarrow E = \sqrt{1400 \times 376.6} = 726.1 \text{ V/m}$$

And, $\frac{EH}{\frac{E}{H}} = 1400/376.6 \Rightarrow H = \sqrt{1400/376.6} = 1.928$ A/m

Hence, the amplitude of electric field,

$E_0 = E\sqrt{2} = 726.1\sqrt{2} = 1027$ V/m,

the amplitude of magnetic field per turn,

$H_0 = H\sqrt{2} = 1.928\sqrt{2} = 2.73$ A-turn/m.

Problem 6. Assuming that all the energy from a 1000 W lamp is radiated uniformly, calculate the average values of the intensities of electric and magnetic fields of radiation at a distance of 2 mt from the lamp.

Solution: The total energy flux over a sphere drawn around the lamp at center is given by

$1000W = 1000J/\sec$

This energy flux falls on an area = $A = 4\pi R^2 = 4\pi \times 2^2 = 16\pi \text{ m}^2$

Hence, the energy flux per unit area per sec

$$= S = EH = \frac{1000}{16\pi} \text{ W/m}^2 \quad ...(52)$$

Also, impedance is given by $Z = \frac{E}{H} = \sqrt{\frac{\mu_0}{\varepsilon_0}} = 376.6\ \Omega$ ...(51)

Using eqns. (51) & (52),

$$E = \sqrt{EH \times \frac{E}{H}} = \sqrt{\frac{1000}{16\pi} \times 376.6} = 86.57 \text{ V/m}$$

And, $H = \sqrt{\frac{EH}{\frac{E}{H}}} = \sqrt{\frac{1000}{16\pi}/376.6} = 0.229$ A-turn/m

Problem 7. A coaxial cable has a outer and inner conductor diameters 3.87 mm and 0.6 mm respectively. If $\varepsilon_r = 2.22$, calculate the characteristic impedance of the cable.

Solution:

The characteristic impedance of the cable is given by

$$Z_0 = \frac{60}{\sqrt{\varepsilon_r}} \log_e\left(\frac{D}{d}\right)$$

Given data: dielectric constant $= \varepsilon_r = 2.22$, outer diameter $= D = 3.87$ mm, inner diameter $= d = 0.6$ mm; hence,

$$Z_0 = \frac{60}{\sqrt{\varepsilon_r}} \log_e\left(\frac{D}{d}\right) = \frac{60}{\sqrt{2.22}} \log_e\left(\frac{3.87}{0.6}\right) = 75\ \Omega$$

Problem 8. A lossless coaxial cable has a capacitance 70 pF and inductance 0.39 μh. Calculate characteristic impedance of the cable at radio frequencies.

Solution: Given data: cable capacitance $= C = 70$ pF, cable inductance $= L = 0.39$ μH; hence,
the characteristic impedance of the cable is also given by

$$Z_0 = \sqrt{\frac{L}{C}} = \sqrt{\frac{0.39\times10^{-6}}{70\times10^{-12}}} = 74.64\ \Omega$$

Problem 9. The velocity factor of a coaxial cable is 0.62. Calculate the dielectric constant of the insulation used.

Solution: Given data: velocity factor of a coaxial cable $= VF = 0.62$; hence, we have,

$$VF = \frac{c}{v_p} = \frac{1}{\sqrt{\varepsilon_r}} = 0.62 \Rightarrow \varepsilon_r = \frac{1}{(0.62)^2} = 2.6$$

Chapter 8. Electromagnetic Theory

Problem 1. The dielectric constant of helium at 0°C and 1 atmospheric pressure is 1.000074. Find the dipole moment induced in each helium atom when the gas is under an electric field intensity of 100 V/mt.

Solution: The relationship between dielectric constant k and electric susceptibility X, polarization P, electric field E is given by

$$k = 1 + X = 1 + \frac{P}{\varepsilon_0 E} \Rightarrow P = (k-1)\varepsilon_0 E$$

Given data: $k = 1.000074, E = 100$ V/m; hence,

$$P = (1.000074 - 1)\times 8.854\times10^{-12}\times100 = 0.0655\times10^{-12}\ \text{Coul/m}^2 \quad \text{...(53)}$$

If p is the dipole moment of helium atom and n is the total nos. of atoms per cubic meter of volume, then,

$P = np$

Using eqn.(53), we have now, $p = \frac{P}{n} = \frac{0.0655\times10^{-12}}{n}$ Coul-m ....(54)

Under STP condition, 1 gram atom of helium contains 6.023×10^{23} helium atoms in 22.4 litres of volume, hence, the nos. of helium atoms in 1 m^3 of volume is given by

$$n = \frac{6.023\times10^{23}}{22.4\times1000\times(10)^3} = 0.268\times10^{26} \text{ atoms/m}^3 \qquad ...(55)$$

Using eqns.(54) & (55), we have,

$$p = \frac{0.0655\times10^{-12}}{n} = \frac{0.0655\times10^{-12}}{0.268\times10^{26}} = 0.244\times10^{-38} \text{ Coul-m}$$

Problem 2. For helium the value of dielectric constant at at 0°C and 1 atmospheric pressure is 1.000074. Calculate its electrical susceptibility.

Solution: We have, $k = 1 + X \Rightarrow X = k - 1$

Given data: $k = 1.000074$; hence,

$X = k - 1 = 1.000074 - 1 = 0.000074$

Problem 3. An isotropic material of ε_r is placed normal to the uniform external electric field with an displacement vector 5×10^{-4} mt^2. If the volume of the slab is 0.5 mt^3 and the magnitude of polarization is 4×10^{-4} mt^2, find the value of ε_r and the total dipole moment of the slab.

Solution: We have, the displacement vector $= D = 5\times10^{-4}$ coul/m^2, volume of slab $= V = 0.5$ m3, polarization $= P = 4\times10^{-4}$ coul/m^2; then,

$D = E + P \Rightarrow E = D - P = 5\times10^{-4} - 4\times10^{-4} = 1\times10^{-4}$ coul/m^2

However, the dielectric constant is given by

$$\varepsilon_r = \frac{D}{E} = \frac{5\times10^{-4}}{1\times10^{-4}} = 5$$

The net dipole moment is given by

$= P\times V = 4\times10^{-4}\times0.5 = 2\times10^{-4}$ coul-m

Problem 4. Two parallel plates have equal and opposite charges and separated by a dielectric 5 mm thick having a dielectric constant 3. If the electric field intensity in the dielectric is 10^6 V/mt, calculate the polarization in the dielectric, displacement and energy density.

Solution: We have, dielectric constant $= k = 3$,

electric field intensity in the dielectric $= E = 10^6$ V/mt; the polarization is given by

$$P = \varepsilon_0 (K-1)E = 8.854\times10^{-12}(3-1)\times10^{6} = 17.71\times10^{-6}\ \text{coul/m}^2$$

The displacement current density is given by

$$D = \varepsilon E = \varepsilon_0 kE = 8.854\times10^{-12}\times3\times10^{6} = 26.55\times10^{-6}\ \text{coul/m}^2$$

The energy density is given by

$$E_d = \frac{1}{2}\varepsilon E^2 = \frac{1}{2}\varepsilon_0 kE^2 = \frac{1}{2}\times8.854\times10^{-12}\times3\times10^{12} = 13.83\ \text{J/m}^3$$

Chapter 9. Special Theory of Relativity

Problem 1. The earth will appear shortened along its diameter due to its orbital motion around the sun to an observer at rest relative to the sun. Calculate the change in length in the diameter. The orbital velocity of the earth is 30 km/sec and radius is 6371 km.

Solution: Let the diameter of the earth is $= l = 2\times6371$ km, the apparent diameter observed by a stationary object outside the earth is $= l'$, velocity of the earth $= v = 30$ km/sec, velocity of the light $= c = 3\times10^5$ km/sec, hence, using length contraction formula,

$$l' = l\sqrt{1-\left(\frac{v}{c}\right)^2} \approx l\left[1-\frac{1}{2}\left(\frac{v}{c}\right)^2\right]$$

Hence, the contraction in length is given by

$$\Delta l = l - l' = l - l\left[1-\frac{1}{2}\left(\frac{v}{c}\right)^2\right] = \frac{l}{2}\left(\frac{v}{c}\right)^2$$

$$= \frac{2\times6371}{2}\times\left(\frac{30}{3\times10^5}\right)^2 = 6.37\times10^{-2}\ \text{m}$$

Problem 2. On her 16th birthday a young lady decides that she will like to remain 16 for at least another 10 years. She decides to go to a journey into outer space with uniform velocity. What is the minimum speed she must have relative to the laboratory so that when she returns after 10 years she can still say that she is 16?

Solution: Here the time of 10 years at earth will appear as 1 day for the lady. Hence, using the time dilation formula, we have

$$\Delta t = \frac{\Delta t'}{\sqrt{1-\left(\frac{v}{c}\right)^2}}$$

The time Δt is the time duration at earth, *i.e.*, $\Delta t = 10$ years and time Δt is the time duration at space, *i.e.*, $\Delta t = 1$ day, hence,

$$\Delta t = \frac{\Delta t'}{\sqrt{1-\left(\frac{v}{c}\right)^2}} \Rightarrow 10 = \frac{(1/365)}{\sqrt{1-\left(\frac{v}{c}\right)^2}} \Rightarrow v = 0.999c$$

Problem 3. A certain star is 20 light year away. How long would it take for a spaceship traveling with a velocity 0.95c to reach the star from the earth as measured by the observer (*i*) on earth, (*ii*) on the spaceship?

Solution: (*i*) The distance of the star = L_0 = 20 light-year

Hence, the time taken to cover the distance with a velocity $v = 0.95c$ is

$$t = \frac{L_0}{v} = \frac{20\times c}{0.95c} = 21.05 \text{ year}$$

(*ii*) For the observer in spaceship, the distance is given by

$$L = L_0\left[1-\sqrt{\left(\frac{v}{c}\right)^2}\right] = 20\times\left[1-\sqrt{\left(\frac{0.95c}{c}\right)^2}\right] = 6.2 \text{ light-year}$$

Hence, time measured by the observer in the space ship is

$$t' = \frac{L}{v} = \frac{6.2\times3\times10^8}{0.95\times3\times10^8} = 6.53 \text{ years}$$

Problem 4. A friend of yours travels in his rocket car of 5 mt long and 1.5 mt high with a speed of 0.6c.

(*i*) What will be its length and height at rest?

(*ii*) How many seconds elapse on your friend's watch when 10 sec is passed in your watch?

Solution:

(*i*) There will be no change in height of the person as there is no relative motion in that direction.

At rest, length becomes

$$L_0 = \frac{L}{\sqrt{1-\left(\frac{v}{c}\right)^2}} = \frac{5}{\sqrt{1-\left(\frac{0.6c}{c}\right)^2}} = 6.25\text{m}$$

(*ii*) Time in my watch $= t = 10$ sec and time in my friend's watch $= t'$

Hence, we have,

$$t = \frac{t'}{\sqrt{1-\left(\frac{v}{c}\right)^2}} \Rightarrow t' = t\sqrt{1-\left(\frac{v}{c}\right)^2} = 10\times\sqrt{1-\left(\frac{0.6c}{c}\right)^2} = 8 \text{ sec}$$

Problem 5. Show that, $S^2 = x^2 + y^2 + z^2 - c^2t^2$ is invariant under Lorentz Transformation.

Solution: For a moving frame, S', the co-ordinates are (x', y', z', t)

We have to prove that $S^2 = (S')^2$

We know, from Lorentz transformation between co-ordinates of fixed frame of reference S and fixed frame of reference S' is given by

$$x = \frac{x'+vt'}{\sqrt{1-\left(\frac{v}{c}\right)^2}}; y = y'; z = z'; t = \frac{t'+\frac{vx'}{c^2}}{\sqrt{1-\left(\frac{v}{c}\right)^2}}$$

Given

$$S^2 = x^2 + y^2 + z^2 - c^2t^2 = \left[\frac{x'+vt'}{\sqrt{1-\left(\frac{v}{c}\right)^2}}\right]^2$$

$$+(y')^2 + (z')^2 - c^2\left[\frac{t'+\frac{vx'}{c^2}}{\sqrt{1-\left(\frac{v}{c}\right)^2}}\right]^2$$

$$= \frac{1}{\left[1-\left(\frac{v}{c}\right)^2\right]}\left[[x'+vt']^2 - c^2\left[t'+\frac{vx'}{c^2}\right]^2\right] + (y')^2 + (z')^2$$

$$= \frac{c^2}{c^2-v^2}\left[x'^2 + 2x'vt + (vt')^2 - \left[c^2t'^2 + 2x'vt + \left(\frac{vx'}{c^2}\right)^2\right]\right]$$

$$+(y')^2 + (z')^2$$

$$= \frac{c^2}{c^2 - v^2}\left[x'^2 + (vt')^2 - \left[c^2 t'^2 + \left(\frac{vx'}{c^2}\right)^2\right]\right] + (y')^2 + (z')^2$$

$$= \frac{c^2}{c^2 - v^2}\left[\frac{c^2 x'^2 + c^2 (vt')^2 - \left[c^4 t'^2 + (vx')^2\right]}{c^2}\right] + (y')^2 + (z')^2$$

$$= \left[\frac{\left(c^2 - v^2\right)x'^2 - c^2 t'^2 \left(c^2 - v^2\right)}{c^2 - v^2}\right] + (y')^2 + (z')^2$$

$$= x'^2 - c^2 t'^2 + (y')^2 + (z')^2 = x'^2 + (y')^2 + (z')^2 - c^2 t'^2$$

$$= S'^2$$

Proved.

Questions for Assignment

Part-A

Interference

1. Explain the formation of interference fringes by means of Fresnel's biprism when a monochromatic source of light is used and derive the expression for the fringe width. How will you measure the wavelength of monochromatic light using biprism method.
2. What are Newton's rings? Explain the formation of Newton's ring in reflected light ?
3. Explain with theory, Newton's ring method to determine the wavelength of monochromatic light and discuss which source is preferred, point or extended?
4. Describe the construction and working of Michelson interferometer to find the thickness of a thin transparent plate.
5. (*a*) Explain the working of Michelson interferometer. How will produce circular fringes?
   (*b*) How will measure the difference in wavelength between D lines of sodium light by it?

Diffraction

6. Differentiate between Fraunhofer and Fresnel's diffraction .Explain the phenomenon of Fraunhofer diffraction through a single slit.
7. Give the construction and theory of plane diffraction grating and explain the formation of spectra by it?
8. Derive an expression for resolving power and dispersive power of plane diffraction grating.

Polarization

9. (*a*) Explain the phenomenon of double refraction in uniaxial crystal
   (*b*) What are quarter and half wave plate? Explain their use in the study of different types of polarized light.
10. Describe the construction and working of Nicole's prism. Explain its limitations.
11. Give the construction and working of Lorentz Half shade device.
12. Describe the action of Bi-quartz plate in Bi-quartz polarimeter.
13. Write note on
    (*a*). Photoelasticity (*b*). Polarized and unpolarized light

Laser

14. Explain the term spontaneous and stimulated emission, population inversion and various pumping process.
15. What are Gas Lasers ?Describe the principle, construction and working of He-Ne Laser.How it is superior than Ruby Laser?
16. Discuss the characteristic of laser beam. What are necessary condition for laser beam. Give description of semiconductor laser.

Fiber Optics

17. Give the principle of optical fibre? How the terms angle of acceptance and numerical aperture are used in optical fiber? Discuss in detail the various modes in fiber optics.

18. What is meant by Graded index fiber and step index fiber? How optical fiber can be used in medical and communication field.

Part-B

Simple Harmonic Motion

1. Define damped harmonic oscillations. Solve the differential equation and discuss the case of oscillatory motion. What is quality factor and how do you define it? What is logarithmic decrement.
2. Discuss the theory of forced harmonic oscillators. How does sharpness of resonance depend on damping?

Dielectrics

3. Define (*i*) Polarization vector (*ii*) Displacement current (*iii*) Susceptibility (*iv*) Dielectric constant (*v*) Permittivity.
   Obtain various relationships connecting them.
4. Write notes on
   (*a*) Frequency dependence of Dielectric
   (*b*) Dielectric losses in a.c.field.

Electromagnetic Theory

5. State and derive Maxwell's equations and give their physical interpretation.
6. Derive simple plane wave equation in free space and prove that light is an electromagnetic wave.
7. Write notes on :
   (*a*) Poynting vector, (*b*) Coaxial cable
   (*c*) Equation of continuity

Special Theory of Relativity

8. What was the aim of Michelson-Morley experiment? Describe the experiment. How is this negative result of the experiment interpreted?
9. Derive Lorentz transformation equations of relativity.
10. Deduce an expression for the variation of mass with velocity.

Nuclear Physics

11. What is nuclear fusion? Explain in brief.
12. Explain nuclear fission on the basis of liquid drop model.
13. Describe the construction and working of a nuclear reactor and explain the terms neutron cross-section and reactor criticality.
14. Write notes on
    (*a*) Scintillation Counter (*b*) Ionization Chamber
    (*c*) Cloud Chamber (*d*) Geiger Muller Counter

Superconductivity

15. (*a*) What is superconductivity? Describe Meissner's effect .Distinguish between Type-I and Type –II superconductor.
    (*b*) Derive London's equation and discuss how it's solution explain Meissner's effect.

Index

A

A frictional force 117
A restoring force 117
Absent spectra 50
Acceleration 114
Acceleration transformation 172
Acceptance angle 102
Acceptance angle 101, 102
Acceptance cone 102
Activation energy 96
Active centers 84
Active medium 84
Amplitude 7, 113
An external periodic force 117
Analyzer 65
Anisotropic medium 60
Anti-reflecting coatings 15
Attenuation in the optical fiber 232
Audible waves 233
Acoustic Grating method 236
Application of ultrasonic waves 237

B

Basov 81
Batholinus 62
Beats 6
Bi-quartz polarimeter 77
Biaxial 60
Biot's law 73
Blunt corners 61
Brewster angle window 93
Brewster's law 58

C

Calcite 60, 63
Calcite 64
Canada balsam 64
Carbon–nitrogen cycle 196
Central maxima 40
Chain reaction and reactor criticality 192
Chemical pumping 86
Circular fringes 24
Cladding 100
Cleavage 61
Cloud chamber 211
Co-axial cable 168
Co-axial cables 99
Coherence 81
Coherence time 82
Coherent 4
Coherent length 82
Collimator 48
Compensating plate 29
Compton effect 199
Constructive interference 6
Controller 195
Cooper pair 228
Coordinate transformation 171, 177
Core 100
Core radius 104
Core size 101
Corning 99
Corpuscular theory 1
Cosine square law 59
Crest 2
Critical angle 101
Critical temperature 218
Critically damped oscillations 124
Crossed nicol prism 66
Crytal detector 209
Curl of a vector field 148
Cut-off parameters 103
Concepts of holography 238

D

D-lines of sodium 64
Damped vibration 123
Dark central spot 23
Destructive interference 6
Dextro-rotatory crystal 72
Detection of Ultrasonic Waves 236
Dielectric constant 132
Dielectric in a.c. Field 141
Dielectric losses 142
Dielectrics 129
Diffraction 1
Diffraction losses 91
Diffused junction type detector : 210
Direct pumping 86
Directional selectivity 87
Directionality 97
Dispersion 108
Dispersion in an optical fiber 231
Dispersive power 51
Displacement 113
Displacement 4
Displacement current density 158
Displacement density 147
Displacement vector 133
Divergence 97
Divergence of a vector field 148
Double refraction 60
Double slit 7
Doubly refractive crystal 60
Dispersion effect 231

E

Einstein 81, 83
Electric 55
Electric displacement 147
Electric field strength 147
Electric field vector 59
Electric susceptibility 134
Electrical pumping 86
Electromagnetic wave 55
Electromagnetic waves 145
Electron collisional thermalisation 85
Electronic communications 99
Electronic polarization 139
Elements of bcs theory 227
Elliptically polarized 67
Endoscopy 107
Energy level diagram 95
Energy of damped harmonic oscillator 126
Energy of the forced oscillator 121
Energy stored in dielectric 138
Equation of continuity 151
Equivalence of mass and energy 187
Ether 170
Excitation energy 92
Extra-ordinary ray 60
Eye piece 48

F

Fabry prot resonator 91
Faraday's law of electromagnetic induction 155
Fiber optic communication 108
Fiberoscope 107
Flash lamp 91
Forced oscillation 116
Four level laser 86
Fractional change of refractive index 105
Frames of reference 169
Fraunhofer 34
Free space 161
Free space wavelength 103
Frequency 115
Fresnel 1
Fresnel biprism 11
Fresnel's diffraction 33
Fringe width 9
Fringes 30

G

Gain medium 85
Gain region 93
Galilean transformation 170
Gallium arsenide 95
Gas filled detectors 201
Gaussian surface 136
Gauss's divergence theorem 151
Gauss's law 136, 149
Gauss's law of electrostatics 154

Geiger-muller counter 204
Glass fibers 99
Graded refractive index type (grin) 106
Gradient of scalar field 148
Grating 41

H

Half shade 75
Half wave plate 71
Half wave plate 75
Heavily doped 96
Helium neon laser 92
High voltage electrode 93
Homogeneous medium 160
Host 84
Human hair 99
Huygens 1
Holography 238
Hologram 238

I

Incoherent light 87
Index profile 104
Induced birefringence 78
Induced dipole 130
Induced dipole moment 130
Inertial 170
Inter-modal dispersion 232
Interaction of radiation with matter 197
Interference 1, 4
Interference filters 15
Interference in wedge shaped thin film 229
Ionic polarization 140
Ionization chamber 202
Ir spectroscopy 107
Isotope effect 220
Infrasonic waves 233

J

Josephson effect 225

K

Kundt's tube method 236

L

Laevo—rotatory crystal 73
Laser 99
Laser cavity 93
Laser medicine 107
Laser transitions 94
Lasing efficiency 86
Lateral 82
Left circularly polarized light 67
Length contraction 182
Level lasers 86
Light 1
Light amplification 88
Light emitting diode 108
Light generator 87
Lines of flux 147
Lines of force 147
Localized 30
Logarithmic decrement 127
London equations 221
London equations 223
London penetration depth 224
Longitudinal coherence 82
Longitudinal modes 90, 93
Lorentz half shade polarimeter 74
Lorentz transformation 177

M

Magnetic field 55
Magneto motive force 159
Mainman 81
Malus law 59
Maser 81
Material dispersion 231
Maxwell 66
Maxwell's equations 153
Mechanical strain 78
Meissner effect 220, 221
Metastable states 94
Mica 60
Michelson – morley experiment 172
Michelson interferometer 7

Michelson's interferometer 27
Microwave communication 98
Mode number 90
Modeartor 194
Monochromatic 7
Monochromaticity 97
Monochromaticity 83
Multimode fiber 100
Material dispersion 231
Magnetostriction effect 233

N

Negative absorption 85
Negative absorption coefficient 85
Negative crystals 63
Negative temperature state 86
Neutron cross section 189
Newton 1
Newton's ring 20
Newton's rings 7
Nicol prism 60
Nicol's prism 63
Non-inertial 170
Non-polar dielectric 130
Non-relativistic case 181
Normal incidence 22
Normalized frequency 103
Nuclear fission 190
Nuclear fission and fusion reactions 188
Nuclear fission reactor 193
Nuclear fuel 194
Nuclear fusion 195
Nuclear radiation detectors 200
Numerical aperture 101

O

Objective 48
Optic axis 61
Optical activity 72
Optical bench 12
Optical cavity resonator 87
Optical cavity resonator 88
Optical communication 81
Optical fiber 99
Optical medium 4
Optical pumping 86
Optical resonator 87
Optically flat 96
Ordinary ray 60
Orientational polarization 140
Oscillations 124
Over damped 124

P

Pair production 200
Path difference 37
Perfect diamagnetism 221
Persistent current 219
Phase and phase difference 115
Phase change 15
Phase difference 37
Photoelasticity 78
Photoelectric effect 199
Physical significance of maxwell's equations 158
Plane of polarization 57
Plane of vibration 56
Plane wavefront 36
Plastic 100
Plastic conduits 99
Polar dielectric 129
Polarization 130, 132
Polarization 1
Polarized light 55
Polarizing angle 58
Polymer 100
Population inversion 84
Population inversion 92
Positive 63
Positive feedback 87
Postulates of special theory of relativity 176
Poynting vector 165
Poynting's theorem 163
Principal plane 61
Principal section 61
Prism table 48
Prokhorov 81
Proportional counter 203
Proton-proton cycle 196
Pumping 86
Piezo-electric effect 233

Piezoelectric method 233

Q

Quality factor 128
Quarter wave plate 70
Quartz 60, 63
Quartz 75
Quenching 206

R

Radiant losses 90
Radiation trapping 85
Radiative transitions 94
Radio waves 99
Ray 3
Rayliegh's criterian 52
Reactor coolants 194
Reactor shield 195
Recombination 96
Refractive index 58, 100
Relaxation time 128
Replica grating 42
Representation of electromagnetic *wave 166
Resolution 52
Resolving power 51
Resonance 118
Right circularly polarized light 67
Rowland 42
Ruby laser 91
Ruled grating 42
Ruling 42

S

Sapphire 91
Scintillation detectors 206
Semi-silvered 28
Semi-transparent mirror 87
Semiconductor laser 95
Semitransparent mirror 88, 92
Sharpness of resonance 120
Sheath 100
Silica 109
Simple harmonic motion 111
Single mode fibers 100
Single slit 36
Signal attenuation 232
Snell's law 65
Solid state detectors 209
Spatial coherence 81
Specific rotation 73
Specific rotation of sugar 76
Spectra 47
Spectrometer 48
Spirit level 12, 48
Splicing 109
Spontaneous emission 83
Standing wave 89
Stimulated emission 83
Stoke's theorem 151
Superconductivity 218
Superposition 4

T

Tangent screw 48
Telescope 48
Temporal coherence 81
The bubble chamber 213
The potential function 150
Thin films 14
Three 86
Threshold current 96
Time dilation 183
Time period 115
Topaz 60
Total internal reflections 99
Tourmaline 60, 63
Townes 81
Transducer 108
Transmission axes 59
Transverse coherence 82
Trough 2

U

Under damped oscillations 125
Uniaxial 60
Unpolarized 55
Ultrasonic waves 233

V

V-number 103
Variation of mass with velocity 185
Velocity transformation 171, 180

W

Wave 2
Wavefront 2, 7
Waveguide 167
Waveguide dispersion 232
Wedge 29
Wedge shaped thin film 229

X

Xylonite 79

Y

Young 1, 7, 55

Z

Zigzag path 99